SHAPE MEMORY ALLOYS – PROCESSING, CHARACTERIZATION AND APPLICATIONS

Edited by **Francisco Manuel Braz Fernandes**

INTECHOPEN.COM

Shape Memory Alloys – Processing, Characterization and Applications
http://dx.doi.org/10.5772/2576
Edited by Francisco Manuel Braz Fernandes

Contributors

F.M. Braz Fernandes, K.K. Mahesh, Andersan dos Santos Paula, Radim Kocich, Ivo Szurman, Miroslav Kursa, Tomasz Goryczka, T. Sakon, H. Nagashio, K. Sasaki, S. Susuga, D. Numakura, M. Abe, K. Endo, S. Yamashita, H. Nojiri, T. Kanomata, Fabiana Cristina Nascimento Borges, V. P. Panoskaltsis, Dezso L. Beke, Lajos Daróczi, Tarek Y. Elrasasi, R.J. Martínez-Fuentes, F.M. Sánchez-Arévalo, F.N. García-Castillo, G.A. Lara-Rodríguez, J. Cortés-Pérez, A. Reyes-Solís, Lei Li, Qingbin Li, Fan Zhang, Marjan Bahraminasab, Barkawi Bin Sahari

Published by InTech

Janeza Trdine 9, 51000 Rijeka, Croatia

Notice

Publishing Process Manager Sandra Bakic
Technical Editor InTech DTP team
Cover InTech Design team

Additional hard copies can be obtained from orders@intechopen.com

Shape Memory Alloys – Processing, Characterization and Applications,
Edited by Francisco Manuel Braz Fernandes
p. cm.
ISBN 978-953-51-1084-2

For EU product safety concerns:
IN TECH d.o.o., Prolaz Marije Krucifikse Kozulić 3, 51000 Rijeka, Croatia,
info@intechopen.com or visit our website at intechopen.com.

Meet the editor

Francisco Manuel Braz Fernandes was born in 1951 in Beira, Mozambique. He graduated in Mechanical Engineering from Instituto Superir Técnico (Lisbon, 1976), had the "Diplôme d'Études Approfondies - Métallurgie Spéciale et Matériaux" from University of Orsay (Paris XI, 1982), the PhD degree from École des Mines de Nancy (Nancy, 1985) and the Habilitation in Physical Metallurgy at Universidade Nova de Lisboa (2008). He is currently an Associate Professor at the Department of Materials Science, Faculty of Science and Technology, Universidade Nova de Lisboa. He is also Head of the Structural Materials Research Group from CENIMAT (Centro de Investigação de Materiais). The research activity where he is currently involved is mostly focused on shape memory alloys, from thin film deposition to thermomechanical processing (including severe plastic deformation), laser welding and shape memory/superelasticity applications (medicine, civil engineering and metalworking), using characterization techniques such as X-ray diffraction, namely with synchrotron radiation, thermal analysis (DSC), mechanical testing and electrical resistance. He has more than 100 papers published in international scientific journals, and is responsible for the supervision or co-supervision of 7 PhD thesis and a large number of MSc dissertations already presented. As a hobby, he is a philatelist owning a stamp collection on Materials Science and Engineering.

Contents

Preface IX

Section 1 Processing 1

Chapter 1 **Thermomechanical Treatments for Ni-Ti Alloys 3**
F.M. Braz Fernandes, K.K. Mahesh and Andersan dos Santos Paula

Chapter 2 **The Methods of Preparation of Ti-Ni-X Alloys and Their Forming 27**
Radim Kocich, Ivo Szurman and Miroslav Kursa

Chapter 3 **$Ni_{25}Ti_{50}Cu_{25}$ Shape Memory Alloy Produced by Nonconventional Techniques 53**
Tomasz Goryczka

Section 2 Characterization 77

Chapter 4 **Thermal Strain and Magnetization Studies of the Ferromagnetic Heusler Shape Memory Alloys Ni_2MnGa and the Effect of Selective Substitution in 3d Elements on the Structural and Magnetic Phase 79**
T. Sakon, H. Nagashio, K. Sasaki, S. Susuga, D. Numakura, M. Abe, K. Endo, S. Yamashita, H. Nojiri and T. Kanomata

Chapter 5 **Iron Based Shape Memory Alloys: Mechanical and Structural Properties 115**
Fabiana Cristina Nascimento Borges

Section 3 Mechanical Behavior 129

Chapter 6 **Mechanics of Shape Memory Alloy Materials – Constitutive Modeling and Numerical Implications 131**
V. P. Panoskaltsis

Chapter 7 **Determination of Elastic and Dissipative Energy Contributions to Martensitic Phase Transformation in Shape Memory Alloys 167**
Dezso L. Beke, Lajos Daróczi and Tarek Y. Elrasasi

Chapter 8 **Micromechanical Behavior of CuAlBe Shape Memory Alloy Undergoing 3-Point Bending Analyzed by Digital Image Correlation 197**
R.J. Martínez-Fuentes, F.M. Sánchez-Arévalo, F.N. García-Castillo, G.A. Lara-Rodríguez, J. Cortés-Pérez, A. Reyes-Solís

Section 4 Applications 213

Chapter 9 **Applications of SMA Bundles in Practical Concrete Structures 215**
Lei Li, Qingbin Li and Fan Zhang

Chapter 10 **NiTi Shape Memory Alloys, Promising Materials in Orthopedic Applications 261**
Marjan Bahraminasab and Barkawi Bin Sahari

Preface

Shape memory alloys have become in the past decades a well established research subject. However, the complex relations between properties and structure have created a continuously growing interest for a deeper insight all this time. The complexity of relationships between structure and properties is mostly related to the fact that strong "multidimensional" interactions are taking place: from the early studies focusing on the thermal and/or mechanical induced phase transformations to the more recent findings on the magnetically induced structural changes. On the other hand, these singular behavioral characteristics have driven a great industrial interest due to the innovative aspects that the applications of shape memory alloys may provide. This makes this subject a highly attractive source of continuous studies, ranging from basics crystallography and thermodynamics to mechanical analysis and electrical and magnetic properties characterization.

In this book, a group of recent studies is compiled focusing on a wide range of topics from processing to the relationship between the structure and properties, as well as new applications.

In the processing section, results are presented for special issues concerning the thermomechanical treatments of binary alloys and ternary alloys, namely using innovative techniques. The first paper details the changes of the transformation characteristics of binary alloys resulting from different variants of thermomechanical treatments, including severe plastic deformation. The second paper is focused on the interest of going from binary to ternary alloys and the problems raised by innovative processing techniques. In particular, the Ni-Ti-Cu alloys have attracted a great interest and the possibilities open by advanced technologies such as powder metallurgy and melt spinning are extensively reviewed in the third and last paper of this section.

The relatively more recent interest on the Ferromagnetic Shape Memory Alloys opens new challenges in terms the characterization techniques combining thermal analysis and magnetic methods. This is the topic for the first paper of the section of characterization, where some alloys from the system based on Ni-Mn-Ga are studied. Next, a structural characterization of the stress-induced martensite in Fe-based alloys is presented.

The singular properties of the shape memory alloys are strongly related to the thermomechanical behavior. A clear understanding of this type of situation requires the use of models, such as the one that is proposed on the first paper of the section on mechanical behavior. The second paper presents a model for hysteretic loops providing the basis for a finer analysis of thermal / mechanical cycling. Eventually, the third paper provides useful information on the experimental approach to get a deeper insight on the link between micro and macromechanical behavior and its relation with the structural details. Iron based SMA are becoming very attractive engineering materials and their structural characterization is presented on the second paper.

Last, a section on applications presents two interesting papers focusing on two major engineering areas: civil and biomedical. Based on a constitutive model and thermal and mechanical characterization, a practical application of SMA on a smart concrete bridge structure is presented. Next, the combination of the superelastic and shape memory effect characteristics with the good biocompatility of Ni-Ti alloys is exploited in the field of biomedical engineering, namely for implants.

The second paper presents a model for hysteretic loops providing the basis for a finer analysis of thermal / mechanical cycling. The third paper provides useful information on the experimental approach to get a deeper insight on the link between micro and macromechanical behavior and its relation to the structural details.

Prof. Francisco Manuel Braz Fernandes
Universidade Nova de Lisboa,
Portugal

Section 1

Processing

Chapter 1

Thermomechanical Treatments for Ni-Ti Alloys

F.M. Braz Fernandes, K.K. Mahesh and Andersan dos Santos Paula

Additional information is available at the end of the chapter

http://dx.doi.org/10.5772/56087

1. Introduction

Thermomechanical treatments for shape memory alloys (SMA) are found to be one of the more economical, simpler, and efficient methods adopted for manipulating the transformation properties. The stability of phase transformation has been found to depend upon the thermomechanical treatments, such as hot- or cold-working, heat-treatment and thermal cycling. It has perhaps more important and wide reaching ramifications than many of the other stages in the fabrication of components and structures.

During the stages of preparation of SMA, hot working is adopted as one of processes in the form of rolling or drawing to incorporate the shape memory effect (SME). Such alloys can be directly employed for the applications. However, most of the times, the ingots are finally cold worked in the form of rolling or drawing before delivering to the application purpose. This allows the application engineers to subject the alloys to appropriate thermal/mechanical treatment in order to obtain the SMA with desired phase transformation properties. Hence, a sequence of cold work followed by heat treatment is considered to be a productive method to tailor the SME and superelasticity (SE).

In order to emphasize the various methods of thermal, mechanical, and thermomechanical treatments, the Chapter is divided into the following Sections and Sub-sections.

i. Cold working
ii. Cold working followed by heat treatments
iii. Effect of cooling rate during heat treatments
iv. Hot working
v. Thermal cycling
vi. Severe plastic deformation
    a. High-pressure torsion (HPT)
    b. Equal channel angular pressing (ECAP)
vii. Concluding remarks

2. Cold working

Cold working can induce dislocations and vacancies in the Nickel-Titanium (Ni-Ti) alloys. It is suggested that the possible mechanisms for the martensite stabilization in the Equiatomic Ni-Ti alloys come from deformed structures and deformation induced dislocations/ vacancies. Thermomechanical treatments of Ni-Ti SMA are important for the optimization of the mechanical properties and phase transformation characteristics. An important characteristic in the Ni-Ti SMA is the stability on direct and reverse transformations, related to the sequence and transformation temperatures, and thermal hysteresis **[1-5]**. The transformation temperatures in Ni-Ti SMA have been shown to be related to the presence of lattice defects introduced by cold working **[6, 7]**. Wu *et al.*, (1996) showed that the defects induced during cold working have the effect of suppressing the martensitic transformation and promoting the R-phase transformation **[8]**. The residual internal stress induced by cold-working defects is considered to be responsible for the R-phase transformation **[9]**. The deformation mechanisms and morphologies in polycrystalline martensitic CuZnA1 alloy have been examined by Adachi and Perkins **[10]**. They observed that a variety of deformation morphologies, including variant-variant coalesce, stress-induced martensite to martensite transformation, injection of foreign variants to plate groups, and internal twinning and slip, are all exhibited simultaneously in moderately cold-worked specimens.

Ni-Ti alloys have a wider application in the form of wires. Therefore, an understanding of the wire drawing properties is important. Thin oxide film with a smooth surface on TiNi wires can be used as a lubricant during the drawing process. However, thick oxide films which have cracks and spalling on the surface can be detrimental to the drawing surface and depress the shape memory effect and pseudoelasticity of TiNi SMA. MoS_2 is an effective lubricant for wire drawing of TiNi SMA **[8]**. Also, cold rolling has been one of the widely adopted processing techniques in order to obtain Ni-Ti alloy in the sheet form. In a study of the cold-rolled equiatomic TiNi alloy, it was found that the same phenomena of martensite stabilization appear, as reported in Cu-based shape memory alloys **[9, 10]**. It is well known that the martensite in the $Ti_{50}Ni_{50}$ alloy has 24 variants **[11]**. The variants will accommodate each other under thermal or mechanical stress. It is reasonable to suggest that the stress exerted by cold rolling causes the variants to accommodate, i.e., the stress forces the preferred orientated variants to accommodate the deformation strain in the favorable stress direction. An intensive study of the microstructure of the deformed martensite shows, in addition to the deformed martensite plates, a large number of dislocations and vacancies can also be induced during the cold rolling. These deformation-induced dislocations and vacancies have an important effect on the martensitic stabilization **[9]**.

In Fig. 1, DSC thermograms of the Ni-Ti alloy plate subjected to 40% cold working are shown. When the cold worked specimen is heated from RT up to around 300°C, no phase transformation is observed and on the further heating, a broad upward peak appears around 350°C corresponding to recrystalization process. However, on cooling to RT, a clear exothermic peak appears around 75°C and that is attributed to A→M phase transformation. While heating again from RT, the DSC thermogram shows an endothermic peak around

85 °C corresponding to M→A phase transformation and while cooling the reverse phase transformation, A→M, is observed.

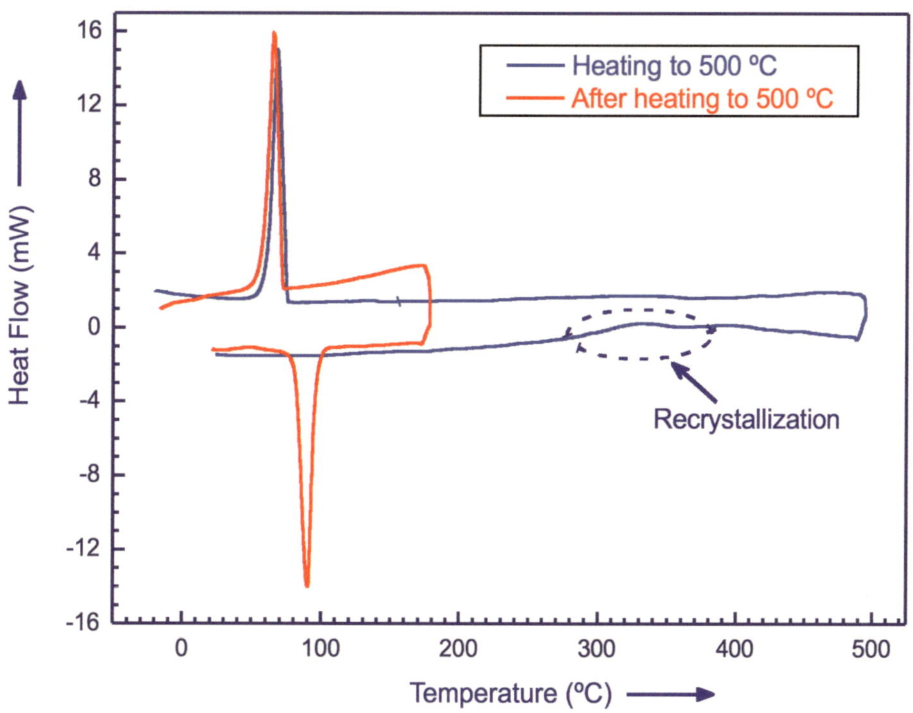

Figure 1. DSC thermograms of the Ni-Ti SMA plate initially cold worked up to 40% and heated to 500°C (in blue), and after (in red).

In Fig. 2, 3-d representation of the XRD profiles obtained at different temperatures from RT to 400°C for the 40% cold worked Ni-Ti specimen is shown. XRD spectra obtained at RT show the peaks corresponding to B19′ structure, which are broad and with low intensity. As the temperature is increased, the peak corresponding to B2 structure starts to emerge around 190 °C and on further heating, the intensity of the peak increases. Broad and low intensity peaks are due to the deformation induced dislocations and vacancies which suppresses the martensitic transformation **[12-14]**.

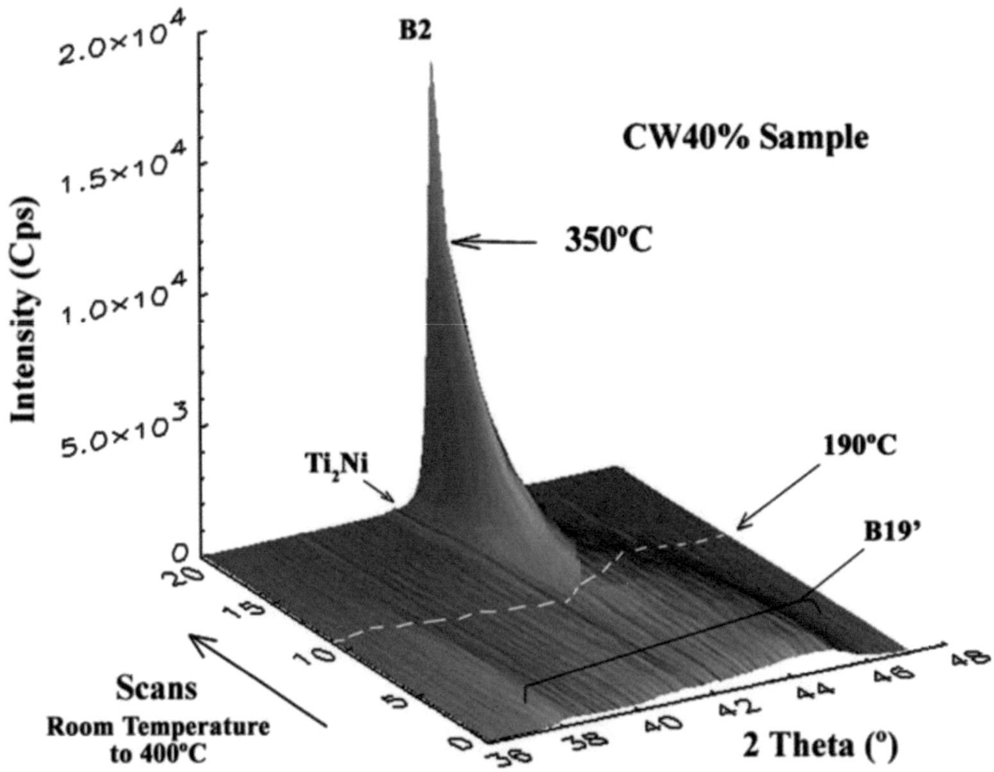

Figure 2. 3-d representation of the XRD profiles obtained at different temperatures from RT to 400°C for the 40% cold worked Ni-Ti specimen.

3. Cold working followed by heat treatments

Heat treatment for metals and alloys has been proved to be an effective and economical process in order to maneuver their properties. Various factors, such as the HTT, annealing time and cooling rate after annealing have their own effects on the final state of the metal/alloy. In the above sub-section, it is mentioned that the defects induced during cold working have the effect of suppressing the martensitic transformation **[8]**. On the contrary, there is a possibility that a reverse phenomenon (restoration) would occur in a rather enhanced manner upon annealing through thermal activation processes of point defects. The migration of vacancies and interstitials could facilitate promotion of the martensitic transformation **[15]**. In this sub-section, the dependence of heat-treatment on the composition and thermal/mechanical history of the alloys has been explained. Heat-treatment plays a crucial role in fixing M_s. The detection of R-phase is found to be critical with the positioning of M_s in relation to R_s. If M_s is above R_s, R-phase is found to be masked by the martensite phase. Earlier, from electrical resistivity measurements, it was shown that while cooling from austenite phase, if R-phase exists, it preceded the martensite phase and it was regarded as the pre-martensitic phase **[16]**. However, later it was shown that both phases coexist at the same temperature, and it has been confirmed by the DSC study on the phase transformation in the 40% cold worked, near equi-atomic NiTi alloy subjected to water quenching from 400°C **[17]**.

Phase transformations associated with SME in Ni-Ti alloys can be one-stage, $B19^{I} \leftrightarrow B2$, two-stage including an intermediate R-phase stage, or multiple-stage depending on the thermal and/or mechanical history of the alloy. In a recent report, it has been highlighted the effect of (i) deformation by cold-rolling (from 10% to 40% thickness reduction) and (ii) final annealing on the transformation characteristics of a Ti-rich NiTi shape memory alloy. For this purpose, one set of samples initially heat treated at 500 °C followed by cold-rolling (10–40% thickness reduction) has been further heat treated at various temperatures between 400 and 800 °C. Phase transformations were studied using differential scanning calorimetry, electrical resistivity measurements and in situ X-ray diffraction. A specific pattern of transformation sequences is found as a result of combination of the competing effects due to mechanical-working and annealing **[18]**.

Fig. 3 (a & b) show the Differential Scanning Calorimeter (DSC) and Electrical Resistivity (ER) curves for (i) as-received (AR), (ii) annealed at 500°C (HT500) and (iii) annealed at 500 °C/cold-rolled to 30%/annealed at 500°C (TMTCR30HT500) samples. For the AR sample, both in the case of DSC & ER techniques, multiple-step (B2↔R, B2↔B19′, R↔B19′, while heating and cooling) phase transformations are observed. For the HT500 sample, in both

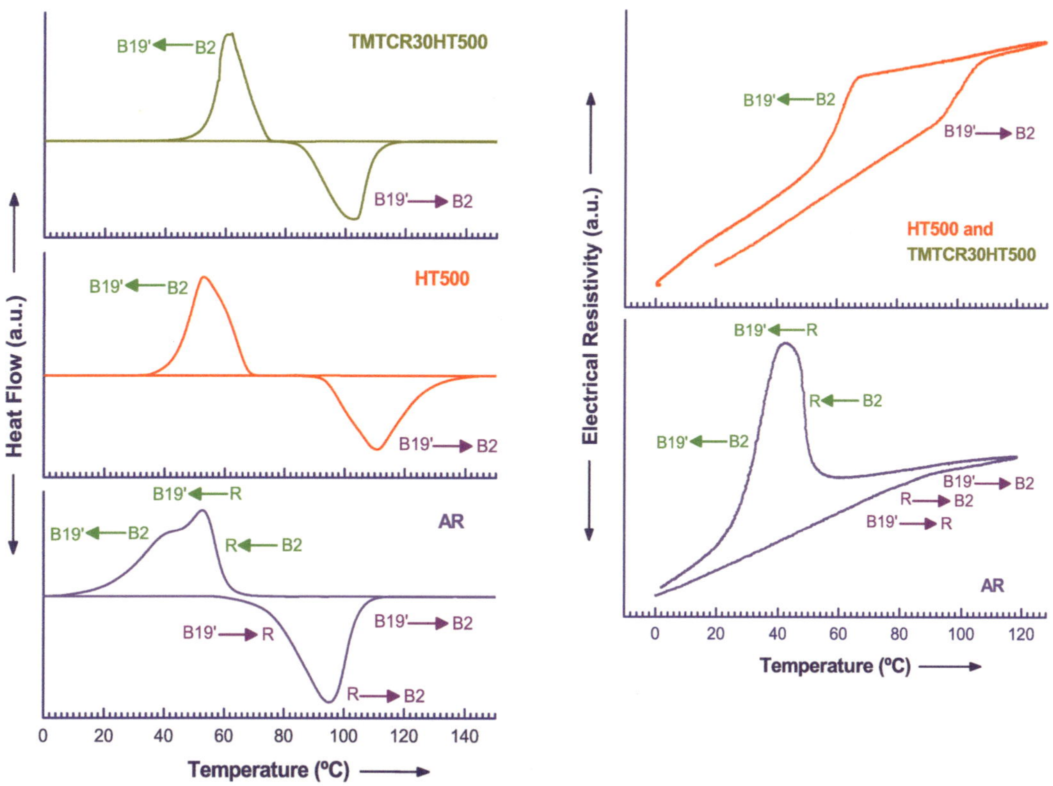

Figure 3. (a) DSC and (b) ER curves for AR, HT500 and TMTCR30HT500 samples

techniques, during heating and cooling, one-step (B19′↔B2) phase transformation is found to be present. Further, in the case of TMTCR30HT500 sample, one-step (B19′↔B2) phase transformation is detected. During heating (for AR samples), a small kink in the DSC and a small hump in ER plots around 60 °C show the presence of R-phase associated to multiple-step, (B19′↔R, B19′↔B2, R↔B2), phase transformation.

The effects of various heat treatment temperatures (HTT) on samples after being cold-rolled to different extents (10 to 40% thickness reduction) are presented in Fig. 4. All the samples were annealed at 500 °C before cold-rolling. Figs. 4 (a to d) show the transformation temperatures (A_f, A_s, R_{fh}, R_{sh}, R_{sc}, R_{fc}, M_s and M_f, obtained from DSC thermograms) as a function of HTT, for the samples annealed after being cold worked up to 10%, 20%, 30% and 40%, respectively. "A", "R" and "M" are the austenite, rhombohedral, and martensite phases; suffixes "s" and "f" are the start (1%) and finish (99%) transformation temperatures; and "c" and "h" refer to cooling and heating, respectively.

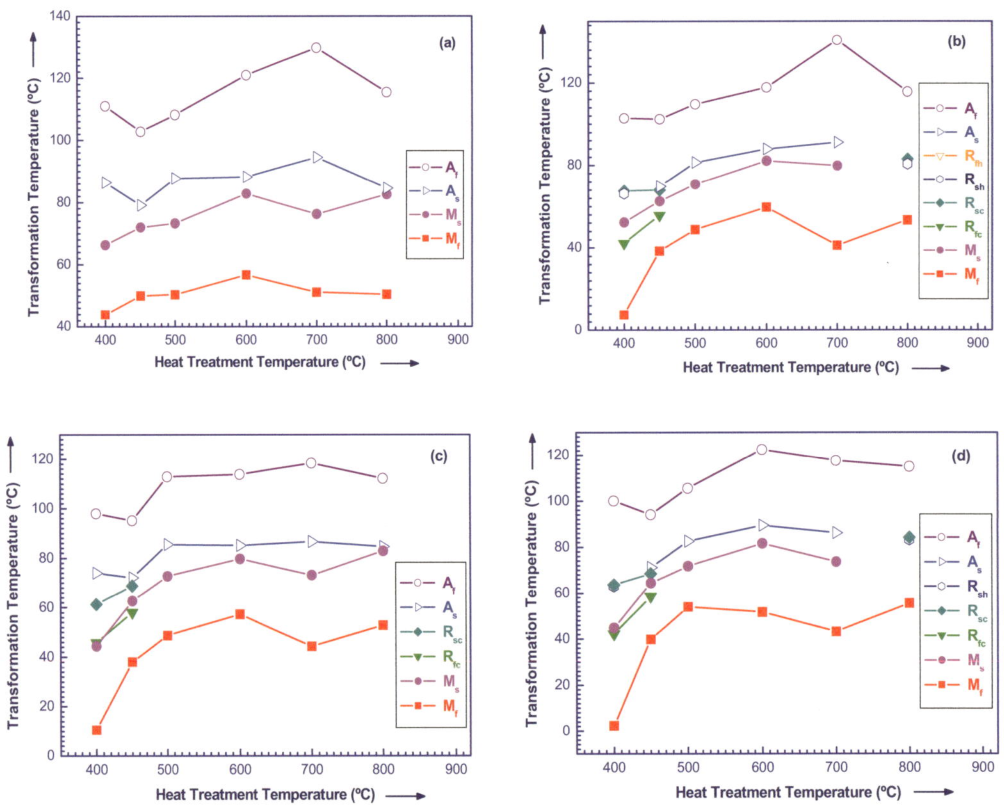

Figure 4. Transformation temperatures of (a) TMTCR10%, (b) TMTCR20%, (c) TMTCR30%, and (d) TMTCR40%

In Fig. 4(a), for the 10% cold worked samples, as the final annealing temperature is increased, M_s and M_f are found to increase gradually up to 600 °C followed by a slight drop up to 800 °C. A_s and A_f are found to decrease as the final annealing temperature is increased from 400 to 500 °C. Further increase up to 700 °C shows gradual increase followed by a decrease for the final annealing temperature of 800 °C.

In Fig. 4(b), it is observed that for the 20% cold worked samples, there is R-phase formation while cooling (R_{sc}, R_{fc}) and while heating (R_{sh}). As the final annealing temperature is increased, R_{sc} and R_{fc} are found to increase till 500 °C. For higher final annealing temperatures, the R-phase formation is no longer detected. M_s and M_f increase with increasing final annealing temperature until it reaches 600 °C, followed by a slight decrease when the sample is heat treated at 700 °C. For the final annealing at 800 °C, M_s is not possible to be determined, but M_f increases slightly. A_s is found to increase with increasing final annealing temperature up to 700 °C along with A_f. For the final annealing temperature of 800 °C, A_s was not possible to be determined and A_f decreases. For this same treatment (800 °C), the R-phase formation is once again detected during cooling and heating.

In the case of samples 30% cold worked, as shown in Fig. 4(c); the R-phase is only present during cooling for final annealing temperatures up to 500 °C (R_{sc} and R_{fc} increase with increasing final annealing temperature). M_s and M_f increase for increasing final annealing temperature up to 600 °C, slightly decrease for 700 °C and then slightly increase for 800 °C. A_s and A_f slightly decrease from 400 to 500 °C and then increase and stabilize after 500 °C.

In Fig. 4(d), it is observed that for the samples 40% cold worked and heat treatment there is R-phase formation only during cooling for the final annealing temperature up to 500 °C. (R_{sc} and R_{fc} are found to increase with increasing annealing temperature). M_s and M_f increase with increasing final annealing temperature till 600 °C. For the final annealing temperature of 800 °C, M_s and A_s were not possible to be determined. For the final annealing temperature of 800 °C, the R-phase formation is once again detected.

The absence of the R-phase formation in the sample annealed at 500 °C (not cold-rolled), may be explained by the annealing out of the structural defects and generation of the strain free crystals **[19]**. The same result is observed for the sample that has been cold-rolled to 10% (very close to the maximum recoverable strain of this class of alloys). With increasing extent of cold-work deformation, the R-phase deformation is only detectable for final annealing temperatures below 500 °C or at 800 °C. The final annealing temperature above 500 °C induces a recrystallization of the marformed matrix that makes the single-step transformation B2↔B19′ more favorable **[14, 20, 21]**. This transformation may be initiated at the coherent interfaces of the very narrow precipitates Ti_2Ni. For the highest final annealing temperature (800 °C) the R-phase formation is once again present and this may be associated to the coalescence of the Ti_2Ni precipitates, making the B2 / Ti_2Ni interfaces incoherent **[22, 23]**. When the DSC and ER results in Figs. 1 and 2 are compared, it is apt to mention that when there is overlap of the phases transformation, ER technique is in a better position to reveal the presence of distinct phases.

Table 1 summarizes the transformation sequences of the samples after the thermomechanical treatments. For the samples cold worked to 10% and subsequently heat treated up to 700 °C, the transformation sequence is found to be clearly one-step (B19′↔B2). On the other hand, no matter the thickness reduction by cold-rolling, when the final annealing temperature is between 500 °C and 700 °C, the transformation is also clearly one-step (B19′↔B2). The two-steps phase transformation while cooling is only observed for the samples cold-rolled to 30 and 40% and for the final annealing temperatures of 400 °C. The multiple-steps phase transformation (with overlap) is only observed in two situations: (i) for the final annealing temperature of 800 °C, no matter the cold-work reduction, both while cooling and heating, and (ii) for the samples cold-worked to 20 to 40%, where the final annealing temperature was 500 °C or below.

HTT (°C)	Reduction by Cold Rolling			
	10%	20%	30%	40%
400	+ / +	⊕ / Ø	++ / +	++ / Ø
450	+ / +	⊕ / Ø	⊕ / Ø	⊕ / Ø
500	+ / +	+ / +	+ / +	+ / +
600	+ / +	+ / +	+ / +	+ / +
700	+ / +	+ / +	+ / +	+ / +
800	Ø / Ø	Ø / Ø	Ø / Ø	Ø / Ø

On Cooling / On Heating: + one-step; ++ two-steps;

⊕ Multiple-steps with overlap; Ø suspect multiple-steps with overlap.

Table 1. Influence of the thermomechanical processing (marforming) conditions on the transformations sequence.

Deformation up to 10% thickness reduction decreases the shape memory effect capability. This behavior is associated with the reorientation of martensite variants and increase of dislocation density, giving rise to a stabilization of martensite at a higher temperature in agreement with previous results **[24]**.

4. Effect of cooling rate during heat treatments

During the heat treatments, one of the parameters which, can be easily controlled is the cooling rate. Otsuka *et al.*, adopted a heat-treatment in which they homogenized the Ni50at%-Ti alloy for 1 h at 1000 °C followed by furnace cooling to eliminate the vacancies and the disorder to some extent. They found that quenched specimen has almost the same transformation temperatures as the furnace cooled one **[25]**. It was found earlier by Saburi *et al.*, that during heat-treatment, M_s and mechanical behavior of Ni-rich off-stoichiometric (>50.7at% Ni) NiTi alloys were sensitive to rate of cooling, whereas, of a near-stoichiometric (50.4 at% Ni) alloys were not **[26]**. Sitepu *et al.*, showed that precipitation of Ni_4Ti_3 particles occurred in a matrix of Ni-rich Ni-Ti SMA of nominal composition Ni50.7at%-Ti, when it was solution annealed at 850 °C for 15 minutes followed by water quenching and aging at 400 °C for 20 h **[27]**. In a more recent study, transformation behavior of NiTi alloys of different composition, heat treated by employing quenching and furnace cooling were investigated **[28]**.

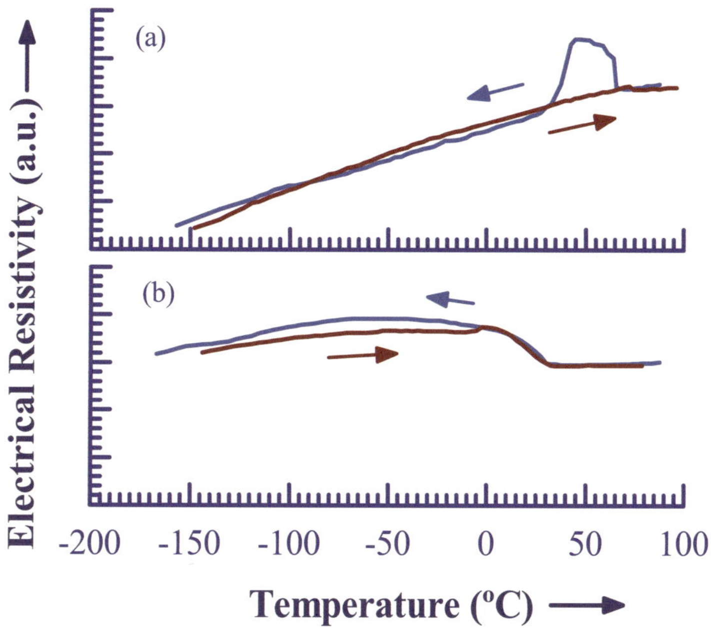

Figure 5. Electrical resistivity profiles for (a) Ni54.76wt%-Ti and (b) Ni56.00wt%-Ti alloys in the as-received condition

Fig. 5 shows resistivity profiles for the 2 samples, (a) Ni54.76wt%-Ti, i.e. Ti-rich and Ni56.00wt%-Ti, i.e. Ni-rich Ni-Ti alloys, in the as-received condition. For Ti-rich alloy, R-phase is found to occur only on cooling and the transformation is confined to a temperature interval of about 60°, above 0°C. In the case of Ni-rich alloy, R-phase is found to appear both while heating and cooling, and its temperature interval is spread over a wide temperature range of more than 150°, below +50°C, and these materials do not undergo the transformation to M-phase in the observed temperature range.

Fig. 6 (a-c) and 6 (d-f) show the resistivity profiles of the quenched and furnace cooled samples of Ti-rich alloy, respectively. In both cases, profiles are similar. R-phase transformation is only present during cooling for all the samples annealed between 100° and 420°C and the transformation region decreases, with increase in annealing temperature due to the increase in M_s temperature. For the annealing temperatures between 420°- 800°C, R-phase is found to be absent.

Fig. 7 (a-d) and 7 (e-h) demonstrate the resistivity profiles of the Ni-rich alloy for the quenched and the furnace cooled samples respectively. For the quenched samples, annealed in the temperature range of 100°- 500°C, two-stage transformation A→R→M during cooling and M→R→A during heating are observed. When annealed between 500° and 600°C, two-stage transformation is observed only in cooling, with decrease in the temperature interval of R-phase. Annealing above 600°C, further suppression of R-phase takes place promoting only M↔A transformation. In the case of furnace cooled alloy, with increase in annealing temperature, a unique discontinuous behavior is observed. With increase of annealing temperature from 100° to 440°C, two-stage transformation is observed both during cooling and heating in the resistivity profile, with reduced R-phase temperature interval. Annealing the sample between 440° and 580°C, R-phase is found only on cooling with further reduction in the temperature interval. For the sample annealed at 590°C, a sudden increase in the temperature interval of R-phase takes place. Hence, annealing around 590°C seems to be very critical. Annealing above 590°C, two-stage transformation is seen both during heating and cooling in the resistivity profile, regaining the initial behavior. The profiles indicate the stabilization of various phases above annealing temperatures of 590°C.

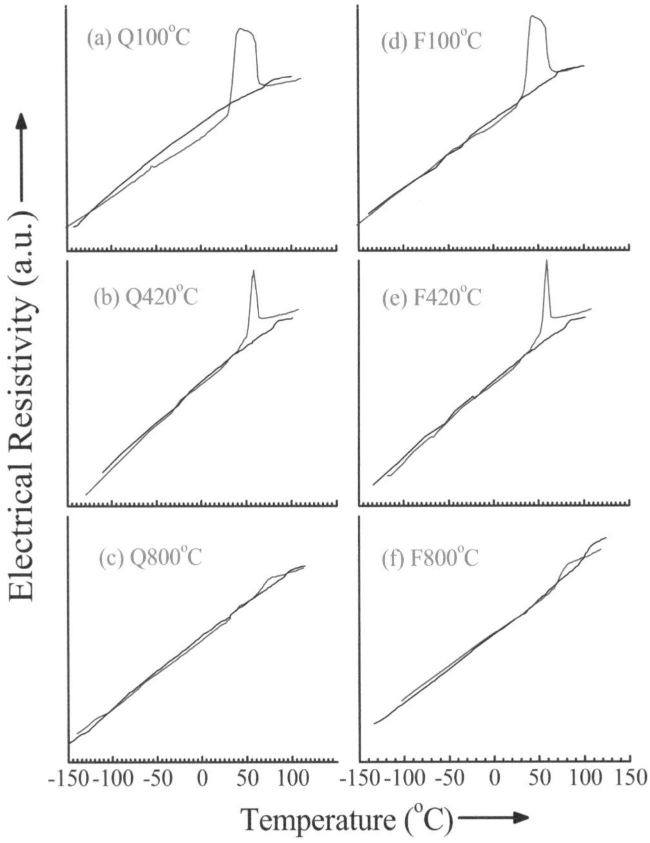

Figure 6. Resistivity profiles for the quenched and furnace cooled Ni54.76wt%-Ti alloys annealed at different temperatures.

For lower annealing temperatures, all the samples of the two alloys, both quenched and furnace cooled, exhibit similar behavior, i.e., M_s increases with increase in annealing temperature, which is attributed to the release of energy stored during the cold work. Cold work introduces high density of lattice defects, residual strain and internal stresses in the materials, which hinders from the movement of martensite interfaces. On annealing such cold worked materials, thermally activated diffusion leads to the annihilation of lattice defects, promoting martensitic transformation [29]. For the quenched samples, at higher annealing temperatures, this trend continues and gradual reduction in R-phase facilitates M↔A transformation. But, the furnace cooled samples, after annealing at higher temperatures, behave differently. A comparison of the resistivity profiles for the quenched and furnace cooled samples, especially annealed at higher temperatures, indicates that Ni-rich alloy is sensitive to the cooling procedure, unlike Ti-rich alloy. There is not much difference in the behavior of Ti-rich alloy either furnace cooled and quenched. In the case of furnace cooled Ni-rich alloy a unique discontinuous behavior is observed, for annealing at 590 °C. This may be due to the microstructural variations, arising as a consequence of two competing processes, viz., annihilation of defects and precipitation. Annealing above this critical temperature, the sample is able to regain and sustain a two-stage transformation, which may be attributed to the dominance of precipitation process over the defect annihilation process. It is proposed that, there is increased chance for Ti_3Ni_4 precipitation while furnace cooling, due to the slow cooling process and the presence of the material at higher temperature for a longer time. As reported by Nishida *et al.*, Ti_3Ni_4 precipitates have rhombohedral structure and are coherent to the matrix having a B2 type structure [**30**].

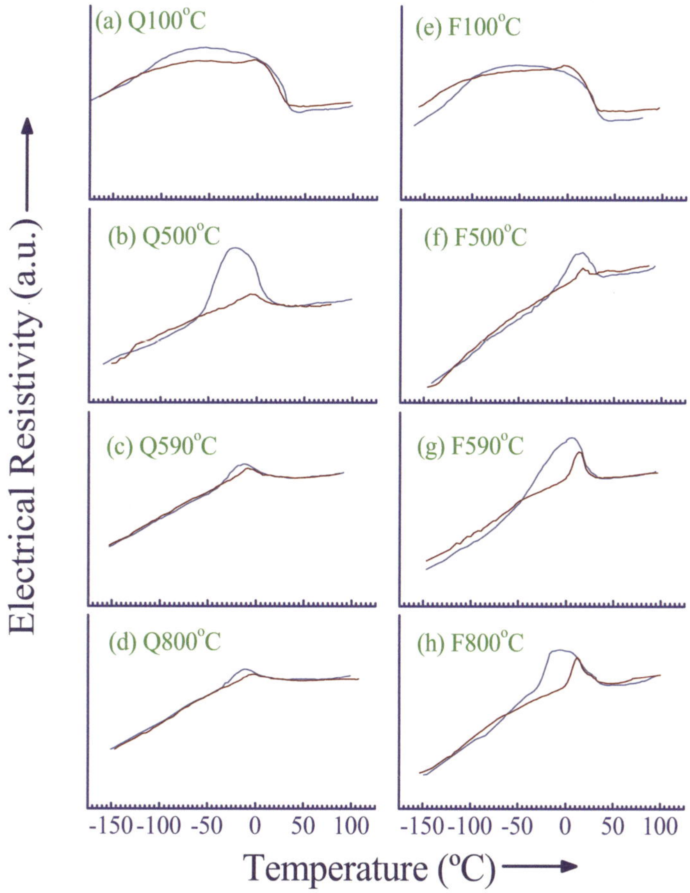

Figure 7. Resistivity profiles for the quenched and furnace cooled Ni56wt%-Ti alloys annealed at different temperatures.

5. Hot working

Both rolling temperature and thickness reduction are important factors that influence the work hardening and hardness of hot-rolled plates. The greater the thickness reduction, the greater the number of dislocations retained, and therefore, the greater the rate of work hardening. At rolling temperatures ≥600 °C, recovery or recrystallization occurs. However, because of the short rolling time and the fast cooling in air, the recovery or recrystallization is incomplete [**31**]. Hot-rolled Ni-Ti materials are found to possess enhanced resistance to low-cycle fatigue (increased pseudoelastic stability) as long as the primary material processing route remains unchanged [**32**]. Paula *et al.*, recently studied Ni-Ti alloys subjected to heat treatment at 767 °C for 300 s followed by hot rolling (50%) after cooling in air to 500 °C and water quenching to room temperature (T_{room}). Phase transformations were studied using differential scanning calorimetry, electrical resistivity measurements and in situ X-ray diffraction [**18**].

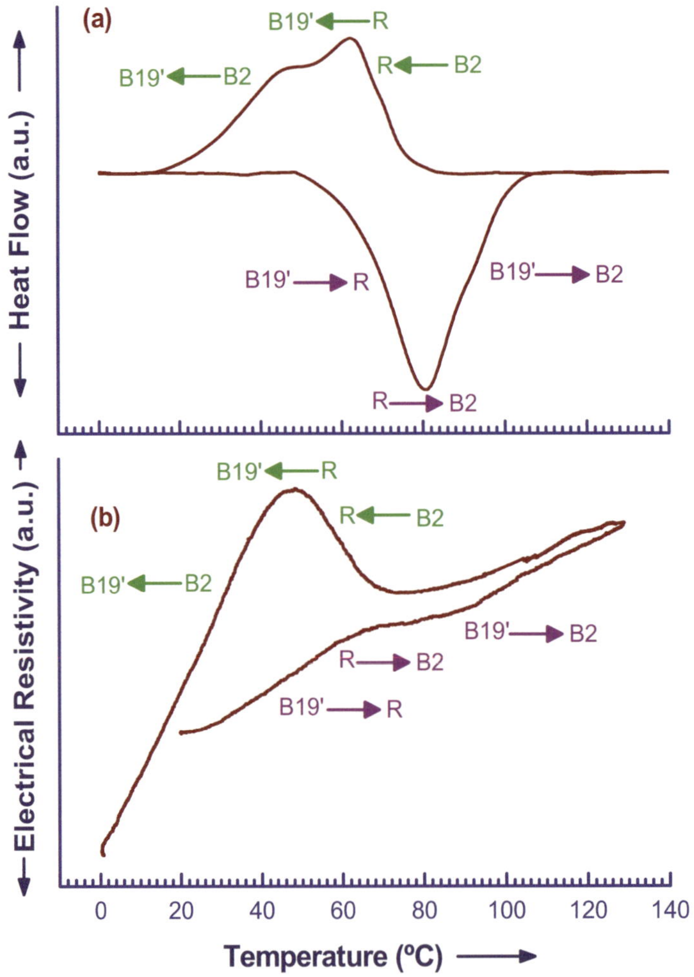

Figure 8. (a) DSC and (b) ER curves for TMTHR500 samples.

Fig. 8 (a & b) shows the DSC and ER curves for the ausformed at 500 °C (TMTHR500) samples. During the cooling and heating stages, multiple-step (B2↔R, B2↔B19′, R↔B19′) phase transformation is clearly detected in both techniques. During heating, a small kink in the DSC and a small hump in ER plots around 60 °C show the presence of R-phase associated to multiple-step, (B19′→R, B19′→B2, R→B2), phase transformation. It was found that the ausforming at 500 °C promotes multiple-step phase transformation on cooling and heating (B2↔R; B2↔B19′; R↔B19′). During the ausforming process at 500 °C, it is not achieved a full recrystallization, in agreement with other authors results **[33]**. Ausforming introduce many defects in the sample, so that R-phase formation becomes necessary to decrease the energy for B2→B19′ or B19′→B2 transformations.

6. Thermal cycling

Thermoelastic martensitic transformation appears to be very sensitive to thermal cycling **[34, 35]**. Also, thermal and mechanical treatments can suppress slip deformation resulting in increase of flow stress and modify the transformation temperatures, recovery stresses and recovery strains **[36]**. These observations indicate that the transformation process is strongly affected by irreversible changes in the microscopic state of the alloy introduced by thermal cycling. Thermal cycling causes a decrease in the characteristic temperatures and heats of transformation **[37]**. Also, thermal cycling is found to promote the intermediate R-phase transformation **[38]**. The effect of training conditions and extended thermal cycling on the two-way shape memory behavior of nitinol has been studied by **Hebda and White, 1995 [39]**. Thermal cycling under constant load was studied by **de Araujo et al., 2000 [40]** and they concluded that the internal stresses created were effective in inducing two-way memory effect.

Below, in Fig. 9, phase transformations are studied during the ab initio 10 thermal cycles by using DSC and ER techniques. In the DSC, thermal cycle was comprised of heating up to 140 °C, holding for 360 s and subsequently cooling down to -30 °C, with heating and cooling rates being 7.5 K/min. ER characterization have been performed by making use of a home made four-probe setup and the thermal cycling is performed by using the temperature controlled silicone oil bath. Ni-Ti (Ti51at%-Ni) alloy has been previously subjected to a series of thermomechanical treatment followed by heat treatment at 500 °C for 30 min. [41].

In Fig. 9 (a & b), during the first thermal cycle, in both the techniques (DSC & ER), it is observed that one-step phase transformation takes place. As the thermal cycling progresses, phase transformation processes are found to shift toward lower temperatures, both while heating and cooling. In Fig. 9(a), DSC thermograms for the first and second thermal cycles, the phase transformation peaks are observed to be symmetrical both while heating and cooling attributing to one-stage M↔A transformation. Also, in the ER profile shown in Fig. 9(b) corresponding to the first and second thermal cycles, it is observed that the specimen undergo one-step M↔A transformation. As the number of thermal cycles is increased, DSC thermogram peaks is found to broaden asymmetrically and shift toward lower temperatures (from the fifth cycle onward), giving rise to increasing evidence of the intermediate R-phase transformation while cooling (Fig. 9b).

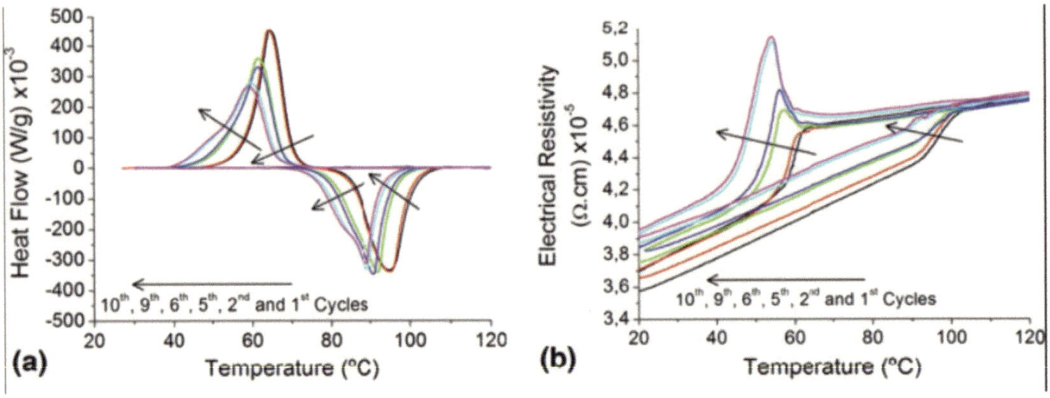

Figure 9. Evolution of phase transformation during thermal cycling up to 10 of the Ni-Ti specimens subjected to series of thermomechanical treatment followed by heat treatment at 500 °C for 30 min. (a) DSC and (b) ER profiles.

This shows that the Ti-rich Ni-Ti alloy under study, when subjected to thermal cycling, after multiple steps of thermomechanical treatments followed by final heat treatments, the stability of the phase transformation is found to sensitive and depend on the final heat-treatment temperatures. Further, the thermal cycling process also found to affect the nature of phase transformation. Further, it can also be inferred that different thermomechanical treatments applied on a specimen are found to have opposing effects on the nature of phase transformations. In contrast to the heat treatments, which tend to increase the phase transformation temperatures, thermal cycling tends to decrease them.

7. Severe plastic deformation

The plastic deformations carried out by cold-working and hot-working presented above have been extended in the recent past, by subjecting these alloys to severe plastic deformation (SPD). It was shown that the effects of high density of grain boundaries on the martensitic phase transformation and the functional properties of SMA became a focus of research investigating the impact of ultrafine and nanograins on the parameters of the SME and SE. Further, methods of SPD, such as high pressure torsion (HPT) and equal channel angular pressing (ECAP) have been successfully applied to achieve ultrafine grained (UFG) and bulk nanostructured SMA **[42–45]**.

a. High pressure torsion (HPT)

Waitz et al. **[44]** showed that martensitic transformation shifts to low temperature when the grain size is less than 150 nm. Initially in their experiments, Ni–Ti alloy was subjected to HPT and later annealed close to recrystallization temperature. By post-deformation annealing at 300°C, it was found that the amorphous structure created by the room-temperature HPT loses its thermomechanical stability and intensively crystallizes **[45]**. The effect of the composition on the phase transformations in Ni–Ti alloys subjected to HPT and followed by heat treatments was recently reported **[46]**.

Bulk Ni-Ti SMA with different compositions have been chosen and subjected to HPT and their phase transformation characterization was carried out. The selected Ni(49.6 to 49.4at%)-Ti (Ti-rich) alloy in the as-received (AR) condition has M_f above RT and Ni(around 50.8at%)-Ti (Ni-rich) has A_f below RT. SPD of Ni-Ti alloys (Ti-rich and Ni-rich) have been performed by HPT at RT. Further, HPT processed separate specimens are subjected to heat treatments at temperatures of 300°C (HPT+HTT300) and 350 °C (HPT+HTT350) for 20 min, and quenched into water at room temperature. Phase transformation temperatures are analyzed by studying the Differential Scanning Calorimeter (DSC) plots. Further, the structural evolution of the samples subjected to SPD in the phase transformation temperature region was studied using in situ X-ray diffraction (XRD) from –180 to +180°C.

The phase transformation temperatures obtained from the thermogram plots of the corresponding sample conditions are presented in Fig. 10. In Fig. 10a, for the Ti-rich alloy in all the conditions, the transformation temperatures correspond to one-step M↔A phase transformation both while heating and cooling. While compared to the transformation temperatures of the AR sample, it is observed that, for the HPT sample, there is a slight decrease in M_f and A_s temperatures, whereas M_s and A_f temperatures increase. As a result, both while heating and cooling, there is a broadening of the temperature intervals in which the phase transformations take place. For the HPT sample after heat treatment at 300°C, designated as HPT+HTT300 in the plot, there is an increase in M_f and A_s temperatures, whereas M_s and A_f temperatures decrease. These results, both while heating and cooling, on narrowing of the temperature intervals where the phase transformations are taking place. After heat treatment at 350°C, designated as HPT+HTT350 in the plot, all the transformation temperatures increase and the phase transformation temperature intervals become narrower.

In Fig. 10b, for the Ni-rich alloy in the AR and HPT conditions, the transformation characteristics show a one-step M↔A phase transformation, both while heating and cooling. It is observed that for the HPT sample, the temperatures corresponding to both phase transformations are higher than those corresponding to the AR sample. However, both while heating and cooling, corresponding to M→A and A→M transformations, respectively, there is a narrowing of the transformation temperature intervals. For HPT+HTT300 sample, M_s decreases, A_s, and A_f increase considerably. M_f decreases to a value below the lower limit of the scanned temperature range. The dashed lines represent the trend of the variation of M_f. Further, R-phase transformations are present both while heating and cooling. On heat treatment at 350°C after the HPT processing, i.e., for Ni-rich HPT+HTT350, it is observed that all the transformation temperatures tend to increase.

AR samples and samples subjected to HPT of both alloys are scanned using XRD technique at different temperatures in the phase transformation temperature range. 3D view of the XRD profiles obtained while cooling and heating are presented in Fig. 11. Miller indices of the diffraction peaks emerging from the corresponding planes of the phases are marked on each peak. In Fig. 11a, for the Ti-rich Ni-Ti AR sample, it might be observed that the recording of the XRD pattern is started at 180°C, where austenite phase exists, followed by cooling and recording the spectra at different temperatures until the martensite transformation is complete, i.e., down to -40°C. Further, the sample is again heated to observe the transformation to austenite, i.e., up to 180°C to complete the thermal cycle. While cooling from 180°C to -40°C, the peak B2(1 1 0) corresponding to austenite (B2 – cubic structure) gradually disappears and peaks associated to martensite (B19′ – monoclinic structure) gradually grow. The diffraction pattern obtained at -40°C, shows the peaks corresponding to martensite. As the temperature is increased from -40 to 180°C, the peak corresponding to (1 1 0) of austenite (B2 – structure) gradually grows and the peaks corresponding to B19′ martensite gradually disappear. In Fig. 11b, for the Ti-rich Ni-Ti sample subjected to HPT, also M↔A phase transformation behavior is observed.

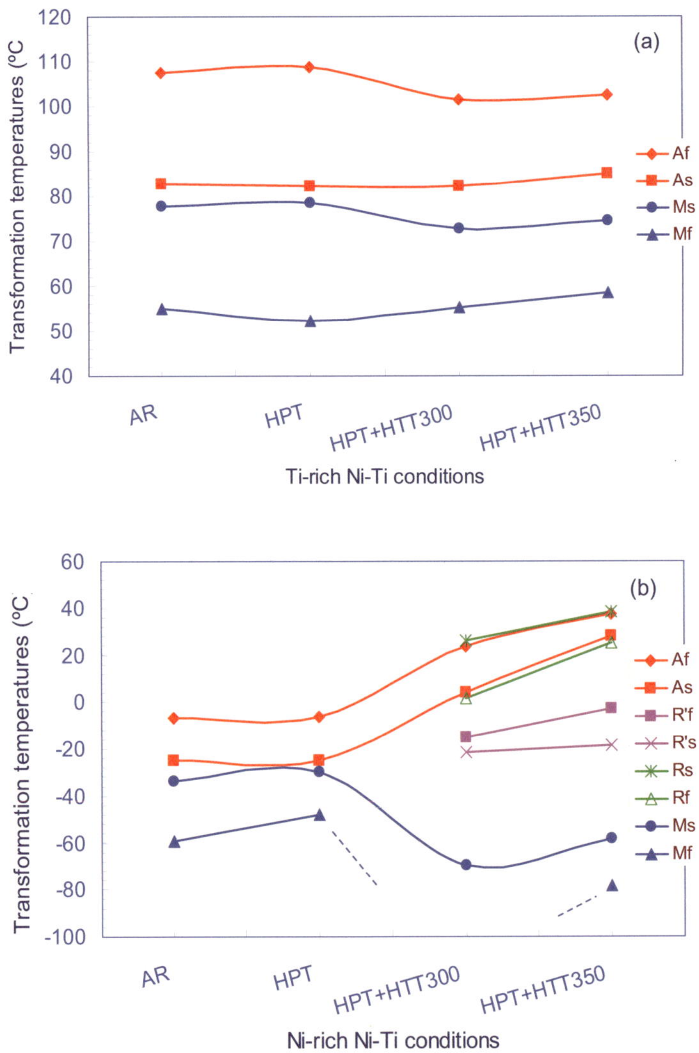

Figure 10. Phase transformation temperatures obtained from DSC plots of (a) Ti-rich and (b) Ni-rich Ni-Ti alloys in different conditions.

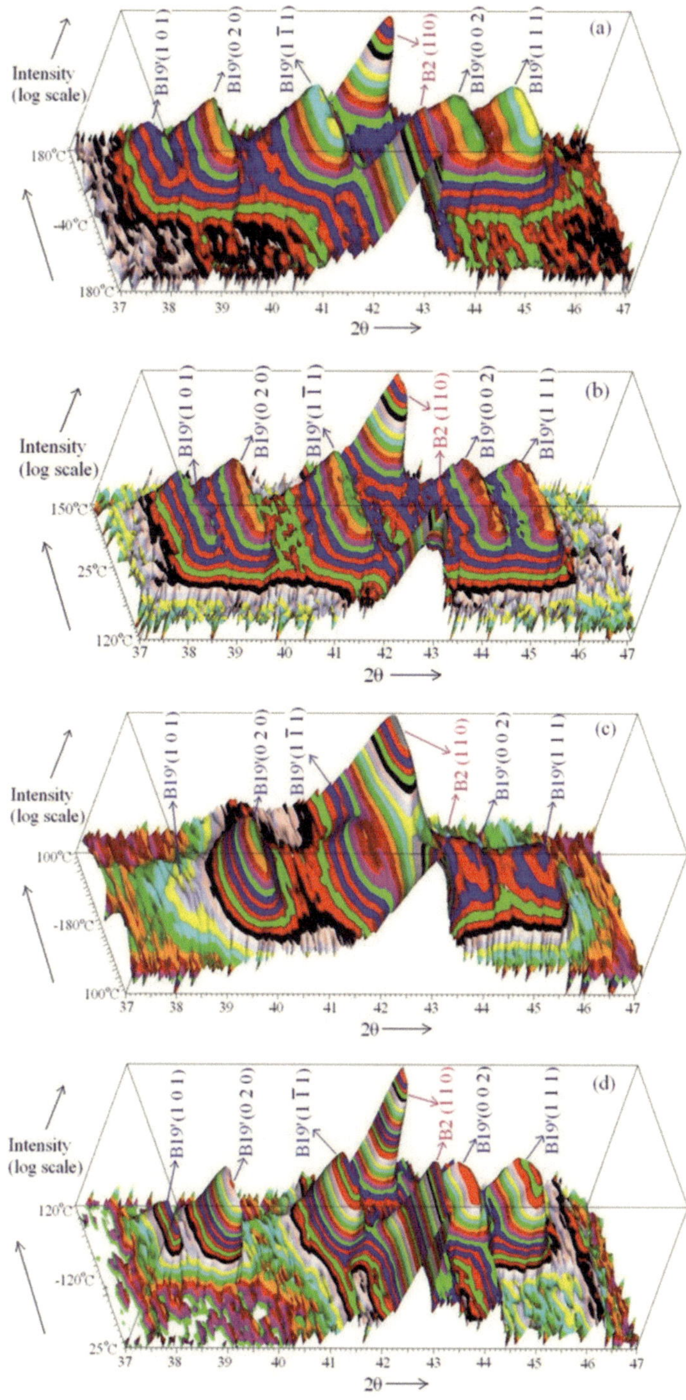

Figure 11. 3-D box layout of the XRD profiles obtained during cooling and heating for Ti-rich Ni-Ti alloy in (a) AR and (b) HPT conditions, and Ni-rich Ni-Ti alloy in (c) AR and (d) HPT conditions.

Fig. 11c shows the phase transformation behavior of Ni-rich Ni-Ti alloy in the AR condition. At 100°C, the sample is found to be in austenite (B2) phase. As the temperature is decreased down to -180°C, the intensity of the peak corresponding to B2(1 1 0) decreases. As the cooling progresses, the diffraction peaks corresponding to B19′ martensite appear. On heating, the peaks related to B19′ martensite disappear and the peak related to B2(1 1 0) appears again. Similar phase transformation behavior is observed for the Ni-rich sample after HPT (Fig. 11(d)). 3D layout of the XRD patterns obtained at selected temperatures during cooling, followed by heating for both Ti-rich and Ni-rich Ni-Ti alloys in HPT+HTT300C and HPT+HTT350C conditions were presented in a recent publication **[47]**. It is clearly observed that the diffraction peaks corresponding to intermediate R-phase are present for the Ti-rich and absent for the Ni-rich Ni-Ti alloys, both while cooling and heating. The result is in agreement with the transformation temperature profiles obtained by DSC thermogram analyses presented in the above Fig. 10.

The results show that for Ti-rich Ni-Ti alloy, after HPT, as well as following the heat treatments, there are no major changes in the phase transformation behavior. But, for Ni-rich Ni-Ti alloy, there is a slight change in the phase transformation behavior after HPT process, and the final heat treatments bring about very significant change, namely, the presence of intermediate R-phase transformation. In the present experiment, during the HPT process, a high speed of rotation of the piston (1,250 rpm) is involved. Initially, when the pressure torque is applied, a very intense and rapid plastic deformation takes place. This causes the specimen to get macroscopically distorted geometrical shape and eventually microscopic disorder. Owing to the process, the specimen gets heated up and might undergo a short duration annealing in the severely strained condition before cooling to room temperature. This situation may lead to accommodate several conflicting processes **[46]**. High speed of rotation during the HPT process might also trigger dynamic recrystallization. Depending on factors, such as the previous condition of the HPT specimen, strain accommodated, temperature attained, and magnitude of the time interval at which the specimen is at high temperature, different final microstructural states will be achieved in the specimen. On one hand, the intense deformation will distort the microstructure and long range order will be broken. On the other hand, the high temperature will have its influence on the recovery of the strains and formation of strain free submicrocrystals.

b. Equal channel angular pressing (ECAP) or Equal channel angular Extrusion (ECAE)

ECAP is an attractive processing technique for several reasons. Processing by ECAP can have a strong effect not only on the mechanical properties but also on the functional properties of materials **[48]**. However, for Ni-Ti SMA, it is difficult to apply ECAP at RT due to their low deformability and accordingly several reports have appeared describing the fabrication of ultrafine-grained alloys using ECAP at elevated temperatures **[49]**. The transformation behavior of TiNi alloy after ECAE process has been reported by **Zhenhua Li**

***et al.*, [50]** by using the experimental material, Ti-50.6at% Ni alloy rods, with a 25 mm diameter, after 850 °C hot forging and 500 °C annealing for 2 h. They concluded that during high temperature ECAE process, there was no dynamical re-crystallization but, most probably, there was dynamical recovery. Annealed for 5 min at 750 °C after two passes of ECAE, grains were refined and became even. After two passes of ECAE, transformation temperatures of the billet of TiNi alloy sharply decreased. Transformation temperature of the sample remarkably increased annealed for 2 h at 500 °C after two ECAE processes, similar to the one of TiNi alloy before ECAE process, which was related to Ni content in the matrix.

Effect of ECAP process on the microstructure and functional properties, such as recovery stress and maximum fully recoverable strain has been reported. The results show that the multipass ECAP of $Ni_{50.2}Ti_{49.8}$ alloy allows one to produce a uniform grain structure with predominantly high-angle grain boundaries with a grain size of about 200-300 nm. ECAP increases strength and insignificantly decreases plasticity as compared to the as-quenched state. The strength increases more than 50% with increasing number of passes; after ECAP using 12 passes. The functional properties of the $Ni_{50.2}Ti_{49.8}$ alloy after ECAP are substantially improved. With increasing number of ECAP passes the maximum recovery stress rises to 1100 MPa and the degree of maximum fully recoverable strain increases to 9.2% **[51]**.

8. Concluding remarks

Phase transformations can be studied by using various characterization techniques, such as DSC, ER, Internal Friction (IF), dilatometry, XRD, and optical/electron microscopy **[5, 14, 16-19, 41, 43, 52, 53]**. Each of these techniques senses different physical phenomena and thus provides information concerning the changes of various physical parameters taking place during the phase transformations. Because of their distinctive nature, when these techniques are employed individually, only partial information about the phase transformation can be delivered.

DSC measures only the sum of all thermal events and, as a result, some important features may be ignored or the results are easily misinterpreted in the cases involving weak and/or complex (overlapping) transformations [5, 16, 18, 19]. ER is the structural sensitive property of a material and it reveals changes during crystallographic phase transformations. In fact, it is found to be more sensitive than DSC in detecting the phase transformations which occur in a narrow temperature range [19, 41]. Dilatometry is capable of sensing small volume changes during phase transformations. Only a limited number of publications report the use of dilatometry to study the phase transformations in Ni-Ti shape memory alloys [17, 19]. These methods have been widely accepted to detect the phase transformations in Ni-Ti SMAs. A combined approach of several characterization techniques would lead to the proper understanding of the phase transformations involved.

Author details

F.M. Braz Fernandes* and K.K. Mahesh
CENIMAT/I3N, Departamento de Ciências dos Materiais, FCT/UNL, 2829-516 Caparica, Portugal

Andersan dos Santos Paula
Post-graduated Program in Metallurgical Engineering, UFF - Universidade Federal Fluminense, Volta Redonda, Brazil

Acknowledgement

The pluriannual financial support (by Fundação para a Ciência e a Tecnologia – Ministério da Educação e Ciência) of CENIMAT/I3N through the Strategic Project - LA 25 - 2011-2012 and the research project Smart Composites (PTDC/CTM/66380/2006) is gratefully acknowledged by KKM and FMBF. KKM gratefully acknowledges the fellowship under the scheme, *'Ciência 2007'* with Ref. No. C2007-443-CENIMAT-6/Ciência2007.

9. References

[1] Effect of Thermal Cycling on the Transformation Temperatures of TiNi Alloys. S. Miyazaki, Y. Igo, K. Otsuka. Acta Metall. 34 (1986) 2045–2051.
[2] Factors Affecting the M_s Temperature and its Control in Shape-Memory Alloys, K. Otsuka, X. Ren. Mater. Sci. Forum 394–395 (2002) 177–184.
[3] Lattice Transformations Related to Unique Mechanical Effects. J. Perkins. Metall. Trans. 4 (1973) 2709–2721.
[4] Effect of Thermal Cycling on the R-phase and Martensitic Transformations in a Ti-Rich NiTi Alloy. V. Pelosin, A. Riviere, Metall. Mater. Trans. A 29 (1998) 1175–1180.
[5] Effect of Thermal Cycling on R-phase Stability in a NiTi Shape Memory Alloy, J. Uchil, K.G. Kumura, K.K. Mahesh, Mater. Sci. Eng. A 332 (2002) 25–28.
[6] Effect of Deformation and Thermal Treatment of NiTi Alloy on Transition Sequence. Morawiec, D. Stroz, D. Chrobak. J. Phys. IV 5 (1995) 205–209.
[7] Effect of Heat Treatment After Cold Working on the Phase Transformation in TiNi Alloy. T. Todoroki, H. Tamura. Trans. J. Inst. Met. 28 (1987) 83–94.
[8] A study on the wire drawing of TiNi shape memory alloys. S.K. Wu, H.C. Lin, Y.C. Yen. Materials Science and Engineering A 215 (1996) 113–119.
[9] The effects of cold rolling on the martensitic transformation of an equiatomic TiNi alloy. H.C. Lin, S.W. Wu, T.S. Chou, H.P. Kao. Acta Metall. Mater. 39 (1991) 2069 - 2080.
[10] Deformation of Martensite in a Polycrystalline Cu-Zn-Al Alloy. K. Adachi, J. Perkins. Metall. Trans. 17A (1986) 945-959.

* Corresponding Author

[11] FEM Simulation of the Martensitic Transformation in NiTi Alloys. S. Zhang, H. Braasch, P.G. McCormick. J Phys IV France 7 C5 (1997) 537-542.

[12] In-situ High Temperature Texture Characterisation in NiTi Shape Memory Alloy Using Synchrotron Radiation. A.S. Paula, K.K. Mahesh, F.M. Braz Fernandes, R.M.S. Martins, A.M.A. Cardoso, N. Schell. Materials Science Forum 495-497 (2005) 125-130.

[13] Textural Modifications during Recovery in Ti-Rich Ni-Ti Shape Memory Alloy Subjected to Low Level of Cold Work Reduction. A.S. Paula, K.K. Mahesh, N. Schell, F.M. Braz Fernandes, Materials Science Forum 636-637 (2010) 618-623.

[14] Study of the textural evolution in Ti-rich NiTi using synchrotron Radiation. A.S. Paula, J.H.P.G Canejo, K.K. Mahesh, R.J.C. Silva, F.M. Braz Fernandes, R.M.S. Martins, A.M.A. Cardoso, N. Schell. Nuclear Instruments and Methods in Physics Research Section B: Beam Interactions with Materials and Atoms 246 (2006) 206–210.

[15] Restoration Phenomena of Neutron-irradiated Ti-Ni Shape Memory Alloys. T. Hoshiya, F. Takada, Y. Ichihashi. Materials Science and Engineering A 130 (1990) 185-191.

[16] Effect of Heat Treatment after Cold Working on the Phase Transformation in TiNi Alloy, Tsunehiko Todoraki, Hirokazu Tamura, Transactions of the Japan Institute of Metals 28(2) (1987) 83-94.

[17] Thermal expansion in various phases of nitinol using TMA. J.Uchil, K.P.Mohanchandra, K.Ganesh Kumara, K.K.Mahesh, T.P.Murali. Physica B (UK) 270 (1999) 289-297.

[18] Thermomechanical behaviour of Ti-rich NiTi shape memory alloys. A.S. Paula, K.K. Mahesh, C.M.L. dos Santos, F.M. Braz Fernandes, C.S. da Costa Viana. Materials Science and Engineering A (UK) 481-482, (2008) 146-150.

[19] One- and Two-step Phase Transformation in Ti-rich NiTi Shape Memory Alloy. A.S. Paula, K.K. Mahesh, C.M.L. dos Santos, J.P.H.G. Canejo, F.M. Braz Fernandes. International Journal of Applied Electromagnetics and Mechanics (Netherlands) 23 (2006) 25-32.

[20] Microstructural evolution kinetics after plastic deformation of equiatomic Ti–Ni alloy during isothermal annealings. F. Khelfaoui, G. Thollet, G. Guenin. Mat. Sci. and Eng. A, 338 (2002) 305-312.

[21] Influence of Work Hardening and Heat Treatment on the Substructure and Deformation Behaviour of TiNi Shape Memory Alloys. P. Filip, K. Mazanec. Scr. Met. et Mat. 32 (1995) 1375-1380.

[22] On the Ti_2Ni precipitates and Guinier–Preston zones in Ti-rich Ti–Ni thin films. J.X. Zhang, M. Sato, A. Ishida. Acta Mater. 51 (2003) 3121-3130.

[23] High-resolution electron microscopy studies on coherent plate precipitates and nanocrystals formed by low-temperature heat treatments of amorphous Ti-rich Ti-Ni thin films. T. Kikuchi, K. Ogawa, S. Kajiwara, T. Matsunaga, S. Miyazaki, Y. Tomota. Philos. Mag. A 78 (1998) 467-489.

[24] Tratamentos Termomecanicos de Ligas do Sistema Ni-Ti. A.S. Paula, Ph.D. Thesis, FCT/UNL, Lisbon – Portugal, 2006.

[25] Experimental test for a possible isothermal martensitic transformation in a Ti–Ni alloy. K. Otsuka, X. Ren, T. Takeda, Scripta Materialia (45)2 (2001) 145–152.

[26] Effects of Heat Treatment on Mechanical Behaviour of Ti-Ni Alloys. T. Saburi, T. Tatsumi, S. Nenno. Journal de Physique Colloques 43(C4) (December 1982) Proceedings of the International Conference on Martensitic Transformations (ICOMAT-82) 261-266.

[27] Neutron diffraction phase analysis during thermal cycling of a Ni-rich NiTi shape memory alloy using the Rietveld method. H. Sitepu, W.W. Schmahl, J. Khalil Allafi, G. Eggeler, A. Dlouhy, D.M. Tobbens, M. Tovar. Scripta Materialia, 46 (2002) 543-548.

[28] Effect of cooling process during heat treatment on martensitic transformation in Ni-Ti and Ni-Ti-Cr alloys. J. Uchil, K. Ganesh Kumara, K.K. Mahesh. Proc. International Conference on Martensitic Transformations (ICOMAT'02) Helsinkin University of Technology, Espoo, Dipoli, Finland, June 10-14, 2002, Edited by J. Pietikainen, O. Soderberg, J de Physique - IV, France 112 (2003) 747-750.

[29] Some aspects of the properties of NiTi shape memory alloy. Y. Liu, J. Van Humbeeck, R. Stalmans, L. Delaey. Journal of Alloys and Compounds 247(1997) 115–121.

[30] Precipitation Processes in Near-Equiatomic TiNi Shape Memory Alloys. M. Nishida, C.M. Wayman, T. Honma. Metallurgical Transactions A 17 (1986) 1505-1515.

[31] Effects of hot rolling on the martensitic transformation of an Equiatomic Ti-Ni alloy. H.C. Lin, S.K. We, Mater. Sci. Eng. A158 (1992) 87-91.

[32] Effect of microstructure on the fatigue of hot-rolled and cold-drawn NiTi shape memory alloys. K. Gall, J. Tyber, G. Wilkesanders, S. W. Robertson, R. O. Ritchie, H. J. Maier. Materials Science and Engineering A 486 (2008) 389–403.

[33] Microstructure and Thermo-Mechanical properties of NiTi shape memory alloys. E. Hornbogen. Mat. Sci. Forum 455-456 (2004) 335-341.

[34] Appearance of an Intermediate Phase with Thermal Cycling on the Transformation of NiTi. H. Matsumoto. J. Mat. Sci. Letts. 10 (1991) 408-410.

[35] Transformation Behavior of NiTi in Relation to Thermal Cycling and Deformation. H. Matsumoto. Physica B190 (1993) 115-120.

[36] P. Filip, K. Mazanec. Influence of Cycling on the Reversible Martensitic-Transformation and Shape-Memory Phenomena in TiNi Alloys. Scripta Metallurgica et Materialia 30 (1994) 67-72.

[37] Kwarciak J., Lekston Z., Morawiec H., Effect of Thermal Cycling and Ti_2Ni Precipitation on the Stability of the Ni-Ti Alloys. J. Mat. Sci. 22 (1987) 2341-2345.

[38] G. Airoldi, B. Rivolta, C. Turco, Heats of Transformations as a Function of Thermal Cycling in NiTi Alloys, in Proc. Int. Conf. of Martensitic Transformations, JIM Ed., 1986, 691-696.

[39] D.A. Hebda, S. R. White. Effect of training conditions and extended thermal cycling on nitinol two-way shape memory behaviour. Smart Mater. Struct. 4 (1995) 298-304.

[40] C.J. de Araujo, M. Morin, G. Guenin. Estimation of internal stresses in shape memory wires during thermal cycling under constant load: A macromechanical approach. J. Int. Mat. Sys. & Structs. 11 (2000) 516-524.

[41] A.S. Paula, K.K. Mahesh, F.M. Braz Fernandes, Stability in Phase Transformation after Multiple Steps of Marforming in Ti-rich Ni-Ti Shape Memory Alloy. Journal of Materials Engineering and Performance, 20 (2011) 771-775.

[42] M. Zehetbauer, R. Grossinger, H. Krenn, M. Krystian, R. Pippan, P. Rogl, T. Waitz, R. Wurschum. Bulk Nanostructured Functional Materials By Severe Plastic Deformation. Adv Eng Mater 12 (2010) 692-700.

[43] K.K. Mahesh, F.M. Braz Fernandes, R.J.C. Silva, G. Gurau. Phase transformation and structural study on the SPD NiTi alloys. Physics Procedia, 10 (2010) 22-27.

[44] T. Waitz, V. Kazykhanov, H.P. Karnthaler. Martensitic phase transformations in nanocrystalline NiTi studied by TEM. Acta Mat 52 (2004) 137-147.

[45] Alloy composition, deformation temperature, pressure and post-deformation annealing effects in severely deformed Ti-Ni based shape memory alloys. S.D. Prokoshkin, i.Yu. Khmelevskaya, S.V. Dobatkin, I.B. Trubitsyna, E.V. Tatyanin,V.V. Stolyarov, E.A. Prokofiev Acta Mater 53 (2005) 2703-2714.

[46] K.K. Mahesh, F.M. Braz Fernandes, G. Gurau. Stability of thermal induced phase transformations in the severely deformed equiatomic Ni-Ti alloys. J Mater Sci 47(2012) 6005–6014.

[47] F.M. Braz Fernandes, K.K. Mahesh, R.J.C. Silva, C. Gurau, G. Gurau. XRD study of the transformation characteristics of severely plastic deformed Ni-Ti SMAs. Phys. Status Solidi C7 (2010) 1348-1350.

[48] R.Z. Valiev, T.G. Langdon. Principles of equal-channel angular pressing as a processing tool for grain refinement. Prog Mater Sci 51 (2006) 881–981.

[49] V.G. Pushin, R.Z. Valiev The Nanostructured TiNi Shape-Memory Alloys: New Properties and Applications, in Proceedings of the Symposium on Interfacial Effects in Nanostructured Materials, Warsaw, Poland, 14-18 September 2002. Solid State. Phenom., 94, (2003) 13–24.

[50] Z. Li, G. Xiang, X. Cheng. Effects of ECAE process on microstructure and transformation behaviour of TiNi shape memory alloy. Materials and Design 27 (2006) 324–328.

[51] E. Prokofyeva, D. Gunderov, S. Prokoshkin, R. Valiev. Microstructure, mechanical and functional properties of NiTi alloys processed by ECAP technique, in Proceedings ESOMAT 2009 - 8TH European Symposium on Martensitic Transformations, Article nr 06028, 5 pgs.

[52] D. Stroz, Z. Bojarski, J. Ilczuk, Z. Lekston, H. Morawiec. Effect of thermal cycling on as-quenched and aged nickel-rich Ni-Ti alloy. Journal of Materials Science 26 (1991) 1741-1748.

[53] M. Piao, K. Otsuka, S. Miyazaki, H. Horikawa. Mechanism of the A_s temperature increase by pre-deformation in thermoelastic alloys. Mater. Trans JIM 34 (1993) 919–929.

Chapter 2

The Methods of Preparation of Ti-Ni-X Alloys and Their Forming

Radim Kocich, Ivo Szurman and Miroslav Kursa

Additional information is available at the end of the chapter

http://dx.doi.org/10.5772/50067

1. Introduction

The continuous development of science and technology in all industrial sectors means connecting and usage of a wide range of new knowledge together with implementation of new modern technologies for production of materials with high functional, specific and special properties. Intermetallic compounds TiNi with shape-memory effect are an interesting group of materials. These materials are used in a wide range of industry, such as electronics, robotics, tele-communication and also in medicine and optics. Shape-memory alloys (SMA) are a group of materials characterized by shape-memory effect (SME) and superelasticity (SE), also called pseudoelasticity.

Ti-Ni binary alloys (sometimes called "Nitinol") are probably the best known from this group of materials. Nevertheless, these alloys are not always the most suitable for the particular purpose. This factor is also the reason for seeking optimized variants of these original binary systems. One of the possible solutions is to modify binary alloys by incorporation of one or more chemical elements into the production process. The resulting materials can be summed up in the term Ti-Ni-(X), where X means presence of another element. Although the best memory characteristics are usually achieved for alloys with Ni content of 49.3 ÷ 51 at. % (Raz & Sadrnezhaad, 2004), by decreasing the content of one element (Ti or Ni) to the prejudice of the third element, modified materials are obtained, while preserving some of original characteristics. Among the main characteristics, surpassing SME and SE, mechanical properties, corrosion resistance and related biocompatibility should be mentioned (Van Humbeeck, 2001) or (Duerig et al., 1999). Intermetallic equiatomic compound of nickel and titanium thus remains as the base of modified binary materials. Nevertheless, it should be stated that characteristics of Ti-Ni SMA may be significantly modified otherwise than by the appropriate choice of chemical composition, namely by forming and thermal (thermomechanical) processing. As will be indicated later, final properties and products made of SMA are significantly influenced not only by the chosen forming technique, but also their mutual sequence. These factors together with the used technique play a major role in the manufacture of products from SMA.

2. Method of preparation

Production of Ni-Ti alloys is mostly done by vacuum melting, whilst various melting procedures are used (electron beam melting, arc melting (Ma & Wu, 2000) and (Meng, 2001), high frequency vacuum melting in a graphite crucible (Noh, 2001) or (Tsai et al., 1994), plasma melting, etc.). When Ni-Ti alloys are melted, there can be unfavourable effects, especially of gases such as nitrogen or oxygen. Other problems consist in the conditions suitable for crystallization and minimalization of micro- and macro-segregation connected with that. Also, contamination of the material by non-metallic intrusions has to be prevented (Schetky & Wu, 2005). Due to the formation of titanium carbide and titanium oxide in Ni-Ti, concentration of individual elements changes and thus so does the transformation temperature. Among other problems arising from the melting of Ni-Ti, there is the formation of low-melting point phase $NiTi_2$, which causes a strong tendency towards hot crack formation.

The basic requirement to metallurgy of these alloys is strict adherence to the chemical composition of the alloy, which is the main condition for obtaining the alloy with the required transformation behaviour. Another condition is obtaining an excellent microstructural homogeneity of the alloy, which is also a condition for functional reliability and guaranteed transformation behaviour. A deviation of about 0.1 at. % from the required chemical composition usually changes the transformation temperature by as much as 10 K. In Fig. 1a you can see the dependence of temperature of martensitic transformation on the nickel content in the alloy. There is a possibility of attenuation of concentration dependence of the martensitic transformation temperature by alloying with other elements, especially Cu, Fe, etc.

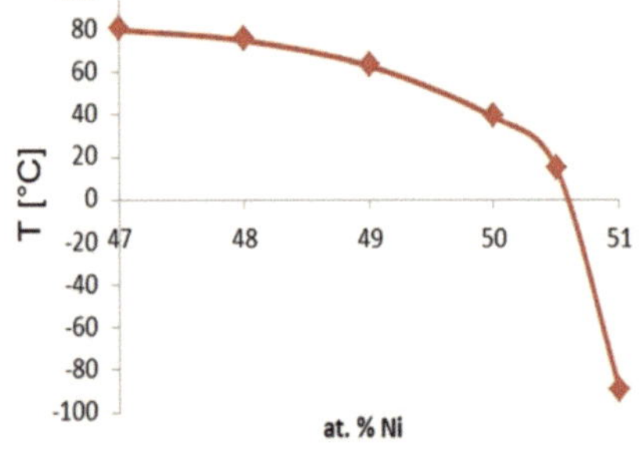

(a)

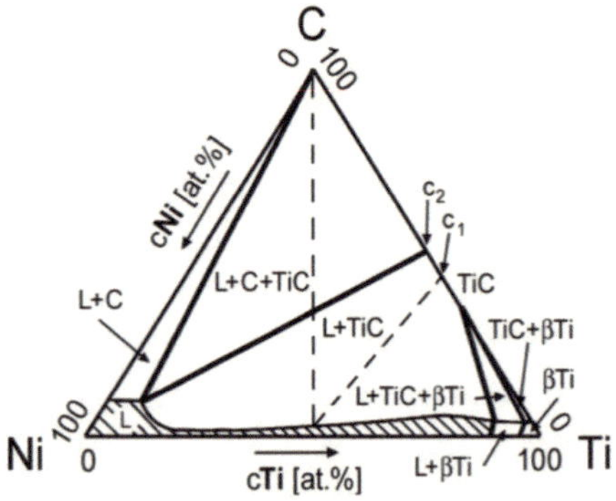

(b)

Figure 1. The dependence of temperature of martensitic transformation (a) Ternary system Ti-Ni-C (b)

Based on specific requirements of applications such as actuators/sensors, temperature control, fatigue properties, etc., various alloys with the addition of a third element giving a ternary alloy were developed (Otsuka & Wayman, 1998) or (Zhang et al. 2006).

There is a certain influence of each alloying element on transformation characteristics of the alloy. For example, the addition of Hf, Zr, Au, Pd and Pt causes the increase of phase transformation temperatures, while elements such as Fe, Co, and V have the opposite effect. Similarly, hysteresis is increased, e.g., by Fe and Nb, and, on the contrary, decreased by Cu (Ramajan et al., 2005). As a consequence of alloying by other elements, the transformation sequence is also changed; e.g., at the content of Cu below 7.7 % one-stage phase transformation B2→B19′ occurs (similarly as in a binary alloy). If the content of Cu exceeds 7.7%, two-steps transformation B2→B19→B19′ takes phase (Tang et al., 2000). The alloy properties may also be significantly influenced by alloy impurities from the production process, forming, heat treatment, etc. As it was already stated, there could be an important role of gases (O_2, N_2, H_2) and carbon. In the resulting structure intrusions of the type $Ti_4Ni_2O_x$, TiO_2 etc. connected with the decrease of Ti content in the matrix can be observed. There is significant influence of these composition changes on transformation characteristics of the alloy.

Typical superelastic nitinols contain ca. 350–500 ppm of oxygen and 100÷500 ppm of carbon. The metallurgical purity (grain structure, presence of impurities etc.), of course, greatly depends on the preparation process. Ni-Ti alloys can be called high-purity alloys if they contain <100 ppm of oxygen and <20 ppm of carbon. These alloys are prepared in vacuum induction furnaces in graphite crucibles with the subsequent repeated re-melting in vacuum arc furnaces (Graham et al., 2004).

2.1. VIM – Vacuum induction melting

As has already been stated, VIM is one of the production processes used for the preparation of TiNi alloys. The technology of vacuum induction melting in graphite crucibles represents the existing key preparation method. Chemical homogeneity within this technology can be achieved by appropriate power control (and stirring of liquid alloy connected with that). When using this technology, the quality of the prepared alloy will strongly depend on the material of the crucible. Usually the mentioned graphite crucible is recommended – where the oxygen content can be neglected; nevertheless, carbon absorption must be considered here (there is a significant influence of carbon on microstructural characteristics and transformation behaviour). During the preparation of the material in a graphite crucible it was also found (Frenzel et al., 2004) that in the case of using Ni-pellets and Ti bars/disks the appropriate arrangement of the material in the crucible was important. The authors of this study have shown that although the inner surface of the crucible was covered with Ti disks, the content of carbon in the produced alloy was lower in comparison with the case of random arrangement of the charge. This phenomenon is caused by formation of a TiC layer, which acts as a diffusion barrier. It was also found that the carbon content strongly depends on temperature and time of dwell of the melt in the crucible. For this reason, a more intensive investigation of these effects was carried out (Zhang et al., 2006). It was established that with increasing time of dwell of the melt in the crucible the melt gets enriched in carbon.

In Fig. 1b (Du & Schuster, 1998) it is possible to see more detailed information on the isothermal section (at 1500°C-temperature recommended for melting of Ni-Ti based alloys) of the Ni-Ti-C ternary system. The composition in this system is given in atomic %. It is shown that there exists a single-phase region of liquids, extends from the area of pure Ni to the area of pure Ti. There exists only a narrow two-phase area L+β-Ti which separates the area of melted material from the β-Ti phase. The diagram also shows that the melted material dissolves a certain amount of carbon (this dissolution is limited). Elementary melted Ti and C cannot coexist in equilibrium state, due to this reason a TiC carbidic phase is created. The diagram in Figure 1b also predicts the existence of three phases in thermodynamic equilibrium: pure carbon, TiC carbidic phase and melted Ni-Ti depleted by Ti. The reactions between the melt will result in a melted material with higher carbon content and certain amount of TiC. In practice we cannot expect this equilibrium.

When the molten Ni-Ti enters into contact with the graphite of the crucible, inter-diffusion causes a growth of the TiC layer and the contents of carbon in the melted alloy grows. This process includes the diffusion of carbon through a thin layer of TiC which grows on the boundary between TiC/melted Ni-Ti. On the boundary between graphite / TiC and the boundary of TiC / melted material we expect local thermodynamic equilibriums. If using a pure (unused, new) crucible for preparing the alloy, the first prepared ingot will have a higher content of carbon than the next one. This fact is in accordance with the creation of the above-listed TiC diffusion barrier. It is also recommended to perform rinse-melting before melting alloys in an unused crucible.

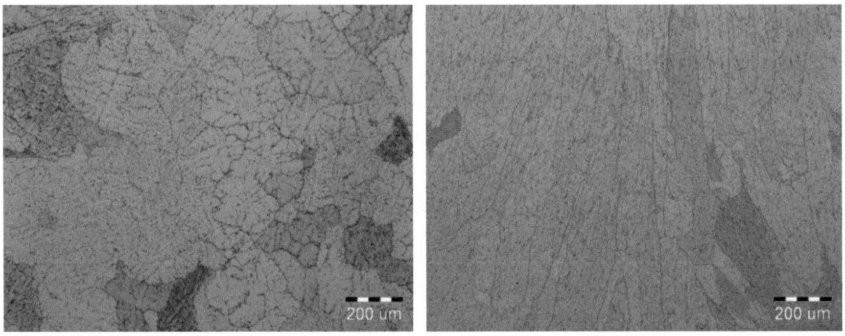

(a) (b)

Figure 2. As cast state of alloy: Ni50.6-Ti(at.%) (a), Ni46-Ti50-Co4 (b)

In order to define the exact influence of the used technique, an experimental study (Szurman & Kursa, 2010) with the aim of monitoring the influence of the preparation process on microstructural characteristics of Ni-Ti-(X) alloys was performed. The examples of microstructures of Ni50.6-Ti (at. %) and Ni46-Ti50-Co4 cast alloys are presented in Figs. 2a and 2b. As a consequence of the preparation of alloys in a graphite crucible, TiC type carbide phases are visible in the alloys' microstructure. A TEM image of the TiC phase (Fig. 3a) with the appropriate diffraction is presented in Fig. 3b. Similarly as with carbides, oxide phases can also be seen in microstructures of Ni-Ti alloys. A specific example is presented in Figs. 4a, b where particles of Ti_4Ni_2O can be seen.

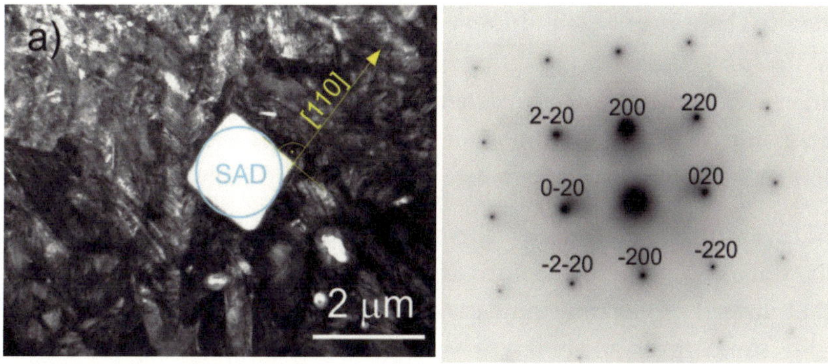

(a) (b)

Figure 3. TiC phase: TEM image (a), corresponding diffraction pattern (b)

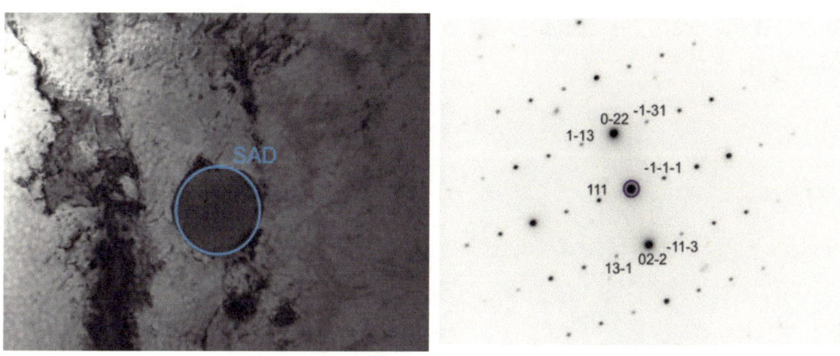

(a) (b)

Figure 4. Particle of Ti_4Ni_2O (a), corresponding diffraction pattern (b)

2.2. Plasma melting – Plasma furnace with horizontal crystallizer

This is another possible preparation process; there are, however, serious drawbacks. During this process, input elemental metals are placed in the copper water-cooled crystallizer. The crystallizer is carried by the screw below the plasma burner. Argon is used as a plasma-forming gas. For the melting as such it is necessary to use the cleanest available argon due to high affinity of titanium to oxygen. The plasma temperature during this process reaches 6500 K (Dembovský, 1985) and (Pacholek et al., 2003). The advantage of this process can be seen in the prevention of contamination of melted material by graphite from used electrodes (crucibles); high concentration of energy, high plasma flow velocity and very quick heat transfer on the heated material ensure high speed of melting. Disadvantages of plasma furnaces in comparison with vacuum induction furnaces include lower degassing of the melted metal, which depends on purity of the used argon. The key disadvantage of this process consists in insufficient homogeneity of the prepared alloy.

The development of plasma furnaces takes place in two main directions. Melting units working on the similar principle as common arc furnaces can be added to the first type of plasma furnaces. There is only one difference – that instead of electrodes, plasma burners are used and the furnace used to be equipped with special soil electrode carrying the current into the charge. The working space of furnaces is often designed to be vacuum-tight, which enables maintaining an ideal inert atmosphere. This type of furnace can be equipped with a relatively simple device for electromagnetic stirring of liquid metal.

The second furnace type is plasma furnaces with water-cooled metal crystallizers. As to the arrangement, the concept of these furnaces is similar to electronic furnaces, with the difference that instead of electron guns plasma burners are used and the furnaces mostly work with the pressure of an inert gas varying around 10^5 Pa. Exceptionally, there are furnaces with overpressure. In metallurgy, so-called low-temperature plasma in particular is considered, which is a system comprising a mixture of neutral particles with the prevailing number of electrons and positive ions with temperatures in orders of 10^3 to 10^4 K. The temperature of 10^5 K can be considered as the temperature of totally ionized plasma (Dembovský, 1978).

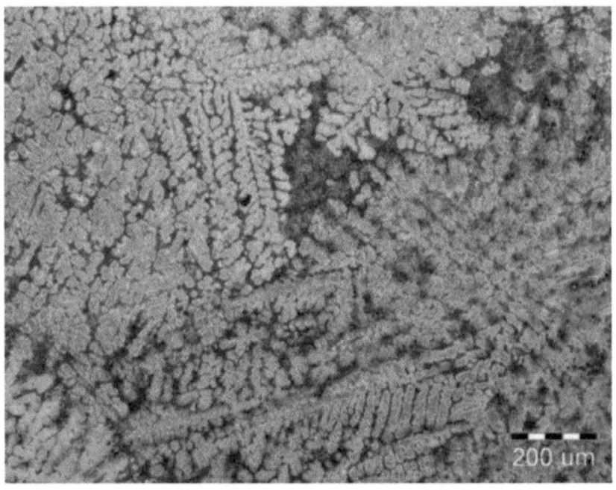

Figure 5. Microstructure of alloy Ni49.5-Ti25.5-Zr10-Nb15 (at. %), plasma

Specific experiments with melting of selected alloys Ti-Ni-(X) are described, e.g., in studies (Szurman & Kursa, 2009). Using this technique, ingots with the weight of 200–1000 g were prepared. In Fig. 5 you can see microstructure of alloy after plasma melting. As you can see, the microstructure of the alloy is highly inhomogeneous. This problem is caused by very high temperature gradients during melting. At the top of the ingot the alloy is heated to a high temperature. On the other hand, the part of the ingot which is in contact with the crystallizer is intensively cooled.

2.3. VAR – Vacuum Arc Melting

VAR technology is widely used to increase metallurgical purity of alloys prepared using standard procedures, e.g., in vacuum induction furnaces. This procedure is also known as "consumable electrodes". Direct current is used for the formation of arc between the electrode (melted material) and a water-cooled copper crucible/crystallizer. As a consequence, the electrode tip is melted and a new ingot is formed within the water-cooled crucible. The melt during the arc vacuum melting therefore is not in contact with the graphite crucible (as it is in the case of the VIM technique with a graphite crucible/mould), thus a "more pure" product can be obtained using this method. The carbon content usually does not exceed 200 ppm (Dautovich & Purdy, 1965) or (Wu, 2001). For this technique a very high vacuum is required.

Nevertheless, there are also drawbacks of this technique – small volume of the alloy and low convection in the melt which may cause inhomogeneity of ingots. That is also the reason why this procedure is usually repeated several times. Often the VIM+VAR process is applied for the preparation of Ti-Ni-(X) alloys. VAR technology is also preferred for preparation of experimental material for basic research of Ti-Ni-(X) alloys. For example, the study (Choi et al., 2005) describes the preparation of experimental Ni-Ti alloys alloyed by Fe. In this study, the ingot prepared using this method is homogenized at 1273 K for 24 hours. In another study (Sakuma et al., 2003) heat treatment at 1223 K for 1 hour after the preparation of the material using this method is proposed. The specific regime is also mentioned in the study (Jung et al., 2003) where heating at 1100°C for 100 h was used.

2.4. EBM – Electron Beam Melting (Vertical Zone Melting)

Crucible-free zone melting – or the floating zone process known as the FZ method represents another specific preparation method. The formation of a narrow melted zone is performed using electron heating. The melting takes place in a vacuum, and values of 10^{-2} Pa are reached in this technology compared to 10 Pa in VIM technology. Using the method of electron zone melting with suitable oriented nuclei, even monocrystals of many high-melting metals – W, Mo, Ta, Nb, V, Zr, Ti, Re – can be prepared. The zone is maintained in the floating condition mainly by forces of surface tension. The zone stability depends on gravitation, surface tension and density of the melt, on material composition and also on direction of zone movement. To maintain the stability of the zone, an outer magnetic field with so-called supporting frequency is used (Kuchař & Drápala, 2000). A circular shape ingot prepared in a vacuum induction furnace is used in this case as input material. There is no risk of other contamination of the material with carbon in this technology (there is no crucible). Carbon contents are usually lower than in the case of the alloy preparation using VIM technology. Contents of gases are usually low as well because of using a high vacuum. The contamination with gases thus depends only on the quality of an input casting and tightness of the vacuum system. The disadvantage of this technology is the control of chemical composition – evaporation of some elements can be expected here during melting. Another drawback is the rather small volume of the prepared material; therefore this technology is not suitable for commercial use (Ramaiah et al., 2005).

Rather integrated results regarding (un)suitability of various methods of preparation are presented, e.g., in a recent study (Szurman & Kursa, 2009). The aim was an intercomparison of techniques selected for the preparation of Ti-Ni-(X) alloys from the point of view of the microstructure and gas contents in the material. Also in this situation a distinct decrease of gas contents after the preparation was observed here. The specific microstructure of the prepared alloy Ni50,6-Ti (at.%) is presented in Fig. 6.

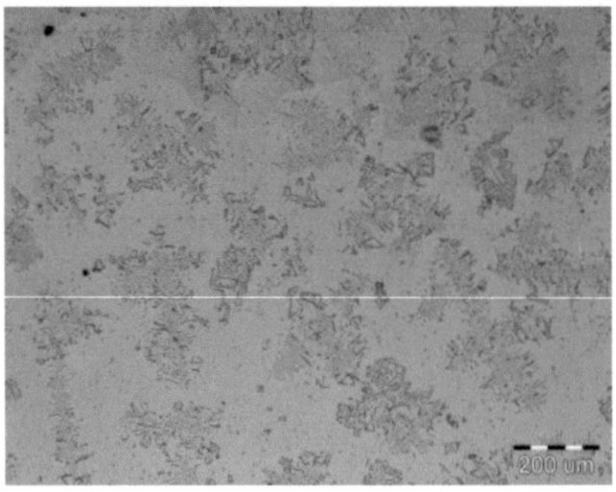

Figure 6. Microstructure of alloy Ni50.6-Ti (at. %), EBM

2.5. Preparation of alloys using powder metallurgy

Powder metallurgy is an important and suitable method for the production of the mentioned alloys. Methods of atomization were developed for preparation of powder metals with precise control of composition. However, the biggest problem with these alloys is oxygen and carbon content. The content of oxygen can be up to 3000 ppm, but it can be decreased by careful treatment to 1500 ppm (Schetky & Wu, 2005). It is well known that with increasing content of impurities (especially oxygen and carbon), transformation temperatures are decreased and a brittle secondary phase is formed (Mentz et al., 2008). Subsequently, the composition of NiTi matrix (depletion by Ti) is significantly influenced by oxide and carbide intrusions and thus can cause degradation of functional and mechanical properties, which was also confirmed, e.g., in the study (Mentz et al., 2006). Other methods are hydridation, pulverization and mechanical alloying (Wu, 2002).

The method for preparation of Ni-Ti alloy using powder metallurgy is described in the study (Mentz et al., 2008). At first, the alloy was prepared by authors using a classical VIM melting from high-purity input raw materials, then atomization with 6N argon followed. The obtained powders were sealed into evacuated capsules made from stainless steel and then compacted using the HIP method. It was also found that each technological step is accompanied by increased content of impurities (oxygen and nitrogen). In the study (Bertheville & Bidaux, 2005) the authors performed the preparation of Ti-rich Ni-Ti alloy from elementary powders – Ni and TiH_2. In another study (Zhu et al., 2005), preparation of Ni-Ti alloy by sintering in argon from elementary Ni and Ti powders is described. Another method of preparation is then described in the study (Mousavi et al., 2008), where a method of mechanical alloying is described. The Ni-Ti alloy was prepared from elementary powders in a planetary ball mill under atmosphere of Ar. During sintering of Ni and Ti, a significantly exothermic reaction takes place so that the heat generated during the process is used for the formation of the intermetallic compound TiNi.

Elemental powders can also be sintered using "combustion" synthesis or explosion. In the first case, a laser can be used as an external energy source (Bertolino et al., 2003). As for explosive sintering, the reaction takes place after explosion during the temperature rise. Another method is based on the passage of an electric current of suitable value under optimal voltage. It was found that the optimal current density is 2822–5290 kA.m^{-2}. The observed sintering times were within the range 5÷40 min (Locci et al., 2003). Ni-Ti materials prepared from powder metals are very porous and contain other intermetallic compounds such as Ti_2Ni and Ni_3Ti.

3. Forming of SMA

Apart from the already mentioned influence of the selected preparation method of SMA, final properties and behaviour of SMA will be also determined by next processing including heat treatment. This means not only the chosen method of forming, but also a sequence of given forming operations. Also the influence of applied regimes of heat treatment should be considered. All the mentioned factors have their own partial effect in formation of final properties of SMA-based materials. With admixture elements, such as Cu contained in binary NiTi alloys, martensitic transformations are considerably shifted and at the same time also mechanical or thermomechanical characteristics of SMA are changed, as can be seen, e.g., in the study (Liu, 2003). Hand-in-hand with forming or heat processing, strengthening and healing processes are also important. Several studies are known, e.g. (Gili et al., 2004) or (Morgiel et al., 2002), which confirm the importance of chemical composition (Cu content) during dynamic recrystallization, esp. near borders of grains. On the other hand, other studies (Nam et al., 1990) bring information on formability of SMA when Ni was substituted by Cu. If Ni is substituted by Cu up to the content of ca. 10% (i.e. binary NiTi alloy will be modified to Ti-40Ni-10Cu), then the phase transformation will take place in two steps and these alloys will be much more deformable in the martensitic state than original NiTi alloys. Despite these findings it is still valid that when Cu content exceeds the limit 10 at.%, Ni-Ti-Cu alloys exhibit a rather low formability.

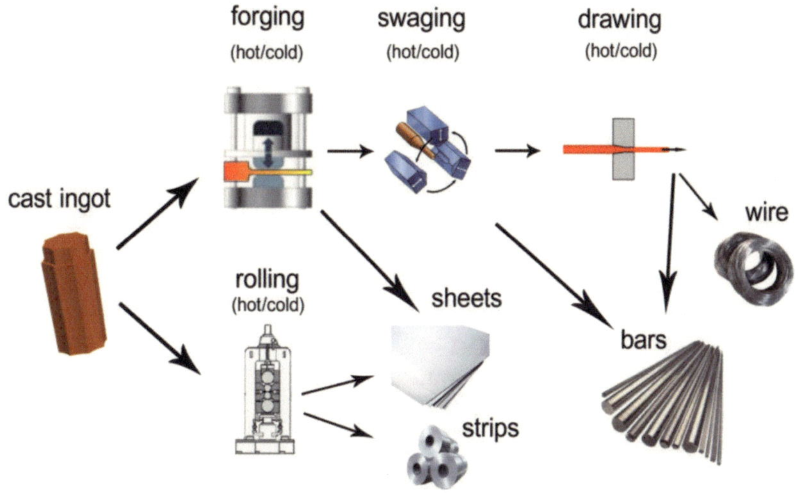

Figure 7. Scheme of basic forming operations used for plastic deformation of SMA

The chosen method of forming together with the method of heat processing is directly proportional to achieved characteristics. Although SMA are mostly used in the form of thin belts, wires or pipes (Kursa et al., 2005), all these products are produced by forming from original cast ingots. Ingots in the cast state are characterized by a very low formability and usually only a small or no memory or superelastic effect. With subsequent hot or cold forming these properties are modified. In Fig. 7 you can see a scheme of basic distribution of forming operations which are usually used for plastic deformation of SMA. Similarly as in other materials, the main aim of hot forming is to change the dimensions and shape of cast ingot, together with modification of its unfavourable microstructure. For the "destruction" of the original dendritic structure, some deformation depending on the cross section of the treated ingot should be applied. It is not unusual for the real size of the applied strain which will provide the required changes to reach values of around 90%. High degrees of deformation performed as hot forming are often also conditioned by requirements arising from the consecutive cold forming, during which such significant reductions of cross section cannot be realized (Ramaiah et al., 2005). Nevertheless, especially recently, when an explosion of unconventional forming techniques occurred, it can be said that intensities of applied strain can reach, and in practice do reach, much higher values than 100%. It should be stated that in these cases it is a shearing deformation where no significant changes in cross section occur. The main aim of these unconventional forming techniques is to achieve structural modification with the effort to deform materials at temperatures as low as possible.

The conventional treatment (forming) of SMA is usually divided into more stages. A frequent sequence of individual operations consists in melting, casting, hot swaging, cold rolling and drawing. Especially during cold forming techniques it is common to insert heat treatment between partial operations. So it is obvious that the transformation behaviour of a particular alloy will be influenced by each of these mentioned operations. In the first stages (melting, casting), there is an already mentioned factor of chemical composition. However, the production process itself can be performed in several various ways with different influence on the studied characteristics, which is documented by a high number of performed studies (Frenzel et al. 2004 and Zhang et al., 2005 and Frenzel et al., 2007a). Regardless of the chosen technique of melting, increased attention should be paid in all techniques to minimization of additional elements, especially oxygen and carbon. These elements have negative influence on the memory effect and also on the brittleness of the particular alloy, which is not without perceptible consequences, especially during the stage of forming.

During alloy forming, which is usually performed in the temperature range 300÷900°C, in addition to the present admixtures, defects in the crystal lattice also begin to come to light. To be more specific, both point defects and changes in the dislocation density begin to activate, which will significantly influence healing and precipitation processes (Frenzel et al., 2007b) or (Kocich et al., 2007). In binary NiTi alloys at around temperatures of 400°C the softening process begins, while at temperatures of 900°C formability (elongation) of alloys determined by tensile tests can exceed 100%. Although SMA at these temperatures exhibit relatively good formability, during alloy forming some cracks may appear, especially near the edges. Usually bars or plate slabs are prepared by forming (forging, rolling). The deformation behaviour of SMA can probably be considered optimal in the temperature range near 800°C. Just these temperatures lie in the range where the alloys are workable and at the same time oxidation of their surface is not as massive as at higher temperatures (Wu, 2001). If forming temperatures are too high, the mentioned oxidation takes place and with increasing temperature the degradation of the material increases as well. The consequence of these processes is formation of very stable oxide layers which are often a part of the surface and cause destruction of the material due to the formation of cracks. When SMA is heated to the temperature 900°C, the alloy tends to be brittle because of occurrence of the Ti_4Ni_2O phase.

After hot forming of alloys, heat treatment is very often applied. The influence of annealing temperatures or cooling rate after annealing is already known very well; more detailed information on partial modes can be found, e.g., in the study (Standring et al., 1980). Generally it can be said that longer annealing times cause higher A_f temperatures (austenite finish). It is known that an increase of transformation temperatures (A_s, A_f) depends on the technology used for the preparation (machining) of SMA and is caused by temperature-induced stresses and defects. High M_s temperature is attributed to incomplete transformation during heating. The mentioned heat treatment is used for optimization of physical and mechanical properties together with maximization of shape memory effect and pseudo-plastic behaviour. The main reason for performed heat treatment after forming is thus modification of transformation temperatures for specific applications. When compared to the relatively wide interval of forming temperatures, the range for annealing temperatures is considerably narrower (300° – 525°C). The times used for annealing are usually in the order of minutes (5–30). Relatively stable resulting transformation temperatures are documented for annealing at temperatures of about 500°C and times shorter than 10 minutes (Liu et al., 2008). With increasing time of annealing A_f temperature increases and stress decreases. Increase of A_f is usually accompanied by depletion of the NiTi matrix by Ni, which is precipitated in the form of precipitates.

3.1. Swaging

As has already been mentioned before, one of the first forming procedures used for plastic deformation of cast ingot is swaging. Swaging is a forming process characterized by a very high rate of deformation (i.e. potential possibilities for the production are 4–6 pieces per minute). It is usually performed on swaging machines. There are many significant advantages of swaging, among them, e.g., possible high reduction of cross section at relatively low energetic costs, significantly dimensionally more precise forged pieces, higher surface quality and considerable improvement of mechanical properties of these products. This process has been known about for a relatively long time, which is also documented by many published studies. One of the main aims of these studies is to find a method of detection of pressure distribution on contact surface during individual deformations

(Zhang, 1984). Individual approaches consider various assumptions related to the parameter of applied energy (Choi et al., 1997 and Canta et al., 1998) or to the course of metal flow (Standring et al., 1980) or (Wang et al., 2005).

Swaging can be divided into two main groups. The first group is hot swaging; the second one cold swaging. Both named groups are procedures characterized by high efficiency and they can be used both for the production of full bars and differently shaped pipes. Swaging is a process which can be described using gradual (incremental) deformations, and is widely applicable for the production of engineering parts such as disks, rings, gear shafts, etc. The main difference between cold and hot swaging is in the working temperature of the process and also the used lubrication (in cold swaging).

However, it is generally known that high-temperature swaging, when compared to the original state of SMA after homogenization, need not necessarily cause a significant difference in reached transformation temperatures. On the other hand, e.g., after cold rolling, there is a distinct (usually full) suppression of phase transformation. It should be noted however that even short annealing of such a deformed state will be enough to restore the memory effect. The annealing will unblock obstacles which prevent the mentioned transformation process and the transformation can proceed again. Although the values of transformation temperatures in the state after hot forming and in the original state are not very different, there are considerable differences in the structural arrangement. During forming, a distinct decrease in grain size occurs, which also determines the final mechanical properties of individual states. Based on the size and type of applied strain, this value after forming can be increased to twice the original value.

3.2. Rolling

During hot rolling, individual grains of SMA forming the structure are deformed and simultaneously recrystallized, which preserves their equiaxed microstructure. At the same time the deformation reinforcement is compensated. The source material is usually in the form of cast semiproducts such as plate slabs, bars or ingots, or semiproducts after forming (e.g., by previous swaging, etc.). Heating of SMA is usually performed in electric furnaces, since SMA semiproducts are generally of small dimensions. In contrast to swaging, there is not such a massive generation of deformation heat to maintain the working temperature of formed components or its slight increase. Although during rolling the temperature as a consequence of deformation changes depending on the deformation rate, it is not enough to compensate the heat loss to the environment or into the tools. That is why for most SMA products (semiproducts small in dimensions) there is a serious danger of going under the recrystallization temperature during the hot rolling itself. The temperature of formed semiproduct after rolling should be near the range from 50 to 100°C above the recrystallization temperature to ensure sufficient heat for the process. In the event of failure to comply with this condition, intermediate heating must be performed before the subsequent reduction. There are relatively many published experiments in the field of rolling focused on the effect of ausforming (forming in the region of austenite) or marforming (forming in the area of martensite). For example, the deformation behaviour of binary NiTi alloys during hot pressure or hot tensile tests is mapped (Dehghani & Khamei, 2010 and Morakabati et al., 2010). Suzuki in (Suzuki et al., 1999) states that significant increase of hot formability can be achieved in NiTi alloys by forming at temperatures of 900° – 1000°C, but there is the drawback of the above mentioned surface oxidation.

When hot rolling is carried out, cold rolling usually follows. Cold rolling of SMA is a process which is much more difficult than the same hot process. The main factor that complicates this procedure is the absence of healing processes which are activated at increased temperatures. During cold rolling, in the course of deformation solidification of SMA takes place and consequently formability of materials decreases. Other influences include a higher value of deformation resistance when compared to hot rolling. Just as a consequence of the limited formability of SMA, microcracks may appear during cold rolling in rolled products. It should be stated that it can occur even at low reductions of height (~20%). In cases when the final wire is produced by cold rolling in calibres (with diameters usually lower than 5 mm) the forming should take place in more stages. That is why a higher amount of passes is necessary to obtain the final wire. Probably the main reason for usage of cold rolling is to obtain dimensional accuracy and also the high surface quality of products prepared in this way. A side effect of cold forming (rolling) is suppression of the shape memory effect, while there is an increase of strength properties and a decrease of plastic properties.

With increasing content of Ni, rolling is more and more difficult and when the limit of 51 at.% of Ni is exceeded, any rolling of NiTi alloys is extremely difficult. The main reason is considerable deformation reinforcement. As is obvious from the more experimental results, yield strength of annealed NiTi alloys is usually lower than 100 MPa, but already after deformation of 40% this limit increases to values of about 1000 MPa. If the deformation continues, it would be very probable that cracks in the material would appear or even destruction of the material would take place. Intermediate annealing must therefore be performed before the next forming. This will cause a decrease of strength properties and a partial recovery of plastic properties. The just mentioned combination of deformation-healing (annealing) causes refinement of the final structure as a result. The temperature of intermediate annealing and its length will accordingly be the key parameters for the microstructure development. Generally it can be said that the temperature of intermediate annealing is lower than the temperature necessary for hot forming and usually is about 600°C (Wu, 2002).

It is known that during cold forming (rolling) the increase of the volume fraction of martensite is much higher than during hot rolling. Relatively many experimental studies focused on thermomechanical processing of SMA and mapping the effects of rolling and subsequent annealing, e.g. (Kurita et al., 2004), confirm this knowledge. It can be briefly stated that the higher the annealing temperature is, the lower content of retained martensite you will find. However, this fraction can be present in SMA even at relatively high temperatures, which is also documented in numerous studies (Brailovski et al., 2006 and Lin

& Wu, 1994). The reason is the presence of dislocations which slow down its conversion to austenite during post deformation annealing. The potential softening during annealing can be considered in three ways (mechanisms): a) dislocation recovery after which retransformation of martensite to austenite takes place (at temperatures of about 400°C), b) particle stimulated nucleation (PSN) at middle temperatures (~500°C), c) recrystallization of the matrix at high temperatures (~600°C). The needed level of deformation ensuring stable shape memory and superelastic behaviour of SMA (during cold rolling) is usually above 90% (Kim et al., 2006).

Among other drawbacks of rolling, there is only a small possibility of controlling the grain morphology or texture at adequate refinement of the microstructure. These are substantial effects which can influence the shape memory characteristics. Generally it can be stated that in the case of the requirement to maximize the shape memory effect, the best solution is to use the NiTi alloy in the state after hot rolling. In the case of the requirement to obtain high strength and hardness at acceptable reversible deformation, the NiTi alloy should be used after cold rolling.

These were not the only reasons for the impulse for researching the effect of unconventional forming techniques on deformation behaviour, or transformation characteristics of SMA. Especially in recent times there is a significant effort during research regarding the application of Severe Plastic Deformations (SPD) in memory materials. Among other applied techniques within the group of SPD, there is e.g. High Pressure Torsion (HPT) technology, or the Equal Channel Angular Pressing (ECAP) process. It was already confirmed several times by experiments that these techniques are a very effective tool for influencing the transformation characteristics, cyclic stability SMA and simultaneously relatively easy control of the texture of these formed materials (Kockar et al., 2010).

3.3. SPD processes

As has already been stated, most of the characteristics of SMA are mainly based on reverse martensitic transformation, which is controlled by chemical composition, microstructural parameters and also the method of preparation. The possibility to control functional properties of materials based on Ni-Ti alloys using thermomechanical treatment can be improved through microstructural refinement, which is documented in many studies, e.g. (Sergueeva et al., 2003). It is most desirable to obtain the structure characterized by very small grain (subgrain) size – ultra fine grain (UFG) structure. Nevertheless, it is necessary to simultaneously preserve exceptional properties of memory alloys.

It is known that TiNi-based materials characterized by a very small grain size can be prepared via three methods. The first method is chilling of cast. The second one consists in the preparation using SPD methods and the third one is then the combination of conventional techniques and subsequent annealing. Using the first and the second method enables one to obtain a fine-grained structure with the grain size in the range of 200–600 μm. If conventional forming with subsequent heat treatment will be used, then generally a larger grain size can be achieved compared to both of the previous variants. Nevertheless, a significant advantage of this method is obtaining a relatively equiaxed structure, although with a little larger grain size than in the case of SPD. A distinct difference between the conventional and unconventional method of forming is mainly in the intensity of applied strain. During conventional forming, only limited degrees of deformation can for technological reasons be applied on the formed material during one forming cycle. In addition to that – as has already been discussed above – this process should usually be performed at higher working temperature.

The principle of SPD processes is based on repeated application of high degrees of plastic deformation during individual forming cycles. This fact, together with relatively low applied temperature when these processes occur, is behind the structural refinement during SPD. The result of such deformations is equiaxed structures characterized by a relatively high amount of grains with high angle misorientation. In addition to these features, the structures can also be described by the presence of subgrains with high dislocation density (especially after ECAP). However, it must be noted that the materials prepared using SPD also contain areas with high internal stress, at least when compared to cast materials. In the case of HPT, it is possible to obtain even amorphous states which, after subsequent annealing, are transformed into nanostructural arrangements. Based on the selected regime, the final grain (subgrain) size can be expected then. The applied annealing of SMA is usually in the temperature range of 200÷400°C. This is, at the same time, the interval of relatively stable grain size, nevertheless there is a significant decrease of dislocation density inside grains. With increasing temperature (~500°C) the grain size is increased to twice the original value, while grain boundaries are also much better defined. The reason of this increase is in the course of healing processes.

3.3.1. High pressure torsion

Localization of applied strain can thus be achieved using large plastic deformations. One of the first suggested and so far the most effective SPD method is the high pressure torsion process (HPT) (Fig. 8). It should be stated that in spite of certain limitations, deformation behaviour of a wide range of materials can be studied with this process. The consequence of the application of SPD is the destruction of the crystal lattice with subsequent transformation to the amorphous state. Accumulated dislocations or grain boundaries are the main driving force of the amorphization process. High density of dislocations may cause formation of amorphous bands thanks to shear deformation instability. In SMA, especially in NiTi-based alloys, a strong crystallographic texture is formed (Frick et al., 2004). In the following course of the process, refinement of grains and simultaneously amorphization are observed. This continues up to the full amorphous state of the formed material volume. According to this study and others, even the nanostructure formed like that contains characteristics of the texture, while in this state no directional deformation occurred. Individual nanograins exhibit preferred orientation which corresponds with orientation of nanograins preserved in the structure after HPT. A certain explanation may be knowledge assigning this influence to nanocrystals thanks to which heterogeneous nucleation occurs.

It should be noted that such a structural state is stable only at low temperatures. For these reasons HPT is usually performed at room temperature or even lower. However, findings on the dependence between M_s temperature and formation of nano-crystalline structure in SMA materials based on NiTi are very important. If the deformation temperature during SPD is lower than the corresponding M_s temperature of the formed alloy, then there is a high probability of formation of nano-crystalline structure. If the deformation temperature is between the M_s temperature and the highest temperature of the beginning of martensitic transformation influenced by deformation (M_d), then the probability of the nanostructure formation is lowered. If the deformation temperature is higher than the M_d temperature, the probability of the nanostructure formation is very low. If after SPD post-deformation annealing is applied, then a submicrocrystalline structure is formed.

It was already mentioned that at specific temperature regimes set during annealing the alloy can crystallize into grains with the size (10–40 nm). The specific temperature range where NiTi alloys are stable is 250° – 300°C (Prokoshkin et al., 2005). With increasing temperature of annealing, the final nanostructure will be "coarser". Total amorphization of the structure is usually achievable only in cases when the deformation temperature is lower than the martensite start temperature (M_s). Rate of the structure amorphization is influenced to large extent by the applied pressure during HPT. Higher imposed pressure suppresses the tendency to form an amorphous structure from nanostructure and as well to form nanostructure from deformably reinforced dislocation structure, as confirmed by the study (Prokoshkin et al., 2005). The probable reason is a decrease of M_s temperature, due to higher values of pressure deformations.

Materials processed using HPT and subsequent heat treatments usually reach very high strength properties (strength higher than 2 GPa). Surprisingly, relatively high plastic characteristics are also preserved (elongation at break up to about 40%) (Sergueeva et al., 2003). There are also assumptions to obtain super-plastic behaviour of NiTi alloys. Although the HPT technology appears to be a suitable candidate for positive modification of the properties, it is in principle excluded from commercial use by its main drawback (very small samples)-see Fig. 8b.

(a) (b)

Figure 8. Machine for HPT process (a) detail of processed sample (b)

3.3.2. *Equal channel angular pressing*

It was necessary to process larger volumes of products, which has led to looking for alternative ways of HPT substitution. One possible candidate for meeting the scheduled targets appears to be the equal channel angular pressing process (ECAP). In Fig. 9a you can see the assembly for practical application of this technique. It must be noted that, in contrast to HPT, various shapes of material sections can be processed here (Fig. 9b). As it was already mentioned, there is an effort when using SPD techniques to decrease the working temperature, because it is known that with decreasing deformation temperature the final grain size in the final structure of the treated material also decreases. Since SMA are characterized by a relatively high deformation resistance, the ECAP process – in contrast to HPT – should be performed at temperatures relatively higher. The most suitable range of forming temperatures found in experiments is 400° – 500°C. Amorphization of the structure, however, cannot occur in this temperature interval. This fact also determines higher size of final grains when compared to HPT technique. The size of the grain is usually decreased to a value that lies in the region of 200–300 nm. Although even in this case the strength properties are significantly increased while preserving relatively good plastic properties, these values are lower when compared with the state after HPT. Also the combination of ECAP and thermomechanical treatment is among tested procedures for another reduction of the grain size. To be more specific, cold rolling was applied after previous angular pressing (Pushin et al., 2006).

(a) (b)

Figure 9. Machine for ECAP process (a) detail of assembly and processed sample (b)

NiTi alloys are very sensitive to the exact chemical composition, which is manifested even during their deformation using SPD techniques. For example, in the study (Khmelevskaya et al., 2001) the binary $Ni_{50}Ti_{50}$ alloy absolved 12 passes (ECAP) at the temperature 500°C, or 8 passes at the temperature 400°C; in the case of $Ni_{49.3}Ti_{50.7}$ alloy, only 3 passes could be applied, since the material exhibited relatively low formability and it was destroyed.

The reason for the formation of a fine-grained structure is probably the proceeding recrystallization process which causes gradual increase of disorientation of subgrain boundaries up to the formation of high-angle grain boundaries (HAGB). As to the ability of reversible deformation, its value after ECAP is comparable to the value obtained after application of conventional forming techniques (cold rolling) followed by heat treatment. It is documented, e.g., by the already mentioned study (Khmelevskaya et al., 2001) where the value of such deformation was determined to be ~7%. It should, however, be noted that the temperature interval of the memory effect after ECAP is narrower and also with lower values than after cold rolling. Similarly as after HPT, the temperature stability of SMA by annealing was tested up to temperatures of 500°C. Similarly, also after ECAP, SMA appeared to be stable, but significant reduction of dislocation density occurred. Increasing temperature caused an increase of grain size.

Generally it is known that cooling of NiTi alloys below the M_s temperature leads to higher formation of coarse-grained structure (50÷80) μm of the R phase. As confirmed by the study (Pushin & Kondratjev, 1994), cooling of coarse-grained monocrystals of B2 NiTi alloys below the M_s temperature leads to the formation of R-martensite with rhombohedral (or hexagonal) lattice. In the case of occurrence of other admixtures in binary and multiphase alloys, formation of monoclinic B19′martensite takes place during cooling below the M_s temperature. In particular, occurrence of elements such as Cu, Pd, Pt or Au during cooling of the alloy below the M_s temperature leads to formation of orthorhombic martensite B19 (Pushin, 2000). Generally it can be said that martensitic transformation on the microstructural level is caused by the presence of microtwins, while on the level of internal areas of grains it is caused by the formation of coherent crystals.

Martensitic phase transformation thus usually takes place from cubic B2 high-temperature phase (austenite) to monoclinic B 19′ phase (martensite). This process is accompanied by high deformation. According to the results from the experiment (Waitz et al., 2004), if the grain size in SMA is in the region of nanometers, then the high density of grain boundaries will act as a significant obstacle during mentioned transformations. There are two main reasons why martensitic transformation in nanomaterials is suppressed. The decrease of transformation temperatures depending on the decrease of grain size follows from the mentioned study. It was also proved that full suppression of martensitic transformation takes place when the grain size is smaller than 50 nm. From the point of view of phase stability, these small grains have a significant influence on morphology of martensite (Waltz, 2005).

Thanks to the experiment performed with the Ni50.4Ti alloy, the specific influence of forming on transformation characteristics and also on microstructure development was mapped. The experiment describes the problem of combination of conventional forming techniques together with unconventional ones. For forming, SMA characterized by the content of O_2 (0.0624 wt.%) and N_2 (0.0039 wt.%) was used. The content of carbon (0.055 at.%) was determined using spectrometry. The diameter of the cast was 20 mm and the cast length 350 mm. Then homogenization at the temperature of 850°C is followed by subsequent cooling.

The forming itself was suggested in the first phase using swaging then pressing using the ECAP technique was performed. Swaging was performed at the temperature of 850°C. During swaging, the strain was applied gradually in individual reductions. The total strain applied on the cross section was 66 %. On the contrary, the ECAP technology was applied on SMA at the temperature of 290°C. Because of possible "negative" influence on the temperature fluctuation of the formed material, the pressed samples were placed in the steel "cans". For the ECAP, matrix with the angle 105° between individual canals was used; the extrusion speed was set to 1 mm/s. The extrusion itself consisted of two performed passes, where Bc was chosen as the deformation path. In particular, the influence of intensities of applied strain in relation to mechanical and thermophysical properties and the course of healing processes were studied. To be able to determine the mentioned effect, heat treatment was carried out after the performed deformation. In addition to optical microscopy, RTG diffraction was also used for evaluation of changes. The differences between after swaging followed by ECAP are obvious from the attached photos (Fig. 10).

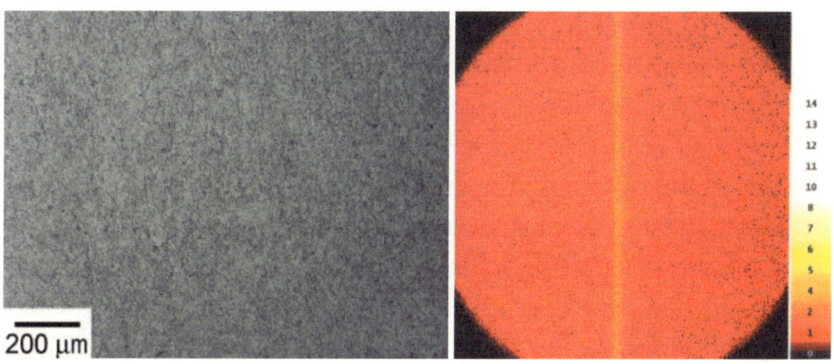

(a) (b)

Figure 10. Ni50.4Ti alloy (swaged+ECAPed) : microstructure (a), diffraction fringe (b)

To be able to specify the differences caused by applied strain, it was necessary to perform heat treatment of the deformed materials. These annealing using selected temperature regimes (550°C/15 min, 600°C/15 min, 650°C/15 min, 900°C/15 min) should provide information on the influence of the accumulated deformation on the beginning of corresponding processes, especially recrystallization.

The azimuthal profile of diffraction lines of the NiTi alloy confirms that all mosaic blocks (DCA-diffraction coherent areas of crystallites) in the structure after ECAP are smaller than 10 μm (Fig. 10b). The DCA are regions that scatter coherently. These areas are defined by borders where is high dislocation density. Within the framework of the mosaic blocks, the moving dislocations face minimal resistance. Inside the mosaic blocks is much less of dislocations than on the borders of blocks, that is why the mosaic blocks are called as dislocationless cells (areas) in the crystallites. Nevertheless, this predication is not quite accurate because one of obstacle in these blocks for dislocation movement is Peierls – Nabarro stress (i.e. dislocation is moving through the mosaic block, so that mosaic block must have at least one dislocation). This labeling (dislocationless blocks) is commonly used, important factor is that inside of the mosaic blocks is considerably less of dislocation than on its boundaries. Due to this matter (low dislocation density inside of mosaic block) can X-ray be diffracted coherently (that is why the mosaic blocks are also often marked as Coherent Scattering Regions – CSR´s). The presence of blocks smaller then 10 μm was confirmed by XRD. Generally may be noticed that mosaic blocks (obtained by XRD) are not the same as grains observed by optical microscopy. Mosaic blocks can be seen by presence of reflections on XRD patterns. In the case of the large block presence will be its reflection large as well. Since cannot into the diffracted volume be much of big mosaic blocks present, there will be their reflections on diffraction lines clearly separated. If will be mosaic block small there will be much of their reflections in diffracted volume. That is why individual reflection are overlapping (they are not separated i.e. reflection is continuous). More detailed information about this technique can be found for example in (Hindelah & Hosemann, 1988).

As a consequence of recrystallization processes after annealing of the extruded alloy at temperatures of 600°C and higher in the structure blocks larger than 10 μm were formed, while after the same length of time annealing at lower temperatures all mosaic blocks remained smaller than 10 μm. It should be noted that the size distribution of mosaic blocks was bimodal (i.e. blocks larger than 10 μm and blocks smaller than 10 μm). This means that coarser crystallites contain more nickel than smaller crystals. The reason is the asymmetry of the area in the phase diagram of titanium – nickel formed by a solid solution of TiNi. Microinhomogeneity of the alloy also influences mechanical properties of the alloy. The dispersion of such elementary composition and thus lattice parameters of individual mosaic blocks in the structure causes so-called microstrain (strain of the 2nd order) influencing the dynamics of dislocations, but also nucleation and growth of microcracks. These strains of the 2nd order also influence shifts of walls of mosaic blocks and thus the course of recrystallization of the alloy. The authors of the experiment performed with the Ni50.6Ti alloy also arrived at similar results (Kocich et al., 2009).

It means that even with very thoroughly conducted preparation of material based on shape memory alloys, microinhomogeneities may be formed. As concerns the forming technologies as such, it can be stated that after completed rotary forging (swaging) the initial grain size and therefore also the ensuing properties were substantially changed. The ECAP process also proved its influence on the final appearance of the micro- and substructure. The differences are even apparent in the number of passes; if material passed through the matrix only once, then no mosaic blocks smaller than 10 μm occurred in the structure, however, influence of the second pass can be seen here, when the samples of the NiTi alloy already contained bi-modal composition. The temperature that appeared to be the starting temperature for initiation of re-crystallisation was 600°C.

4. Conclusions

For some specific purposes, the exclusivity of binary shape memory alloys should be modified by the addition of another element. These prepared Ti-Ni-X based materials are currently very progressive materials. As has already been indicated, their final properties will, to a great extent, be determined by the chosen element. To a great extent, it is thus the factor of chemical composition; nevertheless that is not the only one. A significant influence in the sense of the effect on transformation characteristics together with other properties was also demonstrated for the preparation method. There is an outstanding dependence on used forming technology and their relationship or thermomechanical conditions play an important role. During conventional forming, the deformation behaviour of SMA binary systems is relatively well known. As seems to be the case, multiphase SMA are very sensitive to even small deviations from the required chemical composition. These nuances can be observed through the shifts in recrystallization temperatures, which also means the shifts in their optimum forming temperatures.

Particularly at this time there is intensive research of use of unconventional forming techniques for SMA. Although it is already known that when using these procedures, in contrast to conventional forming, it is possible to obtain structures characterized by grain size of nanometers, the procedure is currently not optimized for practical applications. The effectiveness of procedures based on applications of large plastic deformations consists mainly in a distinct decrease of grain size. There is also an obvious benefit in controlled textures of deformed SMA, which can also be utilized with regard to required final properties.

Although studies monitoring behaviour of multiphase Ti-Ni-X based systems exist, a generally known method for the preparation of SMA is not still available. In the future there should be actions related to preparation of specific alloys with verified behaviour. It is obvious that hand-in-hand with the above mentioned there should also be development focused on the preparation and target modification of properties of these materials.

Author details

Radim Kocich, Ivo Szurman and Miroslav Kursa
VŠB Technical University of Ostrava, Czech Republic

Acknowledgement

This paper was created under the project No. CZ.1.05/2.1.00/01.0040 "Regional Materials Science and Technology Centre" within the frame of the operation programme "Research and Development for Innovations" financed by the Structural Funds and from the state budget of the Czech Republic and project GA 106/09/1573 „Optimisation of chemical composition, structural characteristics, mechanical properties of NiTi alloys for bio-mechanical applications".

5. References

Bertheville, B.; Bidaux, J.E. (2005). Alternative powder metallurgical processing of Ti-rich NiTi shape memory alloys, *Scripta Materialia,* Vol.52, pp. 507-512, ISSN 1359-6462

Bertolino, N.; Monagheddu, M.; Tacca, A.; Giuliani, P.; Zanotti, C.; Tamburini, A.U. (2003). Ignition mechanism in combustion synthesis of Ti–Al and Ti–Ni systems, *Intermetallics*, Vol.11, pp. 41-46, ISSN 0966-9795

Brailovski, V.; Prokoshkin, S.D.; Khmelevskaya, I.Y.; Inaekyan, K.E.; Demers, V.; Dobatkin, S.V.; Tatyanin, E.V. (2006). Structure and properties of the Ti-50.0 at%Ni alloy after strain hardening and nanocrystallizing thermomechanical processing, *Materials Transactions,* Vol.47, pp. 795-802, ISSN 1347-5320

Canta, T.; Frunza, D.; Sabadus, D.; Tintelecan, C. (1998). Some aspects of energy distribution in rotary forming processes, *Journal of Material Processing and Technology,* Vol.80-81, pp. 195-201, ISSN 0924-0136

Choi, S.; Nam, K.H.; Kim, J.H. (1997). Upper-bound analysis of the rotary forging of a cylindrical bilet, *Journal of Material Processing and Technology,* Vol.67, pp. 78-84, ISSN 0924-0136

Choi, M.S.; Fukuda, T.; Kakeshita, T. (2005). Anomalies in resistivity, magnetic susceptibility and specific heat in iron-doped Ti-Ni shape memory alloys, *Scripta Materialia,* Vol. 53, pp. 869-873, ISSN 1359-6462

Dautovich, D.P.; Purdy, G.R. (1965). Phase Transformations in TiNi, *Canadian Metallurgical Quarterly,* Vol.4, pp. 129-134, ISSN 1879-1395

Dehghani, K.; Khamei, A.A. (2010). Hot deformation behavior of 60Nitinol (Ni60 wt%- Ti40 wt.%) alloy: experimental and computational studies, *Materials Science and Engineering A,* Vol.527, pp.684-690, ISSN 0921-509

Dembovský, V. (1978). *Plasma metallurgy*, SNTL, Praha (in Czech) ISBN 04-407-78

Dembovský, V. (1985). *Plasma metallurgy: The principles,* SNTL, Praha, ISBN 0-444-99603-6

Du, Y.; Schuster, J.C. (1998). Experimental investigation and thermodynamic modeling of the Ni-Ti-C system, Zeitschrift fur Metallkunde, Vol.89, pp. 399-410, ISSN 1862-5282

Duerig, T.W.; Pelton, A.; Stöckel, D. (1999). An overview of nitinol medical applications, *Materials Science and Engineering A,* Vol.273-275, pp.149-160, ISSN 0921-5093

Frenzel, J.; Zhang, Z.; Neuking, K.; Eggeler, G. (2004). High quality vacuum induction melting of small quantities of NiTi shape memory alloys in graphite crucibles, *Journal of Alloys and Compounds,* Vol.385, pp. 214-221, ISSN 0925-8388

Frenzel, J.; Zhang, Z.; Somsen, Ch.; Neuking, K.; Eggeler, G. (2007a). Influence of carbon on martensitic phase transformations in NiTi shape memory alloys, *Acta Materialia* Vol.55, pp. 1331-1339. ISSN 1359-6454

Frenzel, J.; Pfetzing, J.; Neuking, K.; Eggeler, G. (2007b). On the influence of thermomechanical treatments on the microstructure and phase transformation behavior of Ni–Ti–Fe shape memory alloys, *Materials Science and Engineering A,* Vol.481-482, pp. 635-642, ISSN 0921-509

Frick, C.; Ortega, A.; Tyber, J.; Gall, K.; Maier, H. (2004). Multiscale structure and properties of cast and deformation processed polycrystalline NiTi shape-memory alloys, *Metallurgical and Material Transactions A,* Vol.35, pp. 2013-2021, ISSN 1073-5623

Gili, F.J.; Solano, E.; Penal, J.; Engel, E.; Mendoza, A.; Planell, J.A. (2004). Microstructural, mechanical and cytotoxicity evaluation of different NiTi and NiTiCu shape memory alloys, *Journal of Material Science,* Vol.15, pp. 1181-1189, ISSN 0022-2461

Graham, R. (2004). Characteristics of high purity nitinol in Shape Memory and Superelastic Technologies. *Proceedings of conferece SMST Society Inc.*, Denver, 16-19.4.2004, pp. 7-12

Hindelah, A.M.; Hosemann R. (1988) Paracrystals representing the physical state of matter, *Journal of Physics C: Solid State Physics* Vol.21, pp. 4155-4162, ISSN 0022-3719

Jung, J.; Ghosh, G.; Isheim, D.; Olson, G.B. (2003). Design of nanodispersion strengthened TiNi-base shape memory alloys, *Proceedings of the international conference on Shape memory and superelastic technologies- SMST 2003,* ISBN 0-9660508, pp. 45-50.

Khmelevskaya, I.; Trubitsyna, I.B.; Prokoshkin, S.D.; Dobatkin, S.V.; Stolyarov, V.V.; Prokofjev, E.A. (2001). Structure and Functional Properties of Ti-Ni-Based Shape Memory Alloys Subjected to Severe Plastic deformation, *Nanomaterials by Severe Plastic Deformation*, pp. 170-176, Wiley-VCH, Weinheim, Germany, ISBN 3527306595

Kim, H.Y.; Ikehara, Y.; Kim, J.I.; Hosoda, H.; Miyazaki, S. (2006). Martensitic transformation, shape memory effect and superelasticity of Ti–Nb binary alloys, *Acta Materialia,* Vol.54, pp. 2419-2425, ISSN 1359-6454

Kocich, R.; Kursa, M.; Greger, M.; Szurman, I. (2007). Deformation behavior of shape memory alloys on ECAP process, *Acta Metallurgica Slovaca,* Vol. 13, pp. 570-576, ISSN 1338-1156

Kocich, R.; Szurman, I.; Kursa, M.; Fiala, J. Investigation of influence of preparation and heat treatment on deformation behaviour of the alloy NiTi after ECAE, *Materials Science and Engineering A,* Vol.512, pp. 100-104, ISSN 0921-5093

Kockar, B.; Atli, K.C.; Ma, J.; Haouaoui, M.; Karaman, I.; Nagasako, M.; Kainuma, R. (2010). Role of severe plastic deformation on the cyclic reversibility of a $Ti_{50.3}Ni_{33.7}Pd_{16}$ high temperature shape memory alloy, *Acta Materialia,* Vol.58, pp. 6411-6417, ISSN 1359-6454

Kuchař, L.; Drápala, J. (2000). *Metallurgy of pure metals-methods of pure materials purifying,* Košice Press. ISBN 80-7099-471-1

Kurita, T.; Matsumoto, H.; Abe, H. (2004). Transformation behavior in rolled NiTi, *Journal of Alloys and Compounds,* Vol.381, pp. 158-166, ISSN 0925-8388

Kursa, M.; Szurman, I.; Drápala, J.; Losertová, M.; Greger, M. (2005). *Shape memory materials Ni-Ti-Me and possibilities of control of their transformation characteristics,* VŠB-TU Ostrava (ed), ISBN80-248-0894-3, pp. 23-46

Lin, H.C.; Wu, S.K. (1994). The tensile behavior of a cold-rolled and reverse-transformed equiatomic TiNi alloy, *Acta Metallurgica Materialia,* Vol.42, pp. 1623-1631, ISSN 1359-6454

Liu, Y. (2003). Mechanical and thermomechanical properties of a $Ti_{50}Ni_{25}Cu_{25}$ melt spun ribbon, *Materials Science and Engineering A,* Vol.354, pp. 286-290, ISSN 0921-509

Liu, X.; Wang, Y.; Yang, D.; Qi, M. (2008). The effect of ageing treatment on shape-setting and superelasticity of a NiTi stent, *Materials Characterization* Vol.59, pp. 402-408, ISSN 1044-5803

Locci, A.M.; Orrù, R.; Cao, G.; Munir Z.A. (2003). Field-activated pressure-assisted synthesis of NiTi, *Intermetallics,* Vol.11, pp. 555-571, ISSN 0966-9795

Ma, j.l.; Wu, k.h. (2000). Effects of Tantalum addition on transformation behaviour of $(Ni_{51} Ti_{49})_{1-x} Ta_x$ and $Ni_{50} Ti_{50-y} Ta_y$ Shape Memory Alloys, *Materials Science and Technology,* Vol.16, pp. 716-719, ISSN 1743-2847

Meng, X.L.; Cai, W.; Wang, L.M.; Zheng, Y.F.; Zhao, L.C.; Zhou, L.M. (2001). Microstructure of Stress-Induced Martensite in a Ti-Ni-Hf High. Temperature Shape Memory Alloy, *Scripta Materialia*, Vol.45, pp.1177-1182, ISSN 1359-6462

Mentz, J.; Bram, M.; Buchkremer, H.P.; Stöver, D. (2006). Improvement of Mechanical Properties of Powder Metallurgical NiTi Shape Memory Alloys, *Advanced Engineering Materials*, Vol.8, pp. 247-252, ISSN 1527-2648

Mentz, J.; Frenzel, J.; Wagner, M.F.X.; Neuking, K.; Eggeler, G; Buchkremer, H.P.; Stöver, D. (2008). Powder metallurgical processing of NiTi shape memory alloys with elevated transformation temperatures, *Materials Science and Engineering A*, Vol.491, pp. 270-278, ISSN 0921-5093

Morakabati, M.; Kheirandish, S.; Aboutalebi, M.; Karimi-Taheri, A.; Abbasi, S.M. (2010). The effect of Cu addition on the hot deformation behavior of NiTi shape memory Alloys, *Journal of Alloys and Compounds*, Vol.499, pp. 57-66, ISSN 0925-8388

Morgiel, J.; Cesari, E.; Pons, J.; Pasko, A.; Dutkiewicz, J. (2002). Microstructure and martensite transformation in aged Ti-25Ni-25Cu shape memory melt spun ribbons, *Journal of Material Science*, Vol.37, pp. 5319-5327, ISSN 0022-2461

Mousavi, T.; Karimzadeh, F.; Abbasi, M.H. (2008). Synthesis and characterization of nanocrystalline NiTi intermetallic by mechanical alloying, *Materials Science and Engineering A*, Vol.487, pp. 46-51, ISSN 1527-2648

Nam, T.H.; Saburi, T.; Shimizu, K. (1990). Shape memory characteristics associated with the B2↔B19 and B19↔B19′ transformations in a Ti-40Ni-10Cu (at.%) alloy, *Materials Transactions JIM* Vol.31, pp. 959-968, ISSN 1073-5623

Noh, J.P. (2001). Phase transformation behaviours and shape memory characteristics of Ti-(45-x)Ni-5Cu-xMo (x=0·3-1·0) alloys, *Materials Science and Technology*, Vol.17, pp. 1544-1550, ISSN 1743-2847

Otsuka, K.; Wayman, C.M. (Eds). (1998). *Shape Memory Materials*, Cambridge University Press, ISBN 052144487X

Pacholek, P.; Szurman, I.; Sklenaříková, I.; Kursa, M. (2003). Preparation specificity of shape memory alloys on the base of NiTi, *Proceedings of 8th International conference Technology 2003*, ISBN 80-227-1935-8, Bratislava, 5. - 7.9. 2003, pp. 38-42

Prokoshkin, S.D.; Khmelevskaya, I.Y.; Dobatkin, S.B.; Trubitsyna, I.B.; Tatyanin, E.V.; Stolyarov, V.V.; Prokofiev, E.A. (2005). Alloy composition, deformation temperature, pressure and post-deformation annealing effects in severely deformed Ti–Ni based shape memory alloys, *Acta Materialia*, Vol.53, pp. 2703-2710, ISSN 1359-6454

Pushin, V.G.; Kondratjev, V.V. (1994). Processing of nanostructured TiNi-shape memory alloys: Methods, structures, properties, application, *Physics of Metals and Metallography*, Vol.7, pp. 497-501, ISSN 1555-6190

Pushin, V.G. (2000). Alloys with a Thermomechanical Memory: Structure, Properties, and Application, *Physics of Metals and Metallography*, Vol.S68–S95, pp. 90-96, ISSN 1555-6190

Pushin, V.G.; Valiev, R.Z.; Zhu, Y.T.; Prokoshkin, S.D.; Gunderov, D.V.; Yurchenko, L.I. (2006). Effect of Equal Channel Angular Pressing and Repeated Rolling on Structure, Phase Transformations and Properties of TiNi Shape Memory Alloys, *Materials Science Forum*, Vol.503-504, pp. 539-540, ISSN 0255-5476

Ramaiah, K.V. (2005). Processing of Ni-Ti Shape Memory Alloys, *Proceedings of conference Smart materials structures and systems*, Bangalore, 4-6.5. 2005, pp. 23-30.

Raz, S.B.; Sadrnezhaad, S.K. (2004). Effect of VIM frequency on chemical composition, homogeneity and microstructure of NiTi shape memory alloy, *Materials Science and Technology*, Vol.20, pp. 593-598, ISSN 1743-2847

Sakuma, T.; Iwata, I.; Okita, K. (2003). Effect of additional elements on erosion and wear characteristics of Ti-Ni shape memory alloy, *Corrosion Engineering, Science and Techology*, Vol.52, pp. 821-832, ISSN 1743-2782

Schetky, L.; Wu, M.H. (2005). Issues in the further Development of Nitinol Properties and Processing for medical Device Applications, *Memry Corporation*. 18.10.2010, Available from http//memry.com/technology/pdfs/ASM03_NitinolDevelop.pdf

Sergueeva, A.V.; Song, C.; Valiev, R.Z.; Mukherjee, A.K. (2003). Structure and properties of amorphous and nanocrystalline NiTi prepared by severe plastic deformation and annealing, *Materials Science and Engineering A*, Vol.339, pp. 159-165, ISSN 0921-509

Standring, P.M.; Moon, J.R.; Appleton, E. (1980). Plastic deformation produced during indentation phase of rotary forging, *Journal of Material Processing and Technology*, Vol.7, pp. 159-167, ISSN 0924-0136

Suzuki, H.G.; Takakura, E.; Eylon, D. (1999). Hot strength and hot ductility of titanium alloys-a challenge for continuous casting process, *Materials Science and Engineering A*, Vol.263, pp. 230-236, ISSN 0921-509

Szurman, I.; Kursa, M. (2009). Processing technologies of Ni-Ti based shape memory alloys. *Proceedings of 8th European Symposium on Martensitic Transformations*, ISBN 978-2-7598-0480-1 Prague, 7-11. 9. 2009, pp. 75-81

Szurman, I.; Kursa, M. (2010). Methods for Ni-Ti based alloys preparation and their comparison. *Proceedings of conference Metal 2010*, ISBN 978-80-87294-17-8, Rožnov pod Radhoštěm, 18. - 20.5. 2010, pp. 861-866

Tang, W.; Sandstro M, R.; Wei, Z.G.; S. Miyazaki. (2000). Experimental investigation and thermodynamic calculation of the TiNiCu shape memory alloys, *Metallurgical and Material Transactions A*, Vol.31, pp. 2423-2429, ISSN 1073-5623

Tsai, J.Ch.; Jean, R.D. (1994). Effects of hot working on the martensitic transformation of Ni-Ti alloy, *Scripta Metallurgica et Materialia*, Vol.30, pp. 1027-1030, ISSN 1359-6462

Van Humbeeck, J.V. (2001). Shape memory alloys: A material and technology. *Advanced Engineering Materials*, Vol.3, pp. 837-850, ISSNXXX 1527-2648

Waitz, T.; Kazykhanov, V.; Karnthaler, H.P. (2004). Martensitic phase transformations in nanocrystalline NiTi studied by TEM, *Acta Materialia*, Vol.52, pp. 137-143, ISSN 1359-6454

Waitz, T. (2005). The self-accommodated morphology of martensite in nanocrystalline NiTi shape memory alloys, *Acta Materialia*, Vol.53, pp. 2273-2280, ISSN 1359-6454

Wang, G.C.; Guan, J.; Zhao, G.Q. (2005). A photo-plastic experimental study on deformation of rotary forging a ring workpiece. *Journal of Material Processing and Technology*, Vol.169, pp. 108-115, ISSN 0924-0136

Wu, M.H. (2001). Fabrication of Nitinol materials and components, *Proceedings of conference SMST –SMM 2001*, ISBN 978-0-87849-896-3, pp. 285-291

Wu, M.H. (2002). Fabrication of Nitinol Materials and Components, *Materials Science Forum,* Vol. 394-395, pp. 284-290, ISSN 0255-5476

Zhang, M. (1984). Calculating force and energy during rotary forging, in proceedings of conference on rotary metal working processes, *Journal of Material Processing and Technology,* Vol.7, pp. 115-124, ISSN 0924-0136

Zhang, Z.; Frenzel, J.; Neuking, K.; Eggeler, G. (2005). On the reaction between NiTi melts and crucible graphite during vacuum induction melting of NiTi shape memory alloys, *Acta Materiala,* Vol.53, pp. 3971-3978, ISSN 1359-6454

Zhang, Z.; Frenzel, J.; Neuking, K.; Eggeler, G. (2006). Vacuum induction melting of ternary NiTiX (X = Cu, De, Hf, Zr) shape memory alloys using graphite crucibles, *Materials Transactions A,* Vol.47, pp. 661-665, ISSN 1073-5623

Zhu, S.L.; Yang, X.J.; Fu, D.H.; Zhang, L.Y.; Li, C.Y.; Cui, Z.D. (2005). Stress-strain behavior of porous NiTi alloys prepared by powders sintering, *Materials Science and Engineering A,* Vol.408, pp. 264-268, ISSN 0921-509

Chapter 3

$Ni_{25}Ti_{50}Cu_{25}$ Shape Memory Alloy Produced by Nonconventional Techniques

Tomasz Goryczka

Additional information is available at the end of the chapter

http://dx.doi.org/10.5772/48764

1. Introduction

Titanium-nickel alloys with a chemical composition close to equiatomic remain in the centre of interest due to the unique properties associated with shape memory effect (SME). In opposite to polymers or ceramic, mechanism of the shape memory effect is strictly correlated to the thermoelastic reversible martensitic transformation (MT). Modification of the course of the martensitic transformation influences the shape memory effect. Generally, course of the martensitic transformation can be influenced, separately or simultaneously, by two ways:

- Modification of chemical composition: addition and/or substitution of the alloying elements (Otsuka at all 1998; Van Humbeeck 1997). Alloying elements such as aluminum, iron, cobalt, cause lowering temperatures of the martensitic transformation even down to -140°C. Addition of copper (Duerig at all 1990), hafnium or zirconium (Li at all 2006) causes an increase of the transformation temperatures. In case of hafnium and zirconium, it is possible to obtain high-temperature shape memory alloys, in which the reversible martensitic transformation occurs at temperatures between 300 and 400°C (Santamarta at all 1999; Monastyrsky at all 2002).
- Modification of microstructure: way of production and/or farther alloy processing. The course of the martensitic transformation and its reversible nature is also affected by the structure of defects formed during the manufacturing or alloy processing. In order to change the transformation temperatures, in fact - the thermal range of the shape memory effect, mostly heat treatment or thermo-mechanical treatment is applied (Besseghini at all 1999; Kima at all 2006; Morawiec at all 1996). However, this requires an additional investments expenditure on devices. In consequence, it increases time of production process as well as its total cost.

Ternary NiTiCu alloy belongs to a large family of Ni-Ti alloys, which reveals shape memory effect (Mercier at all 1978). In the NiTiCu alloys content of titanium is close to 50at.% while nickel does not exceed 20at.%. The rest is copper (Melton at all 1979). Substituting either nickel or titanium results in increasing the characteristic temperatures of the martensitic transformation, when compared to a binary NiTi alloy. In results of that, a thermal range of the shape memory effect moves from the room temperature up to approximately 80^{0}C (Duerig at all 1990). Moreover, copper causes good stability of transformation temperatures as well as prevents from formation of Ti_3Ni_4 precipitation (Fukuda at all 1995). However, the most important feature is appearance of the multistep martensitic transformation and formation of different type of the martensite structures. It results in receiving shape memory alloy, in which a hysteresis of the martensitic transformation can be narrowed from about 30 degrees down to 15 degrees. Dependently on the copper addition, the course of the martensitic transformation can be as follows (Nam at all 1990; Nomura at all 1990; Nomura at all 1992):

- the copper lower than 5-10 at% during cooling the parent phase B2 transforms directly to the B19' monoclinic martensite; thermal hysteresis is about 20 degrees;
- for 10-15at% Cu - transformation occurs in two steps: first, from the B2 parent phase the orthorhombic martensite B19 is formed, next the B19 martensite transforms to the B19' monoclinic martensite; thermal hysteresis is about 30 degrees;
- when the Cu content exceed 20%, again one-step transformation can be observed, however the B2 parent phase transforms directly to the B19 orthorhombic martensite; thermal hysteresis is about 10 degrees;

Special attention was drawn by alloy with nickel content of 25 at.% Ni and copper also 25 at.%. It was due to the wide possibility of its potential application (Gil at all 2004; Grossmann at all 2009; Colombo at all 2006) . The reason for this is the narrow thermal hysteresis loop of the martensitic transformation. Its width does not exceed 15^{0}C.

Recently, for manufacturing of the NiTi-based alloys, intensive effort has been put to adoption of the nonconventional production techniques such as powder metallurgy (PM) (Goryczka at all 2008); Li, 2000), melt-spinning (MS) (Santamarta at all 2004; Morgiel at all 2002; Goryczka at all 2001; Rosner at all 2001) or twin roll casting (TRC) (Goryczka and Ochin 2005; Dalle at all 2003; T. Goryczka, (2004)). This creates a wider possibility of the alloy applications. Despite the fact that these techniques have been successfully applied to metals and alloys production, their application for SMAs manufacturing is quite new.

The subtle nature of martensitic transformation and shape memory effect requires the selection of appropriate process conditions and parameters. In fact not many references about the influence of the processing parameters on the shape memory alloys and transformation behavior in NiTiCu alloy can be found.

The chapter is focused on $Ni_{25}Ti_{50}Cu_{25}$ alloy manufactured by powder metallurgy or rapid solidification techniques. Comparison of transformation behavior, structure, and microstructure is discussed.

2. Powder metallurgy

2.1. From powders to alloy

In order to produce NiTiCu alloy with the nominal chemical composition (25at.% Ni, 50at.% Ti and 5at.% Cu) commercial powders of elements (purity 99,7%) were weighted in proper ratio and mixed in a rotary mixer during 48 hours. Measured average particle size of as-received powders was 6.2 μm, 11.5 μm and 33.7 μm for Cu, Ni and Ti, respectively (Fig. 1).

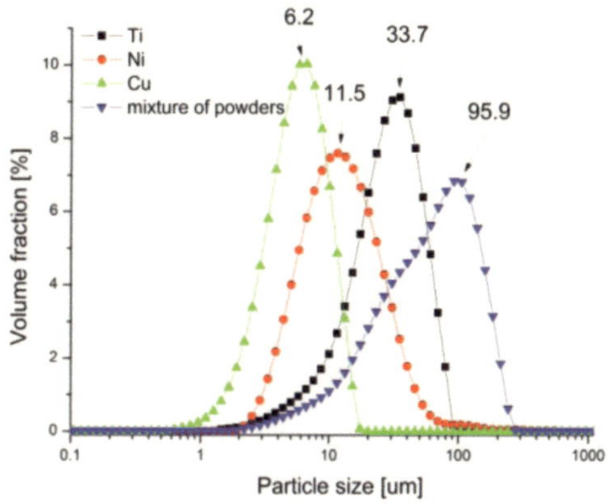

Figure 1. Particle size distribution for powders and their mixture.

During mixing the mixture was homogenized. Simultaneously, powders were agglomerated in mechanical alloying process (Fig. 2a). Due to the significant differences in hardness of the alloying elements (Cu - 369HV, Ti - 970HV, Ni - 638HV), this effect becomes clear between Ti and Cu as well as Ni and Cu. Figure 2b shows distribution of the elements from region marked in Figure 2a . The large particles were composed of Cu and Ti powders as well as Cu and Ni. In consequence, the powders were homogenously distributed in the mixture and its average particle size increased up to 95.9 μm (Fig. 1).

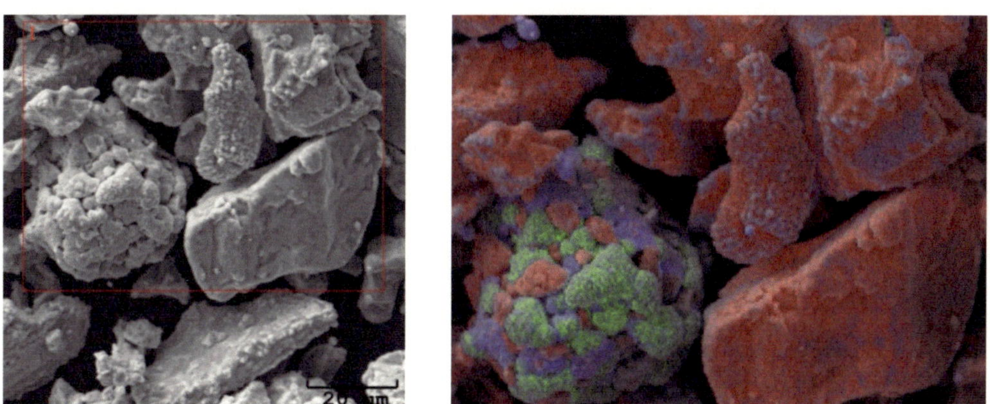

(a) (b)

Figure 2. SEM image of the powder's mixture (a) and distribution of the elements from region marked in left image (red – titanium, green - nickel, blue – copper) (b).

From mixture, compacts were formed at room temperature under pressure of 8 MPa. They were in a shape of cylinder with diameter of 10 mm and 6 mm high. Sintering was carried out in a horizontal tube furnace under argon atmosphere. Sintering conditions were chosen in reference to the melting temperature (1264⁰C). It varied from 850⁰C to 1100⁰C and total sintering time from 7 to 20 hours. After sintering, furnace was slowly cooled down to the room temperature.

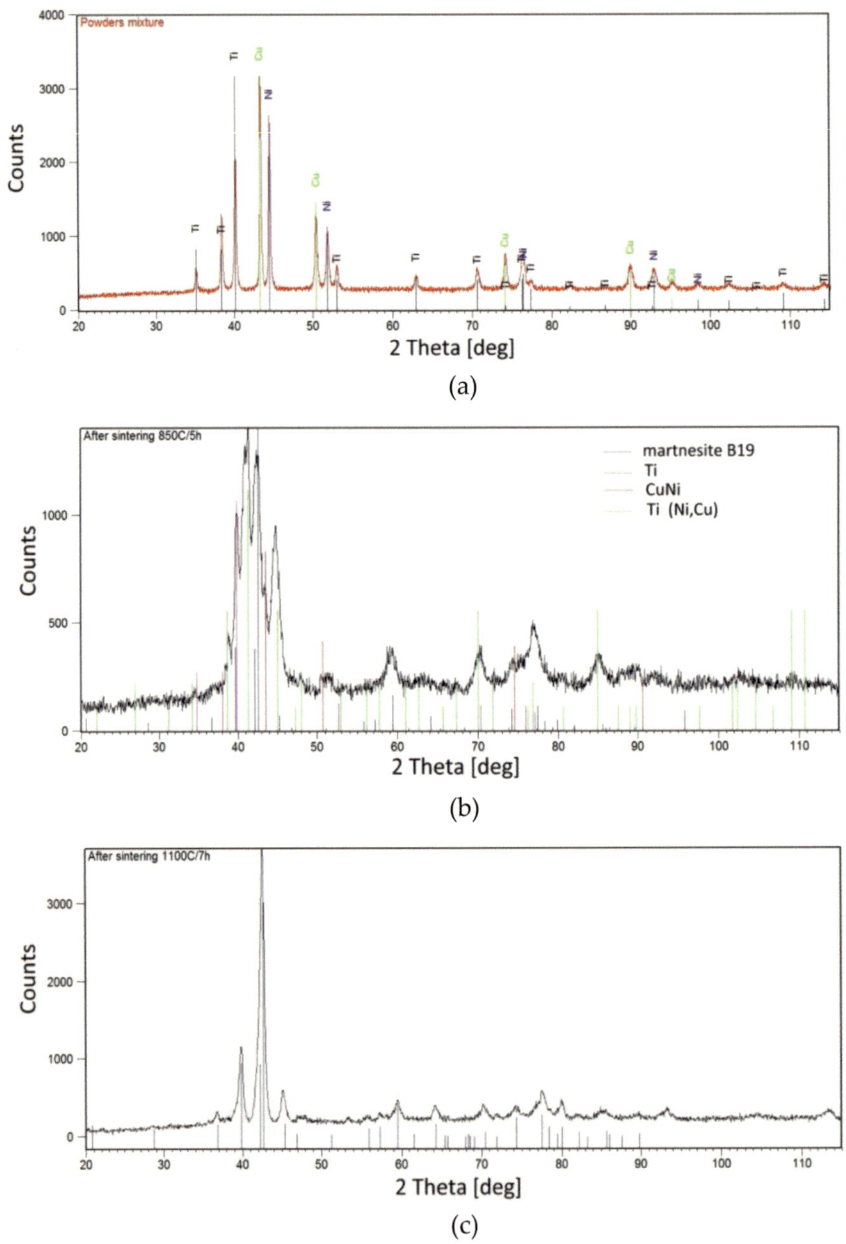

Figure 3. X-ray diffraction patterns registered for powders mixture (a), alloy sintered at 850⁰C for 5 hours (b) and at 1100⁰C for 7 hours (c).

In order to prove, that after sintering the NiTiCu alloys were obtained, X-ray diffraction analysis was done for compacts as well as blends. X-ray diffraction pattern were measured using the X'Pert Pro diffractometer with Cu radiation ($K_{\alpha 1 \text{ and } \alpha 2}$) in the 2θ range: 20 ÷ 140 degrees. Figure 3a shows X-ray diffraction pattern registered for compact before sintering. It can be clearly seen that diffraction pattern contains diffraction lines, which belongs to the pure alloying elements. Separated phases were identified using the ICDD data base PDF-2. It was proved that powders mixture consisted from Ti, Ni and Cu (card no Ti: 065-3362; card no Ni: 04-0850 and card no Cu: 04-0836). Moreover, quantitative analysis done using the Rietveld refinement confirmed that measured chemical composition was comparable to the nominal one. The main goal of sintering was to produce transformable phase, which can be described by nominal chemical composition of the alloy as $Ni_{25}Ti_{50}Cu_{25}$. For such composition, in a bulk alloy, the orthorhombic B19 martensite appears. It undergoes martensitic transformation at temperature above 75^{0}C creating the B2 parent phase (Duerig at all 1990). X-ray diffraction patterns registered at room temperature showed that independently on applied sintering conditions, the orthorhombic B19 martensite was found in all blends. However, sintering took an effect on the phase composition. One has to remember that mechanism of the alloying, which occurs during sintering is different from that, in tradition melting. During sintering alloying elements diffuse each one into other. Thus, from the ternary powder mixture reaction between alloying elements may be combined. In result of that various phases coming from the systems: Ni-Ti, Cu-Ni as well as Cu-Ti may be expected (Gupta 2002). Precursor of that is partial mechanical alloying, which occurred during powders mixing (Fig. 2).

Sintering conditions		Fraction of phase [%]			
Temperature [^{0}C]	Time [h]	B19	Ti_2(Ni,Cu)	NiCu	Ti
850	5	82.7 ± 2.9	13.7 ± 0.6	2.4 ± 0.3	1.2 ± 0.2
	10	83.9 ± 3.7	16.1 ± 0.6	-	-
	20	86.1 ± 2.9	13.9 ± 0.5	-	-
950	7	89.7 ± 1.9	10.3 ± 0.3	-	-
1100	7	100	-	-	-

Table 1. Results of quantitative phase analysis done for blends sintered at various conditions

Lower sintering temperature in combination with short sintering time was not enough to provide a support for diffusion of the alloying elements. In consequence, apart of the B19 martensite also equilibrium phases appeared in sintered blends. Figure 3b shows X-ray diffraction pattern measured, at room temperature, for the blend sintered at 850^{0}C for 5 hours. Qualitative phase analysis revealed that the blend contained also phases such as: NiCu (PDF-2 card no 47-1406), Ti_2(Ni,Cu) – which has crystallographic structure of the Ti_2Ni phase (PDF-2 card no 05-0687) as well as a solid solution of Ni in Ti (PDF-2 card no 89-3073). In order to study of phase contribution in sintered blends the Rietveld refinement was done. Values of reliability factors, which prove goodness of the refinement, were below 8%. Obtained results were compared in Table 1. Quantitative analysis showed that amount of the B19 martensite was about 83% in blends sintered at 850^{0}C for 5 hours. Extending of the sintering time up to 10 hours was enough for nickel and cooper diffusion. Both phases: NiCu and pure titanium were not observed. Extending of the sintering temperature resulted in the limitation of the $Ti_2(Ni,Cu)$ phase formation. The best sintering condition was: temperature 1100⁰C with sintering time of 7 hours. In the sintered blend only the B19 martensite was identified (Fig. 3c).

Obtained results were confirmed by observation carried out using the JEOL JSM 6480 electron scanning microscope. Observations were done on a circular cross-section of the cylindrical blends. Figure 4 shows images taken using back scattered electrons. In places, which varied in a contrast, a chemical composition was determined using energy-dispersive X-ray analysis (EDX) in the JEOL JSM 6480 microscope. Measurements were done in macro scale and 50 points were taken for calculation. Calculated ratio of the alloying elements was in a good correlation with the phases identified with use of X-ray diffraction analysis. SEM image observed for the blends, sintered at 1100⁰C for 7 hours, proved that only the B19 martensite was present at the sample (Fig. 4a).

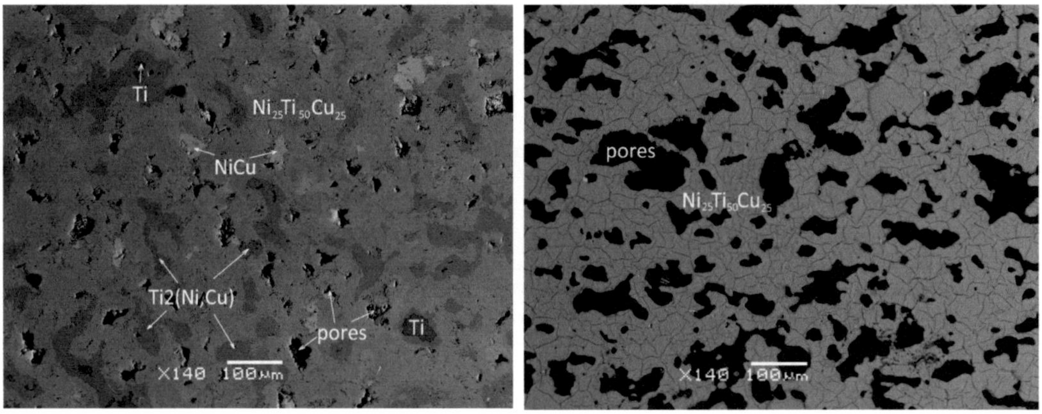

Figure 4. SEM images observed at a cross-section of blends sintered at 850⁰C for 5 hours (a) and 1100⁰C for 7 hours (b)

Sintering, mainly used for ceramic production, also produces pores. Their amount, size and distribution strictly depend on sintering condition as well as particle size of the elemental powders. Dependently on final application of the product, presence of the pores can be considered as a desired element of the microstructure. For example, in medical application pores can be considered as wanted ones, in which the collagen fiber finds way to grow-in. However, their diameter has to be appropriate and varied from 50 to 500 µm (Karageorgiou 2005; Itala 2001; Simske 1995). In case of shape memory alloys, pores do not provide a material, which undergoes the martensitic transformation. For that reason they can be considered as not desired element of the microstructure. In comparison to a bulk material, from the same volume of the sintered alloy lower transformation enthalpy can be received (Yuan 2006). Also, pores form natural barrier for diffusion of alloying elements. It may lead to inhomogeneity even in the frame of grains of the same phase. In order to overcome such inconvenience, sintering time should be extended and/or higher sintering temperature is required.

To bring more light for this problem in the $Ni_{25}Ti_{50}Cu_{25}$ alloy, amount of pores was calculated basing on the SEM images taken over whole cross-section of the sample.

In general, the porosity in alloy sintered at 850⁰C did not exceed 5% (Fig. 5a). The pores were randomly distributed and clearly separated (Fig. 4a). However, pore size varied dependently on the sintering time. The average pore size in the alloy sintered 5 hours was about 8 μm (Fig. 5a). Increase of the sintering time to 20 hours was responsible for increase of the pore size to 38 μm. It is worthy to notice, that calculated values of standard deviation for pore size, in the alloys sintered for 5 or 10 hours, are very close (Fig. 5a). It means that relatively short sintering time did not influence on scattering of the pores size. The pores have similar diameter and were comparable in a shape. Completely different value of standard deviation was obtained for the alloy sintered for 20 hours. It was ±15 μm. Extending of the sintering time up to 20 hours caused formation of the pores, which varied in size from about 8 μm up to 60 μm. In most cases they have irregular shape.

Significant differences in porosity, size of pores and pores distribution can be observed for samples, which were produced applying higher temperatures (Fig. 4b). Increase of the sintering temperature up to 950⁰C or 1100⁰C caused increase of porosity up to 21% or 27%, respectively (Fig. 5). However, alloy sintered at 950⁰C revealed comparable morphology to one observed in alloy sintered at 850⁰C for 5 or 10 hours. The pores, with average diameter of 41 μm, possessed similar regular elliptical shape. Increase of the sintering temperature up to 1100⁰C caused differences in shape and distribution of pores. Average diameter of pores increased to 50 μm. Higher value of the standard deviation indicated that they varied in shape and size. At 1100⁰C diffusion of alloying elements was more intense. In result of that, most of the small pores were joined into large one forming their irregular shape (Fig. 4b).

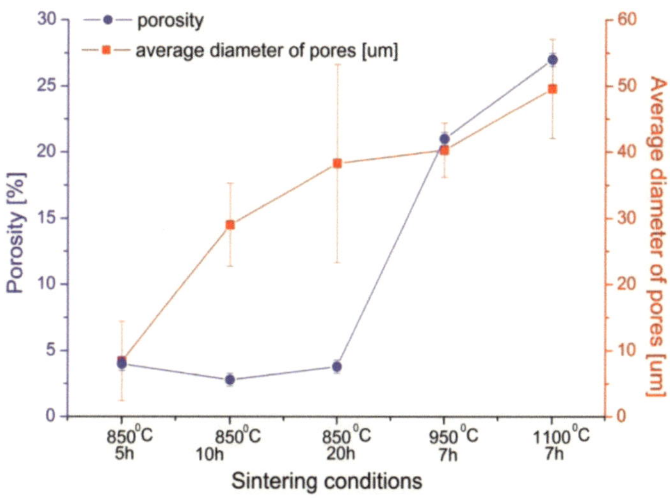

Figure 5. Porosity and average diameter of pores versus sintering conditions

2.2. Martensitic transformation

In order to show influence of the sintering condition on the course of the martensitic transformation DSC cooling/heating curves were registered at temperature range between -100°C and +200°C. Measurements were done using differential scanning calorimeter Perkin Elmer DSC-7 with cooling/heating rate of 10°C /min. Evolution of the DSC cooling/heating curves registered for sintered alloys is shown in Figure 6a. The characteristic temperatures of the martensitic transformation (start M_s, A_s and finish M_f, A_f of the forward and reveres transformation, respectively) were determined using a slope line extension method. Enthalpy of transformation was calculated from the thermal peak (Fig. 6b).

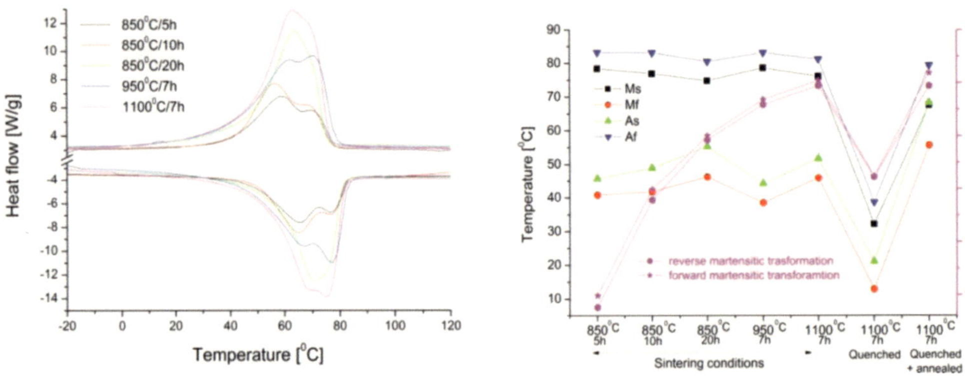

Figure 6. DSC cooling/heating curves (a) and parameters of the martensitic transformation (b) for alloy sintered at various conditions

Generally, all sintered alloys reveal presence of the reversible martensitic transformation. Thus, the main condition for appearing of shape memory effect is satisfied. However, course of the martensitic transformation varies dependently on applied sintering conditions. First, the M_s and A_f temperatures show general tendency for decreasing when the condition for diffusion of alloying elements improves. In opposition to that, the A_s and M_f increases. It takes an effect on thermal ranges of the transformation. Second, the martensitic transformation occurs as one step or two steps transition.

Let's discuss firstly change of the chemical composition as a result of the presence of non-transformable the $Ti_2(Ni,Cu)$ phase. It posses the same type of the crystal structure as the Ti_2Ni phase. In the $Ti_2(Ni,Cu)$ phase, the ratio of Ni atoms to Cu is as 30:1. The presence of this phase decreases titanium content in the transformable phase. Also, one have to keep on mind that the fraction of the $Ti_2(Ni,Cu)$ phase decreases from 16% to 0% dependently on sintering conditions. In consequence, the M_s temperature slightly moved to lower region. Influence of the chemical composition on the M_s temperature has been known and used as key feature in designing of shape memory alloy. However, in ternary alloys this problem is more complex. The thermal behavior of the NiTiCu alloy depends on titanium-nickel ratio as well as nickel-copper ratio. Hanlon et all (Hanlon 1967) shown, that in the NiTi alloy with titanium content around 50-51at% takes a little effect on the M_s temperature. However, when the titanium content lowers from 50 to 49at% the M_s temperature may decrease from 50⁰C to -140⁰C, respectively. In opposition to that increase of copper content, instead of nickel atoms, causes increase of the M_s temperature. For the B2↔B19 transformation in NiTiCu bulk, Mobelry and Melton (Moberly and Melton 1990) showed that increase of copper content from 20 to 25 at% caused increase of the M_s from 48⁰C to 76⁰C, respectively.

The presence of the non-transformable phase takes an effect on the transformation enthalpy (Fig. 6b). Enthalpy of the transformation, at fully transformable bulk alloy, equals about 12 J/g (Hanlon 1967). Presence of non-transformable phases decreases an amount of the phase, which undergoes the martensitic transformation. In result of that, the transformation enthalpy decreases.

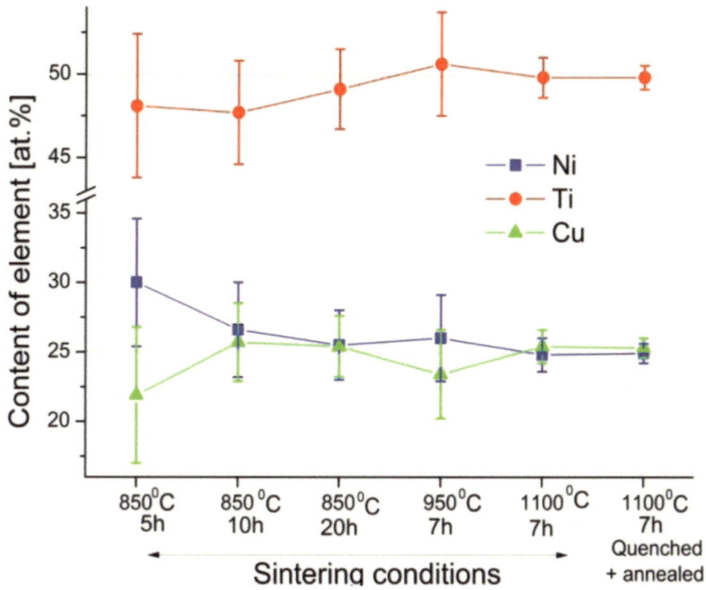

Figure 7. Distribution of the alloying elements, in the transformable phase, versus sintering conditions

The thermal behavior of the sintered alloy may also be a consequence of inhomogeneity of the chemical composition inside of the transformable phase. Figure 7 shows the distribution of the elements versus sintering conditions. Measurements were done in the transformable phase. Calculation of the chemical composition was done over 50 points spread on the cross-section of the sample. It has been clearly seen that a distribution of alloying elements, close to the nominal composition, was obtained in blends sintered at 1100⁰C for 7 hours. It was as follows: Ti - 49.8at.%, Ni - 24.8 at.% and Cu-25.4 at.%. In order to analyze the homogeneity or inhomogeneity, a standard deviation was marked as a vertical bar. The sintering at 1100⁰C for 7 hours causes a small deviation from the average content of the alloying elements and is characterized by relatively low value of the standard deviation. The value of the standard deviation calculated for sintering temperature 850⁰C and time 5 hours or 10 significantly increases to 5% for Ni and Cu as well as for Ti up to 4%. High inhomogeneity of these samples eliminates them as a shape memory material. Lower values of the standard deviation (about 3%) were obtained for sintering at 950^{0}C and 10 hours. However, in this alloy Ti content increased up to 50.6at.%. This was a reason for increase of the M_s temperatures to 78.6^{0}C. In the sample sintered at 850^{0}C for 20 hours the standard deviation was about 2%. However, measured chemical composition slightly varied from the nominal one.

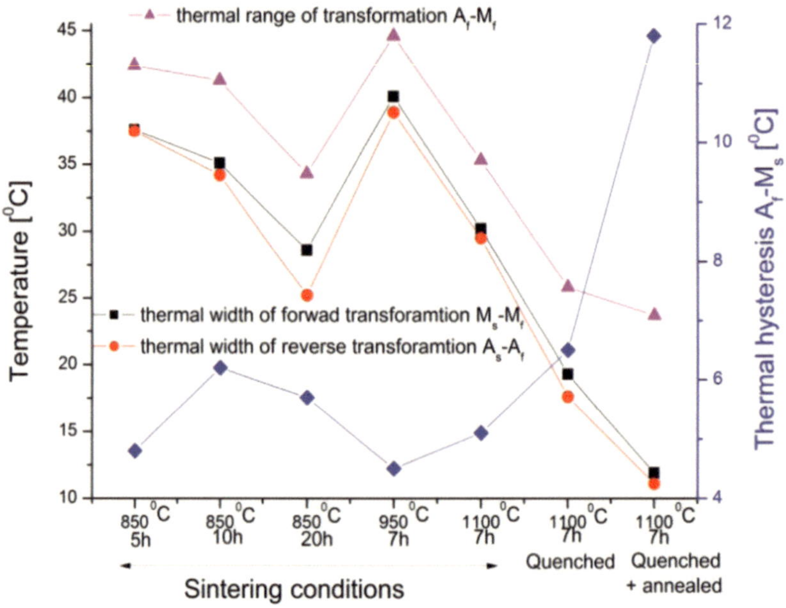

Figure 8. Thermal ranges of the reversible martensitic transformation versus sintering conditions

The local changes of the chemical and its inhomogeneity took an effect on the course of the martensitic transformation. For alloy, sintered at 850^{0}C for 5 hours, the DSC curves reveal two maxima on cooling (at 58.6^{0}C and 69.4^{0}C) as well as two minima on heating (at 65.4^{0}C and 76.8^{0}C). The range of the martensitic transformation (A_f–M_f) equals 42 degrees (Fig. 8 – green line). Thermal hysteresis (A_f–M_s) is about 5 degrees (Fig. 8 – blue line). Difference calculated between position of maximum (on cooling) and corresponding minimum (on heating) keeps comparable value (8^0). The range of the thermal peak on cooling (M_s-M_f) as well as heating (A_s-A_f) is wide and equaled 38^0 (Fig. 8 – red and black line, respectively). Measured X-ray diffraction pattern confirmed that at room temperature the B19 marteniste existed. Taking, all these facts into consideration, it can be concluded that the parent phase transforms to the B19 orthorhombic marteniste. However, in the thermal region of transformation two thermal peaks were overlapped. For such thermal behavior, inhomogeneity of chemical composition may be responsible. Calculated value of the standard deviation (Fig. 7) of measured composition shows significant inhomogeneity in samples. Regions of transformable phase, which differ in chemical composition, undergo martensitic transformation at close temperature, that's why two overlapped peaks were observed.

Neither elongation of the sintering time up to 10 hours nor rising sintering temperature up to 950⁰C did not change course of the transformation – still two overlapped B2↔B19 transformations were observed. A completely different thermal behavior is observed in the alloys sintered at 850⁰C for 20 hours as well as 1100⁰C and 7 hours. Instead of two peaks, one peak was found on cooling as well as on heating. In case of both sintering conditions the range of the martensitic transformation (A_f–M_f) reduces to 34⁰ (Fig. 8 – green line). Also, range of the thermal peak on cooling (M_s-M_f) as well as heating (A_s-A_f) narrows and equals 30⁰ (Fig. 8 – red and black line, respectively). Thermal hysteresis (*Af*–*Ms*) slightly differs and for 850⁰C/20 hours is about 5.7⁰ and for 1100⁰C/7 hours is lower and equals 5.1⁰.

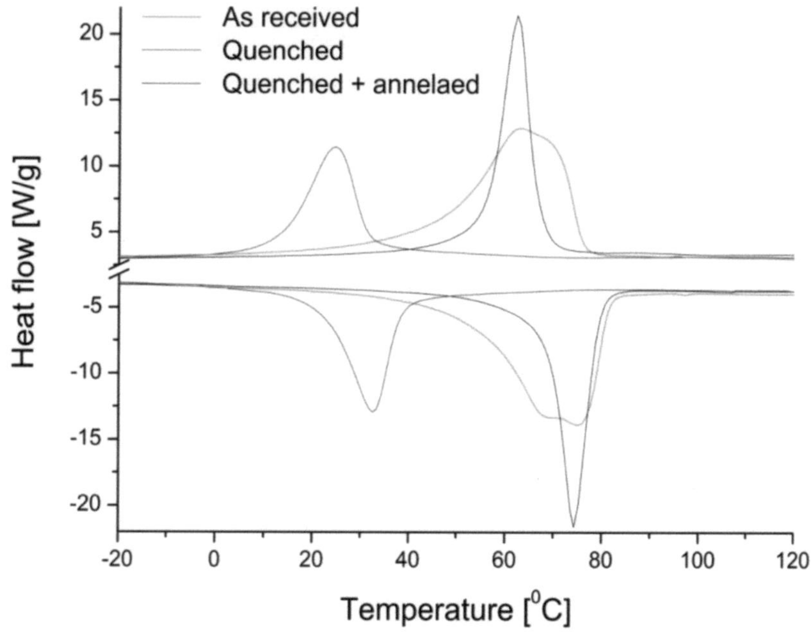

Figure 9. Comparison of DSC cooling/heating curves registered for alloy sintered at 1100⁰C for 7 hours after additional thermal treatment

For both sintered blends irregular shape of the thermal peaks measured on DSC cooling/heating curves was stated. It also may results in slight differences in chemical composition of the transformable phase (standard deviation was 2.2% or 1.2%). In order to improve quality of the alloy sintered at 1100⁰C/7 hours it was quenched from 900⁰C and additionally annealed at 500⁰C for 24 hours. Influence of thermal treatment on a course of the martensitic transformation is shown in Figure 9. First, only one thermal peak was measured during cooling as well as heating. It has symmetrical shape. Second, in comparison to the as-received sample, quenching moved down transformation temperatures of about 34 degrees (Fig. 6b). It is a result of quenched-in structural defects. This effect is better visible in rapidly solidified alloy and will be discussed in next subchapter. Additional annealing at 500⁰C for 24 hours restored thermal region of the reversible martensitic transformation. In consequence, the M_s transformation temperature was comparable to that calculated for the as-received sample (Fig. 8a). Summarizing, additional thermal treatment:

- did not change the chemical composition of the transformable phase: (Ti - 49.8at.%, Ni - 24.9 at.% and Cu-25.3 at.%);
- alloy was stable and did not decompose to equilibrium phases;
- homogeneity of the alloy was improved - calculated standard deviation of the chemical composition was 0.7%;
- the transformation enthalpy is comparable to that determined for bulk alloy (Fig. 6b) .

In result of that, the thermal range of the transformation as well as thermal width of forward and reversible transformation reduced to 23 degrees and 11 degrees, respectively (Fig. 8). However, width of the thermal hysteresis of the transformation increased from 6 (as-received sample) to 12 degrees. It may be caused by remains of the no-annealed point defects. Extending of the annealing temperature to 48 hours caused decomposition of the transformable phase. Again the $Ti_2(Ni,Cu)$ phase appeared.

3. Rapid solidification

Technologies for producing alloys in a finite shape directly from the molten state received a lot of attention. The rapid solidification techniques, realized in melt-spinning or twin roll casting technique, applied to the shape memory alloys offer not only a convenient and economical production technology but also possibilities to improve the alloys' properties as well as to control in a relatively easy way the temperatures of the reversible martensitic transformation. The increase in the wheel speed causes the grain refinement and changes their morphology from equiaxial to columnar (Eucken 1990; Stoobs 1979; Morawiec 2003).

3.1. Melt-spinning

In order to produce melt-spun ribbons, ingots with nominal chemical composition were cast in an induction furnace. Ingots weighted about 10 g. Casting was carried out in a pre-evacuated chamber before melting. Melt was ejected with pressure of 0.02 MPa on a rotating wheel. Increase of cooling rate was realized by increase of wheel speed and/or temperature of melt. Four ribbons were produced with processing parameters shown in Table 2.

Symbol	Melt temp. [^{0}C]	Wheel speed [$m \cdot s^{-1}$]	Thickness [μm]	Width [mm]
MS1	1250	11	83	6,3
MS2	1250	15	64	5,9
MS3	1250	19	48	2,7
MS4	1350	23	46	5,2

Table 2. Processing parameters and dimension of the ribbons

Generally, high cooling rate in the melt–spinning technique (10^6 K/s) causes inhomogeneity of the microstructure on the surface of the ribbons. It is due to fact that the liquid alloy is ejected onto fast rotating wheel. Firstly, the side, which has directly contact with rotating wheel, undergoes solidification. Secondly, until crystallization goes thorough the ribbon, the cooling rate lowers, and finally, the top surface solidify. This procedure produces two different surfaces: one what could be called "frozen" metal from the contact side and second one – top surface.

The combination of the wheel speed (11 m/s) and temperature of melt (1250°C) provided the lowest rate of cooling. It results in elongation of the time, when the liquid alloy contacts surface of rotating wheel. It provides enough time for crystallization of columnar grains, which extends along thickness of the ribbon (Fig. 10a).

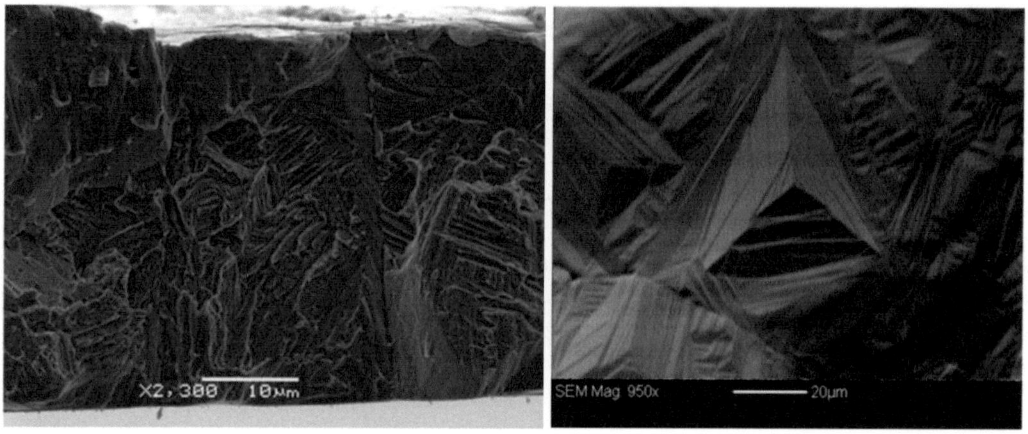

(a) (b)

Figure 10. SEM images observed at the fracture (a) and on the surface (b) of the ribbons MS2

On the surface of the MS1 ribbon self- accommodating plates of the martensite were observed (Fig. 10b). Similar images were observed in the MS2 ribbon, which was produced with higher wheel speed - 15 m/s. X-ray studies confirmed the presence of the B19 martensite. X-ray diffraction patterns registered at room temperature for the MS1 was identified as belonging to the B19 structure (Fig. 11 – "Ribbon MS1"). In case of the MS2 ribbon, apart from the B19 phase, the B2 parent phase was also identified (Fig. 11 – "Ribbon MS2"). It is a result of increase of the cooling rate, which caused increased of the structural defects quenched-in.

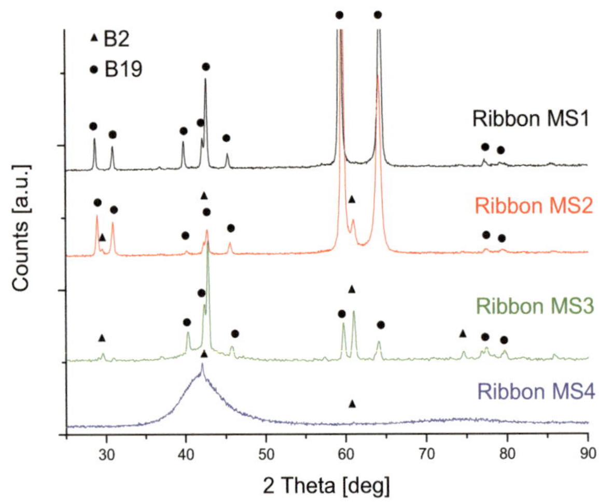

Figure 11. The X-ray diffraction patterns registered at room temperature for melt-spun ribbons

A different microstructure was observed in the MS3 ribbon (Fig. 12). Increase of the wheel speed to 19m/s, while the temperature of melt was the same (1250°C), did not provide a sufficiently time for crystallization of columnar grains. At the thickness of the ribbon three zones were formed (Fig. 12a). First, an amorphous layer, 3-5 μm thick, was formed as a contact side of the ribbon. The second zone, formed during solidification, is the area consisted of spherical grains with an average diameter of about 5-8 micrometers. Finally, the zone composed of the columnar grain is formed, which extends to the top surface of the ribbon. Microstructure of the ribbon is inhomogeneous. Images observed at the fracture, taken from various part of the ribbon, show the columnar grains with width to 35 μm extended along whole thickness of the ribbon. At this area, no amorphous phase was observed. Also in some parts of the ribbon, only small spherical grains were formed along the thickness (Fig. 12b).

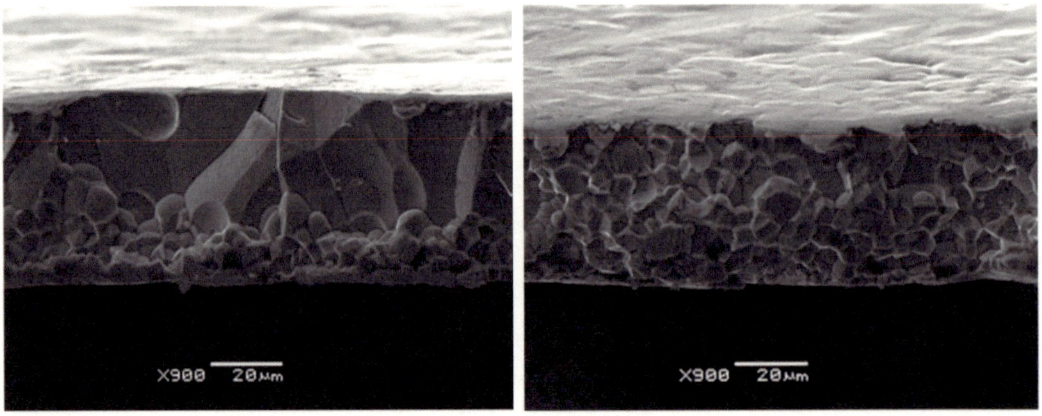

(a) (b)

Figure 12. SEM images observed at the fracture of the MS3 ribbon

Observation of the contact surface shows that this area is not homogeneous. Among the amorphous phase two groups of crystalline "islands" were observed (Fig. 13a). They were randomly distributed and differed in size. Diameter of the "islands" varied from 5 μm to 25 μm. Inside of crystalline areas the martensitic plates were clearly visible. The plates transforms to the parent phase after heating above A_f. After cooling, they reversibly transforms to the martensite.

X-ray diffraction patterns revealed that the MS3 ribbon consists of three phases: amorphous (weak broaden line), the B2 and the B19 (Fig. 11 – "Ribbon MS3"). Increase of the melt temperature to 1350°C and wheel speed to 23 m/s formed almost amorphous ribbon - MS4. The cooling rate realized with these processing parameters does not provide enough time for full crystallization during solidification. The contact surface is completely amorphous. However, at the top surface some crystallized grain can be distinguished (Fig. 13b). The X-ray diffraction patterns shows two broaden maxima characteristic for the amorphous phase. Moreover, weak diffraction lines belonging to the B2 parent phase were identified (Fig. 11 – "Ribbon MS4").

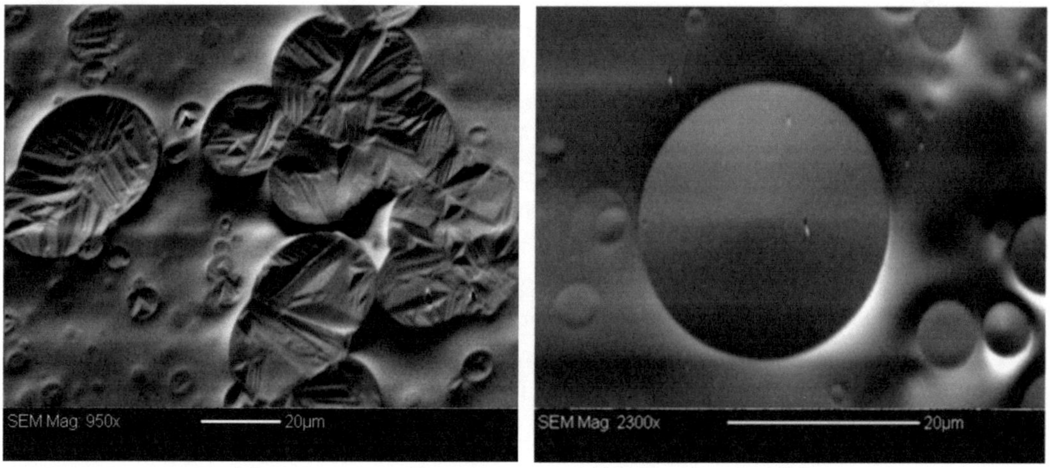

(a) (b)

Figure 13. SEM image of the contact side of the MS3 ribbons (a) and top surface of the MS4 ribbon (b).

From the point of the shape memory effect and the martensitic transformation presence of the amorphous phase is not desirable. The martensitic transformation as a coordinated atom movement at distance proportional to distance to the habit plane occurs only in the crystalline state. Presence of the amorphous phase increases amount of the transformable phase in the ribbons. In order to eliminate influence of the amorphous phase and quenched-in defects on the course of the martensitic transformation crystallization was carried out. Ribbon was heated up to 600°C, and then cooled to 0°C. Heating and cooling was done, inside of DSC calorimeter, with rate of 10 °C/min at argon atmosphere. Observed high temperature peaks correspond to the crystallisation process (Fig. 14). Crystallization of the amorphous phase starts at 445°C and finishes at 466°C. The value of enthalpy was 6.5 J/g. For further studies, the MS4 ribbon was used in crystalline state and denoted as "MS4C".

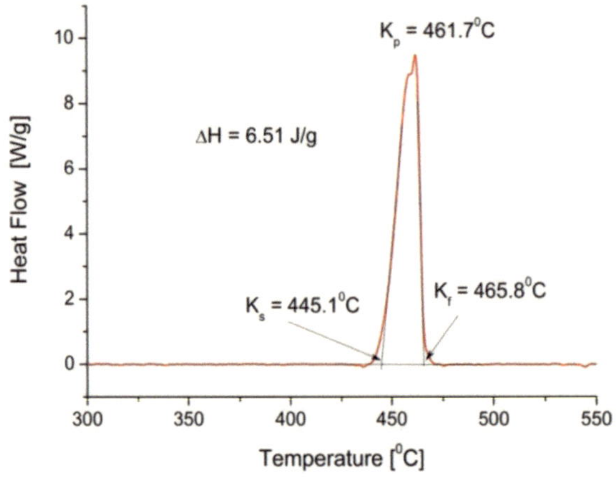

Figure 14. DSC heating curve registered for the MS4 ribbon

Applied casting parameters significantly influenced the course of the martensitic transformation. Figure 15a shows comparison of the DSC cooling/heating curves registered for the first cycle of heating-cooling. In order to obtain the full thermal cycle, the samples were cooled down to 0°C, then heated up to 110°C and again cooled to 0°C. The order and the shape of the thermal peaks occurring at thermograms proved that the martensitic transformation is reversible in all ribbons. However, they differ in the transformation temperatures. The DSC curves measured for the MS1 and MS4C ribbons contain one thermal peak on both heating and cooling curve, respectively. In comparison to that, the DSC curves measured for ribbons: MS2 and MS3 revealed a different course. The thermal peaks measured for the MS2 ribbon show irregular shape, whereas for the MS3 ribbon they were broadened and split. In result of that, for the MS3 ribbon two maxima at temperatures 33°C and 49°C for the forward martensitic transformation as well as two minima for the reverse transformation at 42°C and 55°C were found.

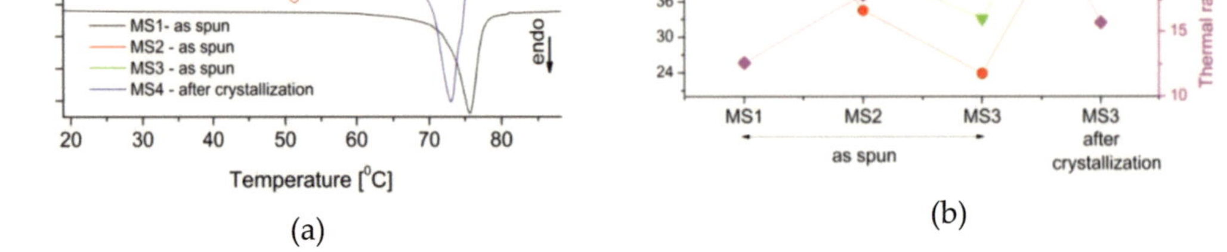

Figure 15. DSC cooling/heating curves measured for the melt-spun ribbons (a) and parameters of the martensitic transformation (b)

The combination of the wheel speed and the temperature of melt, applied in the case of the casting the MS1 ribbon ensured the lowest speed of solidification - similar to the equilibrium condition. The transformation temperatures, determined from DSC curves, are comparable to that determined for the MS4C ribbon (Fig. 15b). For alloys with low concentration of structural defects, the thermal range does not exceed 15 degrees. In the case of the MS1 and MS4C ribbon the thermal range of the martensitic transformation was 12 and 15°C, respectively.

Increase of the wheel speed to 15 m/s caused increase of the cooling rate in the MS2 ribbon. The density of the structural defects increased. Figure 16 shows an example of the image observed with use of the electron transition microscope. Relatively high density of dislocations can be distinguished. Dislocations as well as point defects increase local stress, which has to be overcome when the marteniste plates are formed. It requires additional overcooling and in consequence, the transformation temperatures moves to lower thermal region. In case of the MS2 ribbon transformation temperatures were lower of about 20 degrees in comparison that in the MS1 ribbon. Thermal range of the martensitic transformation increased to 18°C.

Further increase of the wheel speed to 19m/s caused increase of density of the structural defects. In comparison to the MS2 ribbon, the M_f and A_s temperatures decreased of about 8°C whereas the A_f and M_s increased of about 5°C. The thermal range of the martensitic transformation increased to 34°C.

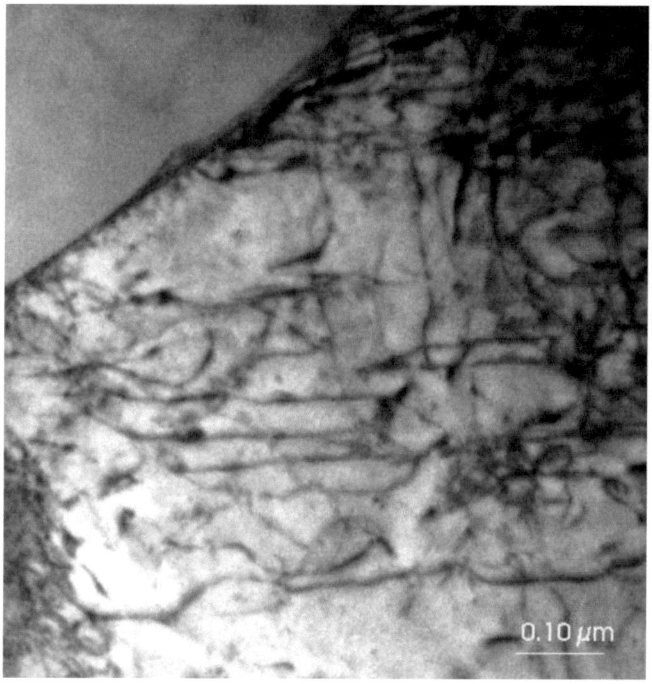

Figure 16. TEM image of the dislocation structure in the MS2 ribbon

3.2. Twin roll casting

Twin roll casting technique (TRC) has been known for almost 30 years in aluminum and iron casting industry (Berg, at all 1995) (Haga 2003). Thin strip, from 10 mm up to 0.5 mm, can be continuously produced replacing conventional processing techniques, which use an ingot and additional thermo-mechanical treatment (Goryczka, at all 2005). The main advantage of TRC is a relatively short processing route which joins casting as well as hot rolling in one step (Haga 2001). Also, low cost equipment, low energy consumption, space saving, strips are free of impurities which come from a crucible and surroundings etc. belong to the most desirable features of TRC technique. In spite of that, also, twin roll casting has some disadvantages. The main one is poorer mechanical properties of the strip in comparison to bulk.

Symbol	Melt temp. [^{0}C]	Rim material	Rollers velocity [m·s^{-1}]	Ejection pressure [MPa]	Pre-set gap [μm]	Nozzle hole [mm]	Thickness [μm]	Width [mm]
T1395	1395	Cu-Be-Co	0.6	0.025	100	3.0	296	45
T1400	1400	Cu-Be-Co	0.6	0.025	100	3.0	306	45
T1430	1430	Cu-Be-Co	0.6	0.025	100	3.0	304	45
T1450	1450	Cu-Be-Co	0.6	0.025	100	3.0	305	45

Table 3. Processing parameters and dimension of the strips

Four strips were produced from a previously cast ingots, with a nominal composition of Ni 25 at.%, Ti 50 at% and Cu 25 at.%. In order to get good quality strips processing parameters such a rolls velocity, a pre-set gap, an ejection pressure and diameter of nozzle hole were optimized and remained the same for all castings (Table 3). Casting parameters differs in solidification temperature, in the range 1395 - 1450^{0}C. Obtained strips were about 40 cm long and revealed smooth surface without any visible cracks and inclusions.

In twin roll technique ingot is placed to the quartz crucible fixed directly over a gap between two rolls. After induction melting, the melt is vertically ejected with argon pressure between two rolls rotated in opposite directions. Thin metal layers solidify on the surface of the rolls simultaneously, when the remaining liquid is spread from central part of the gap to its outer side forming the strip (Goryczka, at all 2005). As the crystallization front proceeds from both rolls the columnar grains are developed. Due to the two-sides solidification an interface between solidified shells is formed (Fig. 17).

However, final microstructure depends on the cooling rate, which is realized by combination between solidification temperature and roll velocity, as well as a gap between rotating rolls. In case of a narrow distance between rolls the strip solidifies over the end-point and is subjected to hot rolling. If the distance is wider the end-point is placed below roll's centre interface consist of pores and discontinuities (Goryczka at all, 2005). The microstructure of surface observed in the strip T1395, is shown in Figure 18a. General, in the strip T1395 solidification on wheel surface causes mountain-like ranges continuously strength along the strip. Each range consists of two zones. First one (A), containing fine

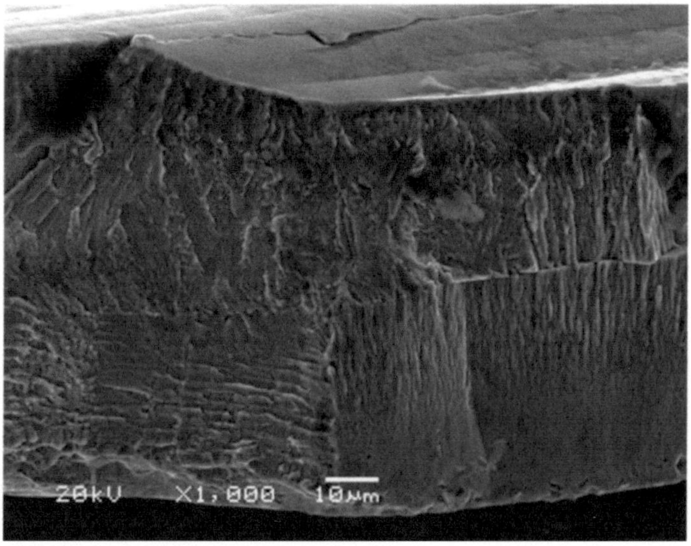

Figure 17. SEM image of fracture in the strip T1395

grains, is associated with hot rolling. Zone B reveals columnar grains growing perpendicular to the hot-rolling direction. Also, the highest applied cooling rate, realized in the strip T1450, causes formation of ranges, however, they consist of separated islands (Fig. 18b). Each island shows presence of zone A and B.

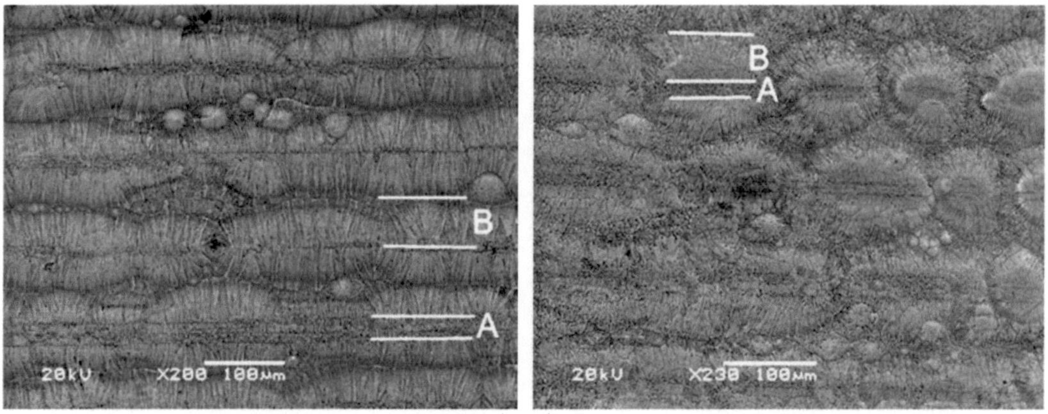

(a) (b)

Figure 18. Surface of the strip T1395 (a) and T1450 (b)

Influence of the solidification temperature on the course of the martensitic transformation was studied using DSC. Figure 19a shows comparison of the DSC cooling/heating curves measured for the as-cast strips. All DSC curves reveal only one exo- and one endothermal peak on cooling and heating, respectively. Similarly, to the sintered blends and melt-spun ribbons, the martensitic transformation occurs in one step. The parent phase transforms to the B19 orthorhombic martensite. From the DSC measurement the transformation temperatures were determined and compared in Figure 19b.

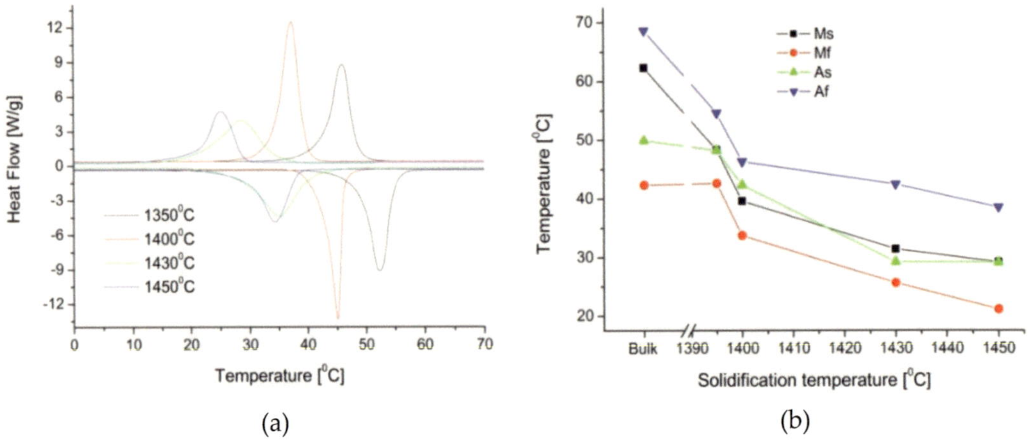

Figure 19. Comparison of the DSC cooling/heating curves (a) and transformation parameters (b) for as-cast strips.

It can be clearly seen, that increase of solidification temperature affects the transformation temperatures. According to a general trend, the martensitic transformation temperatures represented by M_s, M_f, A_s and A_f are lower in the strips than in the bulk alloys. Especially, this effect is enhanced in the melt-spun ribbons, where cooling rate is several times higher than in the strip. A progressive decrease of reversible and forward martensitic transformation temperature with respect to the bulk alloys was observed. For example, decrease of the A_f temperature reached almost 15 degrees when compare to the results obtained for T1395 and T1450. This is a typical behavior of the as-quenched alloy. Decrease of martensitic transformation temperature is caused by increasing numbers of quenched-in vacancies and stress field surrounded precipitates. Figure 20 shows TEM images of precipitates which were observed in the strips T1395 and T1430. EDS analysis revealed that chemical formula of precipitates is $Ti_2(Ni,Cu)$, which is non-transformable phase – stated in the sintered blends.

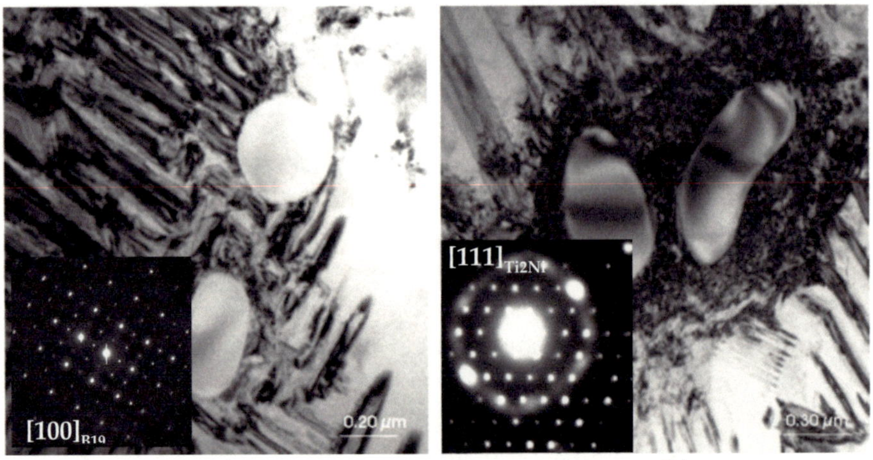

Figure 20. TEM images of precipitates observed in the strips T1395 (a) and T1430 (b)

Phase identification carried out using electron diffraction pattern proved the Ti_2Ni type of structure for the $Ti_2(Ni,Cu)$ phase (Fig. 20b). Moreover, in the strip produced with higher solidification rate precipitates are surrounded by highly deformed zone (Fig. 20b). It affects the course of the martensitic transformation. In order to trigger martensitic transformation some critical energy is required for nucleation. Energy accommodated around vacancies, dislocation and precipitates decreases driving force of the martensitic transformation. The stress generated in these zones requires additional energy for triggering the start of the martensitic transformation. In consequence, further overcooling is required for transformation continuing. In results of that, the transformation temperatures are moved to the lower thermal region (Fig. 19b). Also, the range of the martensitic transformation and the thermal hysteresis increase (Fig. 21). Additional measure of increasing amount of structural defects and non-transformable phase is degradation of transformation enthalpy. Figure 21 (magenta line) shows enthalpy dependence on solidification temperature. Enthalpy determined for forward and reversible transformation reveal similar behavior – increase of the solidification temperature causes decrease of their value.

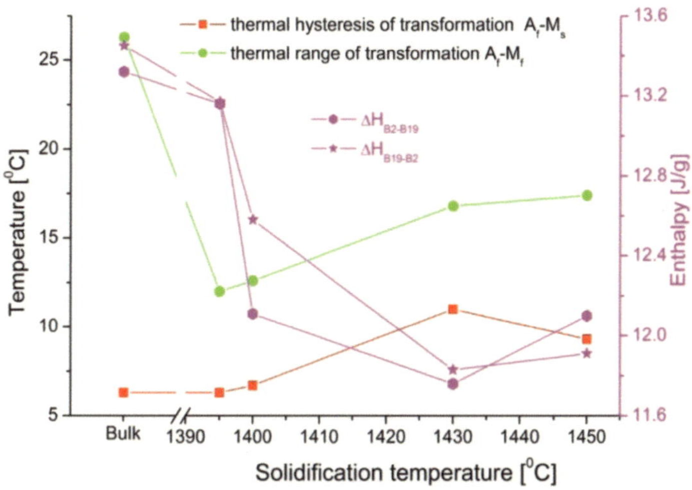

Figure 21. Thermal ranges of the reversible martensitic transformation versus the solidification temperature

4. Summary

This chapter extends the state of knowledge about the technological possibilities of the $Ti_{50}Ni_{25}Cu_{25}$ shape memory alloy manufacturing. It has been proved that, at the stage of alloy production, it is possible to control the reversibility, thermal range and thermal region of the martensitic transformation. Thus, it allows influencing the thermal range of the shape memory effect. Such methods for the $Ti_{50}Ni_{25}Cu_{25}$ alloy manufacture are powders metallurgy, melt-spinning and twin roll casting.

The porous $Ti_{50}Ni_{25}Cu_{25}$ shape memory alloys have been successfully manufactured by powder technology. However, only a correct combination of sintering temperature and time leads to a homogeneous alloy. The best sintering condition for the $Ti_{50}Ni_{25}Cu_{25}$ alloy demands sintering temperature of 1100°C with the sintering time of 7 hours. Additional thermal treatment such as quenching and additional annealing extends possibility for thermal control of the martensitic transformation. Produced alloy reveals a similar reversible martensitic transformation when compared to dense material. The porosity in homogenous samples does not exceed 25%.

Rapid solidification techniques, realized by means of melt-spinning as well as twin roll casting technique, allows producing fully transformable $Ti_{50}Ni_{25}Cu_{25}$ alloy. The advantage of these techniques, in comparison to traditional casting, is wide possibility for controlling thermal region of the martensitic transformation. Moreover, thin ribbons or strips can be produced directly from the melt in thickness varied from 30 μm to 350 μm. It allows avoiding of mechanical treatment needed after traditionally cast alloy. Also, control of the cooling rate allows for steering of the course of the martensitic transformation.

In melt-spinning technique, the martensitic transformation is sensitive to the wheel speed – processing parameter. Applying speed from the range of 11 m/s -19m/s enables receiving $Ti_{50}Ni_{25}Cu_{25}$ alloy, which transforms in the thermal range from 25°C to 80°C. Increase of the cooling rate, realized when melt temperatures is higher than 1350 °C and the wheel speed increases to 23 m/s, leads to obtaining alloy in amorphous state. However, it opens way for production of nanocrystalline $Ti_{50}Ni_{25}Cu_{25}$ shape memory alloy (Ye Xu, 2009).

Application of the twin roll casting technique moves the thermal range of the martensitic transformation to region between temperatures: 20°C and 55°C.

Author details

Tomasz Goryczka
University of Silesia, Institute of Materials Science, Katowice, Poland

Acknowledgement

The author thanks Dr. Patrick Ochin (*ICMPE Institut de Chimie et des Matériaux Paris Est)* for cooperation in shape memory alloys manufacturing with use of rapid solidification techniques.

5. References

B. Yuan X.P. Zhang, C.Y. Chung, M. Zhu *Materials Science and Engineering* A 438–440. - (2006). pp. 585–588.

Berg B.S. *Journal of Materials Processing Technology* 53. (1995). pp. 65-74.

Besseghini S. Villa E., Tuissi A. *Materials Science and Engineering* A273–275. (1999). pp. 390–394.

Ch. Grossmann J. Frenzel, V. Samphath, T. Depka, G. Eggeler, *Metall. Mater. Trans.* 40A. - (2009). - pp. 2531–2544.

F. Dalle G. Despert, Ph. Vermaut, R. Portier, A. Dezellus, P. Plaindoux, P Ochin, *Mater. Sci. Eng.* A 346. - (2003). - pp. 320–327.

F. Fukuda T. Kakeshita, M. Kitayama, K. Saburi, *J. Phys.* IV 5. - (1995). - pp. C8–717.

F.J. Gil E. Solano, J. Penal, E. Engel, A. Mendoza, J.A. Planell, *J. Mater. Sci. Mater. Med.* 15 - (2004). - pp. 1181–1185.

G.E. Monastyrsky V. Odnosum, J. Van Humbeeck, V.I. Kolomytsev, Yu.N. Koval, *Intermetallics* 10 - (2002). - pp. 95–103.

Goryczka T. and Ochin P. *Journal of Materials Processing Technology* 162–163. - (2005). - pp. 178–183.

Goryczka T. Van Humbeeck *J Journal of Alloys and Compunds* 456. - (2008). - pp. 194-200.

Gupta K.P. [Article] // Journal of Phase Equilibria Vol. 23. - 2002. - pp. 541-547.

H. Funakubo *Shape memory alloys*, Amsterdam: Gordon and Breach Science Publisher, 1987.

H. Morawiec T. Goryczka, J. Lelatko, D. Stróż, M. Gigla, P. Tkacz, *Arch. Mater. Sci.* 24. - (2003). - p. 5-21.

H. Rosner P. Schloßmacher, A.V. Shelyakov, A.M. Glezer, *Acta Mater.* 49. - (2001). - p. 1541.

Haga T. Suzuki S. *Journal of Materials Processing Technology* 118. - (2001). - pp. 165-168.

Haga T. Takahashi K., Ikawa M., Watari H *Journal of Materials Processing Technology* 140. - (2003). - pp. 610-615.

Van Humbeeck J. *J. Phys. IV France* 7. - (1997). - pp. C5-3-C5-11.

Itala AI Ylanen HO, Ekholm C, Karlsson KH, Aro HT. *J Biomed Mater Res.* 58. - (2001). - pp. 679-683.

J. Morgiel E. Cesari, J. Pons, A. Pasko, J. Dutkiewicz, *J. Mater. Sci.* 37. - (2002). - pp. 5319–5325.

J.E. Hanlon S.R. Butler and R.J. Wasilewski, *Trans. AIME 239*. - (1967). - pp. 1323-1327.

Karageorgiou V Kaplan D. *Biomaterials* 26. - 2005. - pp. 5474-5491.

Kima H.Y. Kima J.I., Inamura T., Hosoda H., Miyazaki S. *Mat. Sci.e and Eng. A* 438–440. - (2006). - pp. 839–843.

Li B.Y., Rong, L.J., Li, Y.Y., Gjunter, V.E., *Intermetallics* 8. - 2000. - pp. 881–884.

Li C. Zheng Y. F. *Materials Letters* 60 - (2006). - pp. 1646–1650..

Melton: O. Mercier and K. N *Met. Trans. A,*. - (1979). - Vol. 10A. - p. 387.

Mercier: K. N. Melton and O. *Met. Trans. A,*. - (1978). - Vol. 9A. - p. 1487.

Morawiec H. Stróż D., Goryczka T., Chrobak D. *Scripta Materialia* 35. - (1996). - pp. 485-490.

Nomura K. Tsuji and K. *J. Mater. Sci., 27,*. - 1992. - p. 2199.

Nomura K. Tsuji and K. *Scripta Metall., 24.* - 1990. - p. 2037.

Otsuka K. Wayman C. M. *Shape Memory Materials* Cambridge: Cambridge University Press, (1998).

R. Santamarta E. Cesari, J. Pons, T. Goryczka, *Metall. Mater. Trans.* 35A. - (2004). - pp. 761–770.

S. Colombo C. Cannizzo, F. Gariboldi, G. Airoldi, *J. Alloys Compd.* 422. - (2006). - pp. 313–320.

S. Eucken J. Hirsch *Mater. Sci. For.* 56–58. - (1990). - p. 487.

Santamarta R. Segui C., Pons J., Cesari E. *Scripta Materialia* 41. - (1999). - pp. 867–872.

Simske SI and Sachdeva R. *J Biomed Mater Res.* 29. - (1995). - pp. 527-533.

T. Goryczka, M. Karolus, H. Morawiec, P. Ochin *Journal de Physique IV,* 11. - (2001). - pp. Pr8-345.

T. Goryczka P. Ochin, H. Morawiec *Archives of Metallurgy and Materials,* 49. - (2004). - pp. 891-906.

T.H. Nam T. Saburi and K. Shimizu, *Mater. Trans. JIM.,* 31. - 1990. - p. 959.

T.W. Duerig K.N. Melton, D. Stockel, C.M. Wayman *Engineering aspects of sahpe memory alloys* - London: Butterworth-Heinemann, 1990.

W. J. Moberly and K.N. Melton at *"Engineering aspects of shape memory alloys"* by T.W. Duering K.N Melton, D. Stöckel, C.M. Wayman, Butterworth-Heinemann Ltd. 1990, p.52.

W.M. Stoobs J.V. Wood *Acta Met.* 27. - (1979). - p. 575..

Ye Xu Xu Huang, A.G. Ramirez *Journal of Alloys and Compounds* 480. - (2009). - pp. L13–L16.

Section 2

Characterization

Chapter 4

Thermal Strain and Magnetization Studies of the Ferromagnetic Heusler Shape Memory Alloys Ni_2MnGa and the Effect of Selective Substitution in 3d Elements on the Structural and Magnetic Phase

T. Sakon, H. Nagashio, K. Sasaki, S. Susuga, D. Numakura, M. Abe, K. Endo, S. Yamashita, H. Nojiri and T. Kanomata

Additional information is available at the end of the chapter

http://dx.doi.org/10.5772/47808

1. Introduction

Ferromagnetic shape memory alloys (FSMAs) have been extensively studied as potential candidates for smart materials. Among FSMAs, Ni_2MnGa is the most familiar alloy [1]. It has a cubic $L2_1$ Heusler structure (space group $Fm\bar{3}m$) with the lattice parameter a = 5.825 Å at room temperature, and it orders ferromagnetically at the Curie temperature $T_C \approx 365$ K [2,3]. Upon cooling from room temperature, a martensite transition occurs at the martensite transition temperature $T_M \approx 200$ K. Below T_M, a superstructure forms because of lattice modulation [4,5]. For the Ni–Mn–Ga Heusler alloys, T_M varies from 200 to 330 K by non-stoichiometrically changing the concentration of composite elements.

Several studies on Ni–Mn–Ga alloys address the martensite transition and correlation between magnetism and crystallographic structures [6–18]. Ma *et al.* studied the crystallography of $Ni_{50+x}Mn_{25}Ga_{25-x}$ alloys (x = 2–11) by powder X-ray diffraction and optical microspectroscopy [7]. In the martensite phase, typical microstructures were observed for x < 7. The martensite variants exhibit configurations typical of self-accommodation arrangements. The TEM image of $Ni_{54}Mn_{25}Ga_{21}$ indicates that the typical width of a variant is about 1 μm. The interaction between the magnetism and crystallographic rearrangements was discussed in Refs. [1,8,17,18]. The memory strain was observed in single crystal Ni_2MnGa and polycrystalline $Ni_{53.6}Mn_{27.1}Ga_{19.3}$ [10]. As for the magnetism, the magnetic anisotropy constant K_U in martensite phase is 1.17×10^{-5} J/m^3, which is forth larger than that in austenite phase (0.27×10^{-5} J/m^3) [1]. Manosa *et al.* suggested that the martensitic transition take place in the ferromagnetic phase, and the decrease in magnetization observed at intermediated fields ($0 < B < 1$ T) is due to the strong magnetic anisotropy of the martensite phase in association with the multi-domain structure of the martensite state [8]. Likhachev *et al.* stated that the magnetic driving force responsible for twin boundary motion is practically equal to the magnetic anisotropy constant K_U [17]. The magnetization results indicate that the martensite Ni–Mn–Ga alloys have higher magnetocrystalline anisotropy. This is because lower initial permeability or lower magnetization at low fields than the cubic austenite phase. Furthermore, the magnetization results indicate that the coercivity and saturation field at martensite phase are higher than those of the cubic austenite phase [11–15]. Zhu *et al.* investigated the lattice constant change $\Delta c/c$ of -4.8 % by means of X-ray diffraction study around martensite transition temperature [11]. Chernenko *et al.* also studied about the magnetization and the X- ray powder diffractions and clear changes were found at martensite temperature for both measurements [12]. Murray *et al.* studied the polycrystalline Ni–Mn–Ga alloys [18]. The magnetization step at T_M is also observed and this is a reflection of the magnetic anisotropy in the tetragonal martensite phase. In the martensite phase, strong magnetic anisotropy exists. Then the magnetization that reflects the percentage of the magnetic moments parallel to the magnetic field is smaller than that in the austenite phase where the magnetic anisotropy is not strong in the weak magnetic field. Therefore the magnetization step is observed at T_M. NMR experiments indicate Mn-Mn indirect exchange via the faults in Mn-Ga layers interchange caused by excessive Ga [13]. This result indicates the exchange interaction between Mn-Mn magnetic moments is sensitive with the lattice transformation. Then the magnetism changes from soft magnet in the austenite phase to hard magnet in the martensite phase, which is due to higher magnetic anisotropy.

To use Ni–Mn–Ga alloys as advanced materials for actuators, polycrystalline materials are useful because of their robustness. Moreover, in daily use, magnetic actuators should be used around room temperature (300 K). Therefore, we selected the $Ni_{52}Mn_{25}Ga_{23}$ alloy, which shows ferromagnetic transition at the Curie temperature T_C, about 360 K, and the martensite transformation occurs around 330 K.

The purpose of this study is to investigate the correlation between magnetism and crystallographic structures as it relates to the martensite transition of $Ni_{52}Mn_{12.5}Fe_{12.5}Ga_{23}$, $Ni_2Mn_{0.75}Cu_{0.25}Ga$, $Ni_2MnGa_{0.88}Cu_{0.12}$ and $Ni_{52}Mn_{25}Ga_{23}$, which undergoes the martensite transition below T_C [6,7]. Especially, we focused on the physical properties in magnetic fields. We performed in this study that by using the polycrystalline samples, it is possible to provide information on the easy axis of the magnetization in the martensite structure with temperature dependent strain measurements under the constant magnetic fields. In this paper, thermal strain, permeability, and magnetization measurements were performed for polycrystalline $Ni_{52}Mn_{12.5}Fe_{12.5}Ga_{23}$, $Ni_2Mn_{0.75}Cu_{0.25}Ga$, $Ni_2MnGa_{0.88}Cu_{0.12}$ and $Ni_{52}Mn_{25}Ga_{23}$ in magnetic fields (B), and magnetic phase diagrams (B–T phase diagram) were constructed. The results of thermal strain in a magnetic field and magnetic-field-induced strain yield information about the twin boundary motion in the fields. From the permeability and magnetization measurements, the magnetic anisotropy constant K_U can be calculated. The experimental results were compared with those of other Ni–Mn–Ga single crystalline or polycrystalline alloys, and correlations between magnetism and martensite transition were found.

2. Experimental details

The $Ni_{52}Mn_{12.5}Fe_{12.5}Ga_{23}$ and $Ni_2Mn_{0.75}Cu_{0.25}Ga$ alloys were prepared by the arc melting of 99.9 % pure Ni, 99.99 % pure Mn, and Cu, 99.95 % pure Fe, and 6N pure Ga in an argon atmosphere. To obtain homogenized samples, the reaction products were sealed in double evacuated silica tubes, which were annealed at 1123 K for 3 days, and quenched into cold water. The samples obtained for both alloys were polycrystalline.

The $Ni_2MnGa_{0.88}Cu_{0.12}$ alloy was prepared by the arc melting of 99.99 % pure Ni, 99.99 % pure Mn, and Cu, and 99.9999 % pure Ga in an argon atmosphere. To obtain the homogenized sample, the reaction product was sealed in double evacuated silica tubes, which was annealed at 1123 K for 3 days, and quenched into cold water. The obtained sample was polycrystalline. From the x-ray powder diffraction, *14M* (*P2/m*) martensitic structure and $D0_{22}$ tetragonal structure were mixed at 298 K [19]. The lattice parameter of tetragonal structure is a = 3.8920 Å and c = 6.5105 Å.

The $Ni_{52}Mn_{25}Ga_{23}$ alloy was prepared by arc melting 99.99% pure Ni, 99.99% pure Mn, and 99.9999% pure Ga in an argon atmosphere. To obtain a homogenized sample, the reaction product was sealed in double-evacuated silica tubes, and then annealed at 1123 K for 3 days and quenched in cold water. The obtained sample was polycrystalline. From x-ray powder diffraction, the *14M* (*P2/m*) martensite structure and the $D0_{22}$ tetragonal structure were mixed at 298 K. The lattice parameters of the *14M* structure were a = 4.2634 Å, b = 5.5048 Å, c = 29.5044 Å, and β = 85.863°, and those of the $D0_{22}$ structure were a = 3.8925 Å, and c = 6.5117 Å. The size of the sample was 2.0 mm × 2.0 mm × 4.0 mm.

The measurements in this study were performed at atmospheric pressure, P = 0.10 MPa. Thermal strain measurements were performed using strain gauges (Kyowa Dengyo Co., Ltd., Chofu, Japan). Electrical resistivity of the strain gauges was measured by the four-probe method. The relationship between strain, ε, and deviation of electrical resistivity, ΔR, is given by

$$\varepsilon = \frac{1}{K_S} \cdot \frac{\Delta R}{R_0} = \frac{1}{K_S} \cdot \frac{(R - R_0)}{R_0}, \tag{1}$$

where K_S is the gauge factor (K_S = 1.98) and R_0 is the electrical resistivity above T_R. The strain gauge was fixed parallel to the longitudinal axis of the sample.

Thermal strain measurements were performed using a 10 T helium-free cryocooled superconducting magnet at the High Field Laboratory for Superconducting Materials, Institute for Materials Research, Tohoku University. The magnetic field was applied along the longitudinal axis of the sample. The thermal strain is denoted by the reference strain at the temperature just above T_M.

Magnetization measurements were performed using a Bitter-type water-cooled pulsed magnet (inner bore: 26 mm; total length: 200 mm) at Akita University. The magnetic field was applied along the longitudinal axis of the sample. The values of magnetization were corrected using the values of spontaneous magnetization for 99.99% pure Ni. The magnetic permeability measurements were performed in AC fields with the frequency f = 73 Hz and the maximum field B_{max} = 0.0050 T using an AC wave generator WF 1945B (NF Co., Ltd., Yokohama, Japan) and an audio amp PM17 (Marantz Co. Ltd., Kawasaki, Japan) at Akita University with the same magnet we used for the magnetization measurement, having the compensating high homogeneity magnetic field. AC fields were applied along the longitudinal axis of the sample.

3. Results and discussions

3.1. $Ni_{52}Mn_{12.5}Fe_{12.5}Ga_{23}$ and $Ni_2Mn_{0.75}Cu_{0.25}Ga$

Figure 1 shows the temperature dependence of the linear thermal expansion of $Ni_{52}Mn_{12.5}Fe_{12.5}Ga_{23}$ in static magnetic fields. When cooling from 310 K (Ferro-A phase), the alloy shrinks gradually in zero magnetic fields. Small elongation was observed at 288 K. Then, sudden shrinking occurs below 286 K, which indicates transformation from austenite phase to martensite phase. We define the martensitic transformation temperature T_M as the midpoint of the steep decrease in the cooling measurement. The T_M of this alloy is 284 K. The reason of small elongation at 288 K is considered that $L2_1$ and *14M* structures coexist each other. Therefore apertures between $L2_1$ and *14M* structures were originated and small expansion occured. As for $Ni_{2+x}Mn_{1-x}Ga$ alloys, small elongation was observed just above T_M [21]. As shown in reference 20, the phase below T_M is Ferro-M. When heating from 270 K, expansion occurs at about T_R = 288 K, which indicates reverse martensitic transformation. Small elongations just above the temperatures of T_M and T_R were also observed in polycrystalline $Ni_{2+x}Mn_{1-x}Ga$ ($0.16 \le x \le 0.20$) [21].

T_M and T_R gradually changed with increasing magnetic fields. The strain at T_M and T_R was about -2.5×10^{-3} (−0.25 %) and was almost the same as that in magnetic fields. Kikuchi *et al.* performed the x-ray diffraction experiments of $Ni_{50+X}Mn_{12.5}Fe_{12.5}Ga_{25-X}$ [20]. The x-ray patterns at room temperature (T = 300 K, austenite phase) for the samples of $0 \le x \le 2.0$ were indexed with the $L2_1$ Heusler structure. In the x-ray diffraction pattern at room temperature of the sample with x =2.0, a very weak reflection from a γ phase was observed, where the γ phase has a disordered fcc structure. The lattice parameter a of x = 2.0 was found to be 5.7927 Å [22]. On the other hand, for $x \ge 3.0$, the martensite phase appeared at room temperature. The martensitic structure of x = 3.0 was indexed as a monoclinic structure with *14M* (*7R*) structure. The lattice parameters of the sample were determined as a = 4.2495 Å, b = 2.7211 Å, c = 29.340 Å, and β = 93.36° at room temperature.

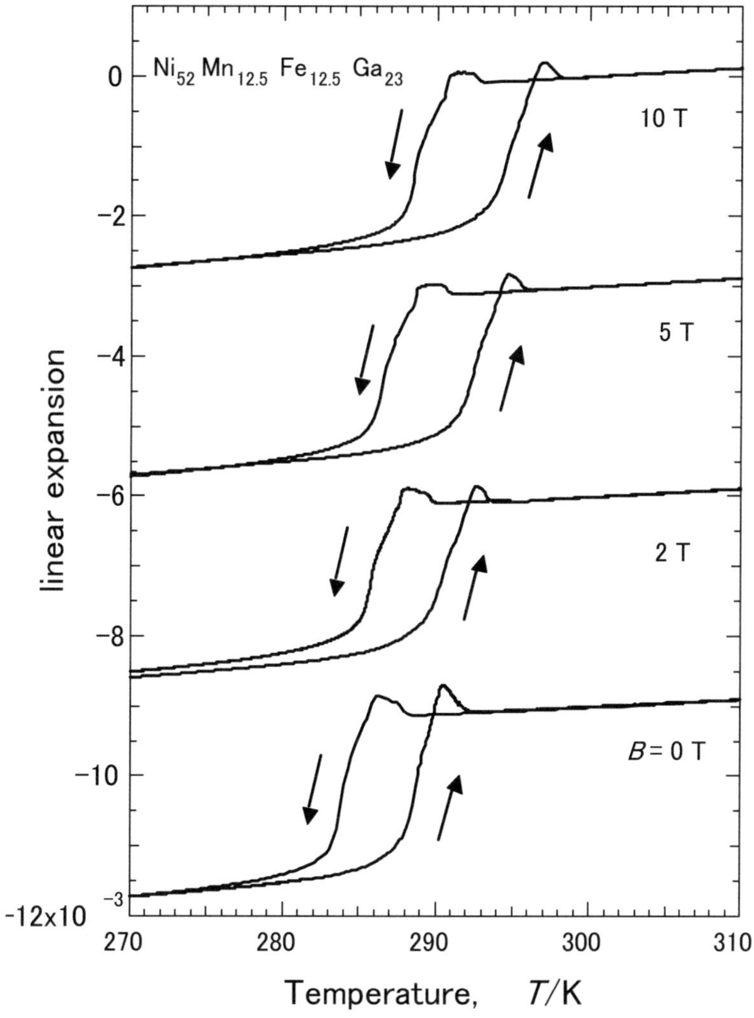

T. Sakon

Figure 1. Temperature dependences of linear thermal expansion of $Ni_{52}Mn_{12.5}Fe_{12.5}Ga_{23}$ in static magnetic fields.

We also estimated the strain of $Ni_{52}Mn_{12.5}Fe_{12.5}Ga_{23}$ (x = 2.0) at T_M using the lattice parameter of x = 2.0 in the austenite phase and that of x = 3.0 in the martensite phase. In the austenite phase, for the $L2_1$ cubic structure, the lattice parameter a was 5.7927 Å [22]. The distance between Mn–Mn atoms was $a/\sqrt{2}$ = 4.0961 Å, and the volume of the unit cell was V_A = $\left(a/\sqrt{2}\right)^3$ = $(4.0961)^3$ = 68.72 Å^3. Furthermore, the volume V_M in the martensite phase was estimated and compared with V_A in the same area. In the *14M* (*7R*) martensite phase, a = 4.2495 Å in the basal plane, is parallel to one of the a axis in the $L2_1$ structure, and is of the same unit. The other axis in the martensite phase corresponds to one of the a axis in the $L2_1$ structure of the Mn–Mn ridge in the basal plane $(\sqrt{2}\times b)$. The c axis is almost normal (β = 93.36°) to the basal plane and the seven Mn–Mn cycles at c = 29.340 Å. Therefore, the volume,

$$V_M = a \times (c/7) \times (\sqrt{2} \times b) \times \sin\beta = 4.2495\text{x}4.1914 \times (1.4142\text{x}2.7211)\text{x}\sin 93.36^{\circ} = 68.55\text{x}0.9983 = 68.43 \text{ Å}^3. \quad (2)$$

The linear strain of a polycrystal is one-third of the volume strain [24]. Therefore, we estimate the linear strain $\Delta\ \varepsilon$ as,

$$\Delta\varepsilon = \{(V_M - V_A)/ V_A\}\times 1/3 = \{(68.43\text{–}68.72)/68.72\}\times 1/3 = (-0.29/68.72)\times 1/3 = -0.14\ \%. \quad (3)$$

This estimated value is approximately comparable to the strain value $\Delta\varepsilon$ = −0.25 % of $Ni_{52}Mn_{12.5}Fe_{12.5}Ga_{23}$ obtained from this experimental study.

Figures 2 (a) and (b) show the temperature dependence of magnetic permeability and linear thermal expansion of $Ni_2Mn_{0.75}Cu_{0.25}Ga$ in zero magnetic fields, respectively. When cooling from a high temperature, it shrinks and the permeability increases at about T_M = 308 K. The permeability at austenite phase is very low as compared with that at the martensite phase. These results indicate that the region above T_M or T_R is Para-A and the region below T_M or T_R is Ferro-M. When heating from a low temperature, the expansion occurs at about T_R = 316 K, which indicates reverse martensitic transformation. The strain at T_M or T_R is about 3.0×10^{-3} (0.30 %). This value is higher than that of $Ni_{52}Mn_{12.5}Fe_{12.5}Ga_{23}$. Kataoka *et al.* studied the x-ray powder diffraction of $Ni_2Mn_{1-x}Cu_xGa_2$ ($0 \leq x \leq 0.40$) [23]. In the vicinity of martensitic transformation, the strain exhibits complicated behavior; when cooling from 342 K, it shrinks gradually and rapid shrinking occurs at T_M = 308 K, subsequently, exhibiting elongation; repetition of small elongation and shrinking was observed between 303 K and 291 K; in addition, it shrinks linearly below 291 K. When heating from 257 K, the repetition of small elongation and shrinking was observed between 307 K and 311 K. Thereafter, it shrinks by 9.0×10^{-4} and exhibits elongation. This sequential phenomenon has been observed in single crystalline $Ni_{2.19}Mn_{0.81}Ga$ [21]. In particular, steep shrinking occurs before elongation due to reverse martensitic transformation during heating. As for polycrystalline $Ni_{2+x}Mn_{1-x}Ga$ ($0.16 \leq x \leq 0.20$), the shape of the small elongation or small shrinking due to the large change of the strain associated with martensitic transformation is broader than that of the single crystalline alloy. In our study, $Ni_2Mn_{0.75}Cu_{0.25}Ga$ showed steep shrinking before elongation during heating from a low temperature, which is similar to that of single crystalline $Ni_{2.19}Mn_{0.81}Ga$. It is possible that the $Ni_2Mn_{0.75}Cu_{0.25}Ga$ crystal is oriented to some extent.

The x-ray diffraction measurement of $Ni_2Mn_{0.75}Cu_{0.25}Ga$ indicates that cubic $L2_1$ phase and the *14M* phase coexist in the martensite phase. The reason for the repetition of small elongation and shrinking in Figure. 2 (b) is supposed to be this complex structure.

Figure 3 shows the temperature dependence of the linear thermal expansion of $Ni_2Mn_{0.75}Cu_{0.25}Ga$ in static magnetic fields. T_M and T_R gradually changed with increasing magnetic fields.

Next, we compared the two samples. The linear thermal coefficients α of $Ni_{52}Mn_{12.5}Fe_{12.5}Ga_{23}$ and $Ni_2Mn_{0.75}Cu_{0.25}Ga$ in zero magnetic fields obtained in this study are shown in Table 1. In the austenite phase, α of $Ni_{52}Mn_{12.5}Fe_{12.5}Ga_{23}$ is much lower than that of $Ni_2Mn_{0.75}Cu_{0.25}Ga$, which means that $Ni_{52}Mn_{12.5}Fe_{12.5}Ga_{23}$ is harder than $Ni_2Mn_{0.75}Cu_{0.25}Ga$. α is higher in the martensite phase than in the austenite phase of $Ni_{52}Mn_{12.5}Fe_{12.5}Ga_{23}$. This is probably due to the *14M* martensitic structure.

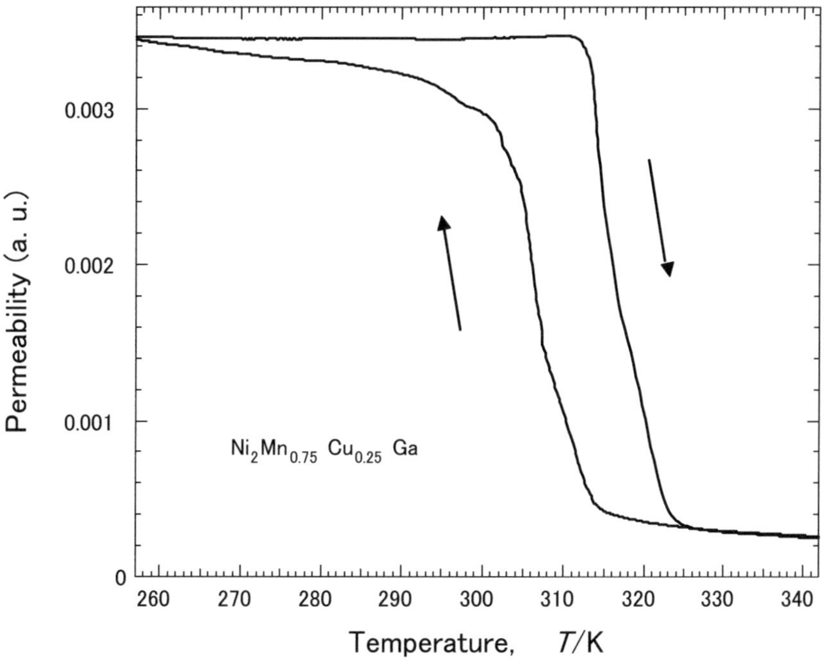

(a)

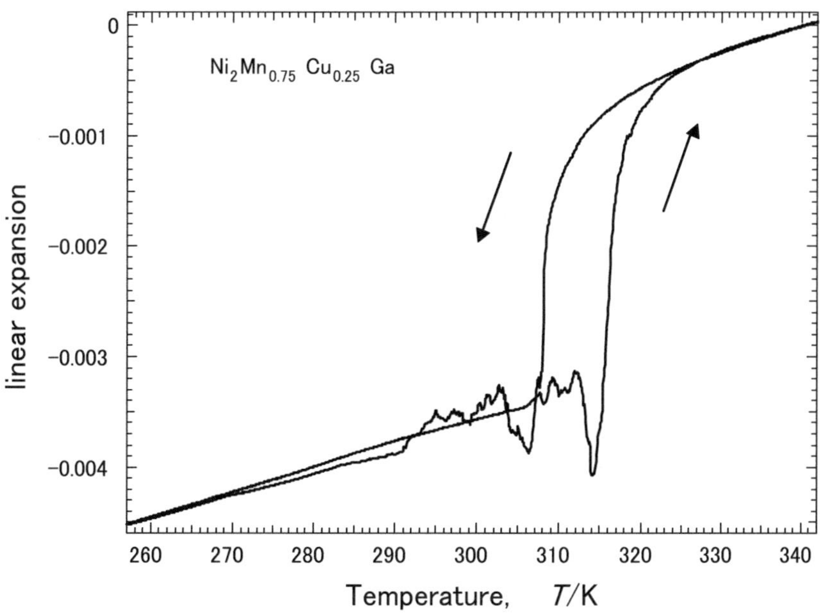

(b)

T. Sakon

Figure 2. (a). Temperature dependence of the magnetic permeability μ of $Ni_2Mn_{0.75}Cu_{0.25}Ga$ in zero magnetic fields. The magnetic permeability measurement was perforemed in AC fields with $f = 73$ Hz and $B_{max} = 1.0$ mT. Zero means the permeability $\mu = 0$. (b) Temperature dependences of linear thermal expansion of $Ni_2Mn_{0.75}Cu_{0.25}Ga$ .The strain was defined by the difference from the sample length at 340 K.

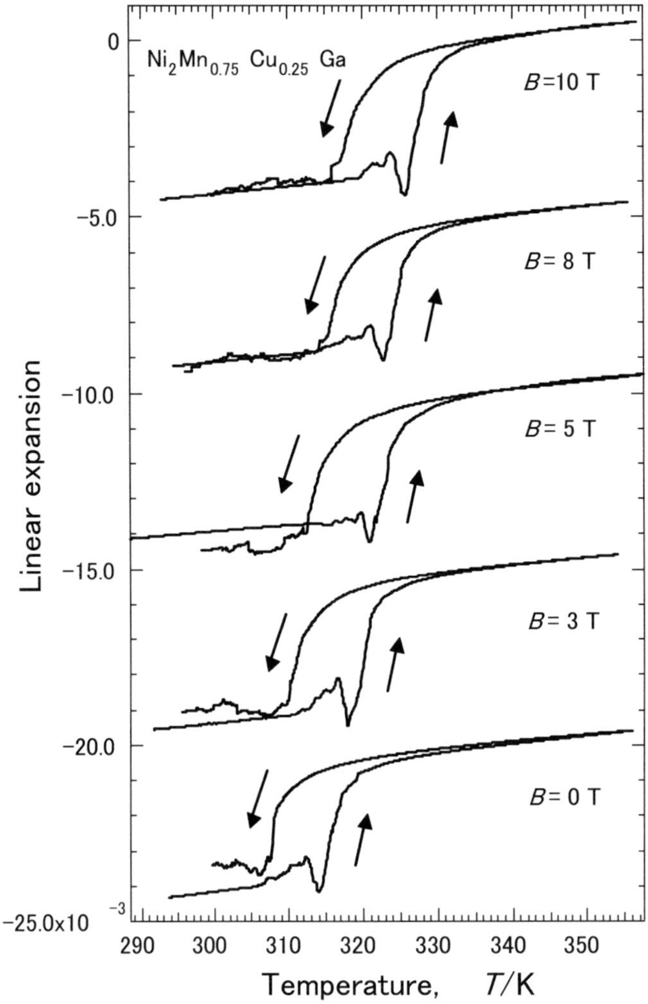

T. Sakon

Figure 3. Temperature dependences of the linear thermal expansion of $Ni_2Mn_{0.75}Cu_{0.25}Ga$ in static magnetic fields.

Figure 4 shows the magnetic phase diagram of the thermal expansion of $Ni_{52}Mn_{12.5}Fe_{12.5}Ga_{23}$ in static magnetic fields. T_M and T_R gradually changed with increasing magnetic fields like $Ni_{2+x}Mn_{1-x}Ga$ alloys. The shifts of T_M and T_R in magnetic fields were estimated as $dT_M/dB \approx 0.5$ K/T and $dT_R/dB \approx 0.5$ K/T, respectively. The shifts of T_M and T_R can be explained by the difference of the magnetization between austenite phase and martensitic phase. Afterwards we discuss about the correlation between magnetization and the shift of T_M.

Figure 5 shows the magnetic phase diagram of the thermal expansion of $Ni_2Mn_{0.75}Cu_{0.25}Ga$ in static magnetic fields. T_M and T_R gradually changed with increasing magnetic fields such as in the $Ni_{2+x}Mn_{1-x}Ga$ or $Ni_{52}Mn_{12.5}Fe_{12.5}Ga_{23}$ alloys. The shifts of T_M and T_R in magnetic fields were estimated as $dT_M/dB \approx 1.2$ K/T and $dT_R/dB \approx 1.1$ K/T, respectively. These ratios are within measurement errors.

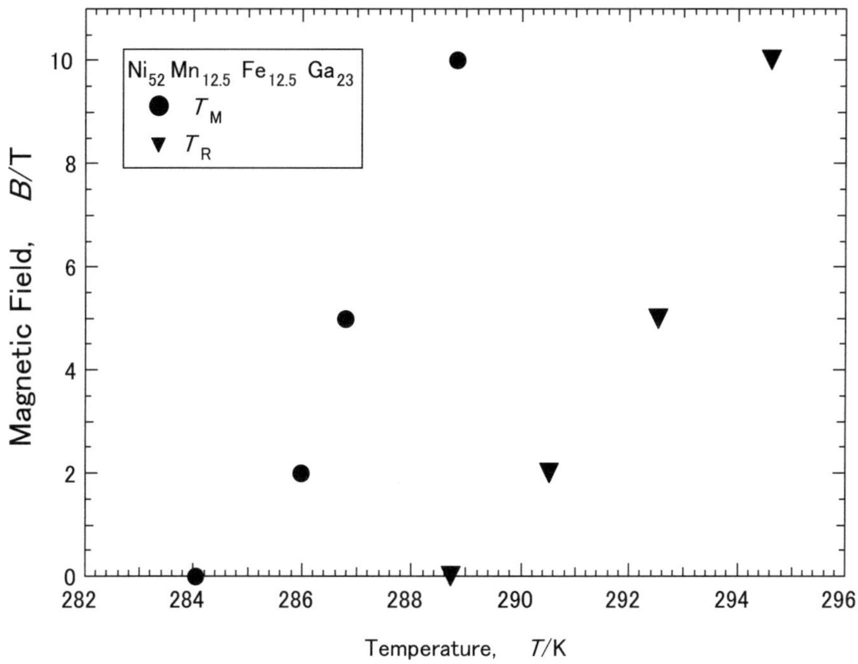

T. Sakon

Figure 4. Magnetic phase diagram of $Ni_{52}Mn_{12.5}Fe_{12.5}Ga_{23}$.

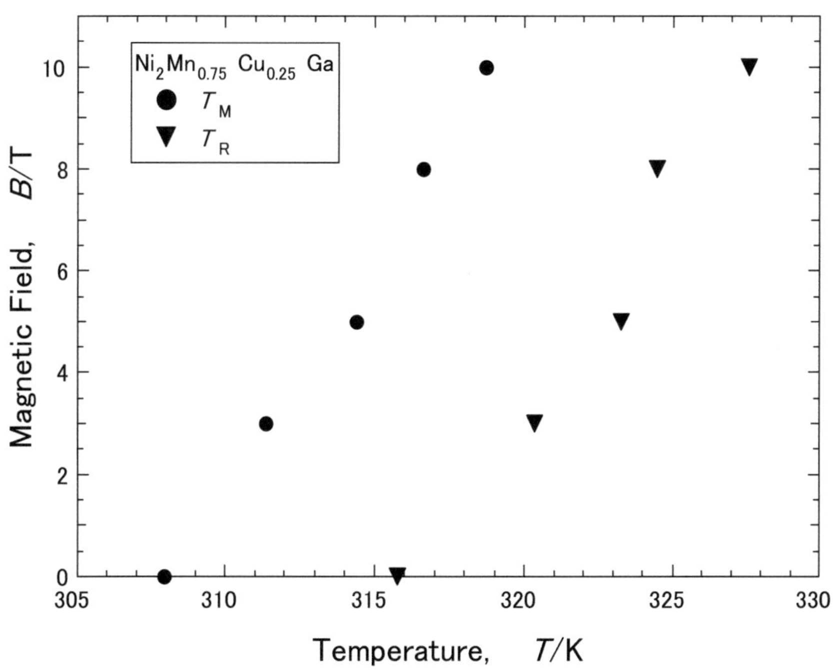

T. Sakon

Figure 5. Magnetic phase diagram of $Ni_2Mn_{0.75}Cu_{0.25}Ga$.

sample	M_M	M_A	$(M_M - M_A)/M_M$	dT_M/dB(K/T)	remarks
Ni_2MnGa	90 J/$\mu_0$kgT at 180 K (*1) Ferro	80 J/$\mu_0$kgT at 220 K (*1) Ferro	0.11	0.20 (*2) 0.40 ± 0.25 (*3)	*1 ref. 2 *2 ref. 35 *3 ref. 36
$Ni_{2.19}Mn_{0.81}Ga$	2.0 (a.u.) (*4) at 300 K Ferro	0 (a.u.) (*4) at 350 K Para	1.0	1.0 (*4)	*4 ref. 38
$Ni_{52}Mn_{12.5}Fe_{12.5}Ga_{23}$	63.1 J/$\mu_0$kgT at 250 K Ferro	52.7 J/$\mu_0$kgT at 300 K Ferro	0.16	0.5	ref. 27
$Ni_2Mn_{0.75}Cu_{0.25}Ga$	42.4 J/$\mu_0$kgT at 300 K Ferro	0 J/$\mu_0$kgT at 307 K Para	1.0	1.2	ref. 27
$Ni_2MnGa_{0.88}Cu_{0.12}$	37.3 J/$\mu_0$kgT at 330 K Ferro	0 J/$\mu_0$kgT at 340 K Para	1.0	1.3	ref. 39
$Ni_{52}Mn_{25}Ga_{23}$	42.2 J/$\mu_0$kgT at 333 K Ferro	34.2 J/$\mu_0$kgT at 335 K Ferro	0.19	0.43	this work

Table 1. Spontaneous magnetization and dT_M/dB of $Ni_{2+x}Mn_{1-x}Ga$, $Ni_{52}Mn_{12.5}Fe_{12.5}Ga_{23}$, $Ni_2Mn_{0.75}Cu_{0.25}Ga$, $Ni_2MnGa_{0.88}Cu_{0.12}$, and $Ni_{52}Mn_{25}Ga_{23}$. M_M and M_A indicate the spontaneous magnetizations in martensite phase and austenite phase, respectively. Ferro and Para mean the ferromagnetic and the paramagnetic phases, respectively.

The magnetic phase diagrams constructed from the thermal expansion measurements of this study are shown in Figures 4 and 5.

Figure 6 (a) shows the magnetization of $Ni_{52}Mn_{12.5}Fe_{12.5}Ga_{23}$ in a pulsed magnetic field. Below 250 K in the Ferro-M state, the *M-B* curves resemble each other, and this is consistent with the results in reference 7. In the Ferro-A state, the magnetization at 300 K is lower than that in the Ferro-M state. Figure. 6 (b) shows the high-field magnetization in a pulsed magnetic field. At 90 K, steep increase in magnetization occurs when magnetic field is applied. Above 2 T, the magnetization increases gradually. The magnetization at 300 K, which is above T_M and T_R, is also ferromagnetic. The magnetization above 5 T is almost flat. This property is quite different from that at $T = 90$ K. The magnetism of the austenite phase appears to be similar to a localized ferromagnetic state, because the magnetization value is constant in high magnetic fields.

Figure 6. (c) shows an Arrott plot, i.e., M^2 vs B/M, of the magnetization of $Ni_{52}Mn_{12.5}Fe_{12.5}Ga_{23}$. The spontaneous magnetizations at 90 K and 250 K in a Ferro-M state are 70.1 J/$\mu_0$kgT and 63.1 J/$\mu_0$kgT, respectively. The spontaneous magnetization at 300 K in a Ferro-A state is 52.7 J/$\mu_0$kgT.

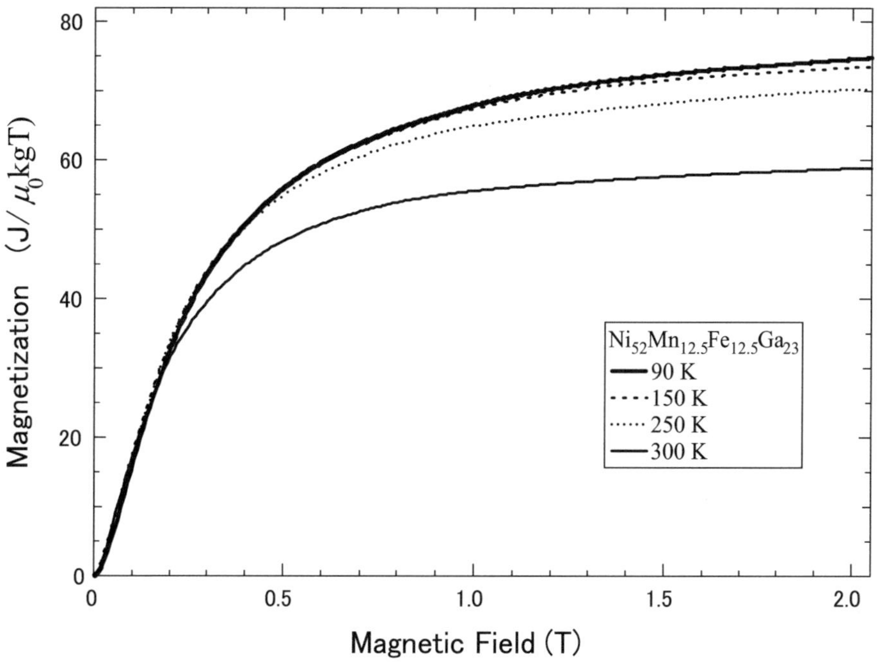
Magnetization (J/ μ0kgT)
Magnetic Field (T)
0
20
40
60
80
0
0.5
1.0
1.5
2.0
Ni52Mn12.5Fe12.5Ga23
90 K
150 K
250 K
300 K

(a)

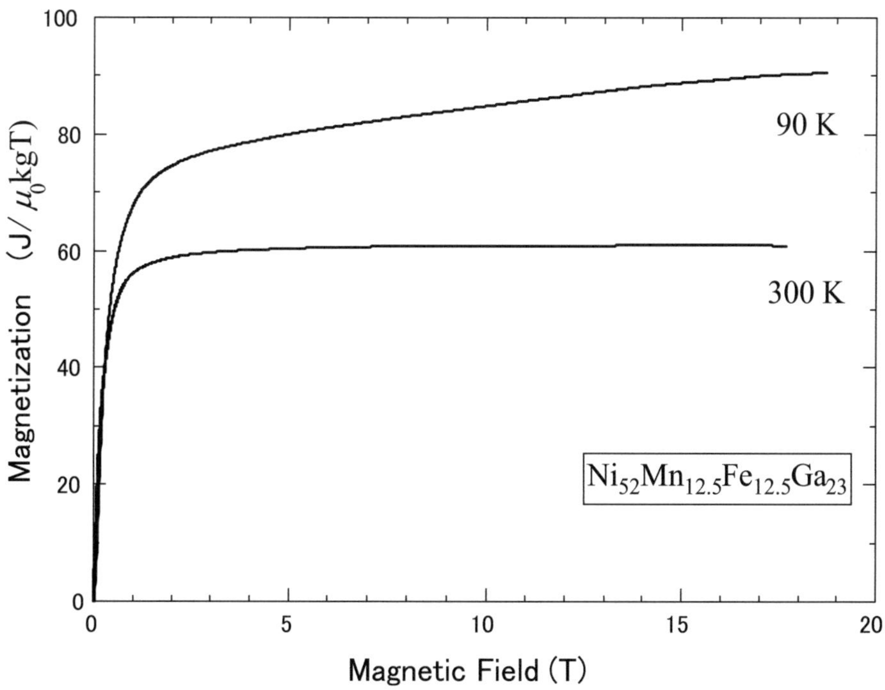
Magnetization (J/ μ0kgT)
Magnetic Field (T)
0
20
40
60
80
100
0
5
10
15
20
90 K
300 K
Ni52Mn12.5Fe12.5Ga23

(b)

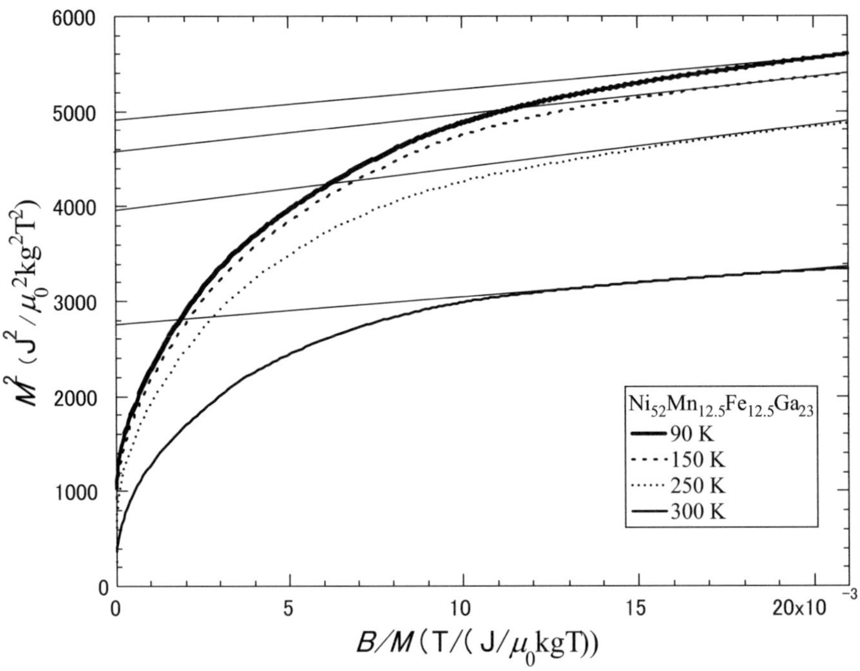

(c)

T. Sakon

Figure 6. (a) Magnetization of $Ni_{52}Mn_{12.5}Fe_{12.5}Ga_{23}$ in a pulsed magnetic field up to 2 T. (b) High field magnetization of $Ni_{52}Mn_{12.5}Fe_{12.5}Ga_{23}$ using a pulsed magnet. (c) Arott plot of the magnetization of $Ni_{52}Mn_{12.5}Fe_{12.5}Ga_{23}$. Fine straight lines are extrapolated lines.

Figures 7 (a) and (b) show the magnetization of $Ni_2Mn_{0.75}Cu_{0.25}Ga$ in a pulsed magnetic field. These measurements were performed after zero-field cooling processes at 323 K in the austenite phase. Below T_M, the magnetization shows ferromagnetic properties, whereas above T_M it exhibits paramagnetic properties. This is consistent with the permeability result shown in Figure. 2. Below T_M, for instance, at 300 K, a steep increase occurred around zero fields and a spin-flop like behavior was shown below 0.06 T. Usually, magnetic alloys such as $FeCl_3$ show spin-flop behavior, and a linear extrapolation line at the canted magnetic moments phase crosses the origin point of the coordinate axis in the *M-B* graph. However, in Figure 7 (a), the *M-B* graph shows that the linear extrapolation line at the canted magnetic moments phase did not cross the origin point at 300 K. It is possible that steep increase just above the zero fields was due to the localized magnetic moments on the Mn atoms, for example, 3.8–4.2 μ_B/Mn atom which was obtained by the neutron scattering experiments of $Ni_{2+x}Mn_{1-x}Ga$ alloys [2, 24-25]. The magnetic moments on Ni atoms are considerably low, such as 0.2 μ_B/Ni atom for Ni_{2+x} $Mn_{1-x}Ga$ alloys [2, 24-25], and therefore, it is possible that the Ni moments that were arranged in a canted-like formation get ordered by the mutual correlations between external magnetic fields and internal magnetic fields due to the Mn moments.

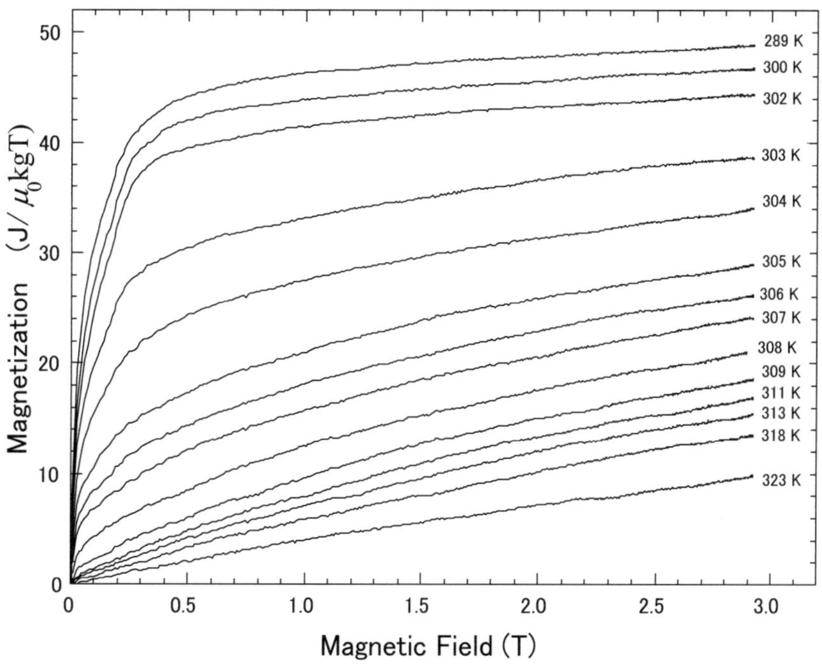
Magnetization (J/ $\mu_0$kgT)
Magnetic Field (T)
289 K
300 K
302 K
303 K
304 K
305 K
306 K
307 K
308 K
309 K
311 K
313 K
318 K
323 K
0
10
20
30
40
50
0.5
1.0
1.5
2.0
2.5
3.0

(a)

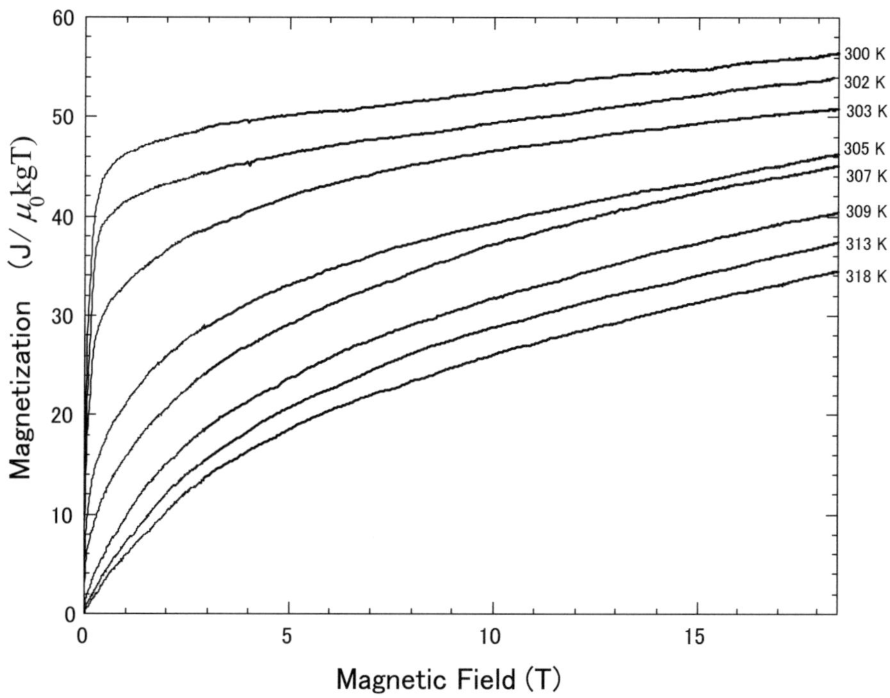
Magnetization (J/ $\mu_0$kgT)
Magnetic Field (T)
300 K
302 K
303 K
305 K
307 K
309 K
313 K
318 K
0
10
20
30
40
50
60
5
10
15

(b)

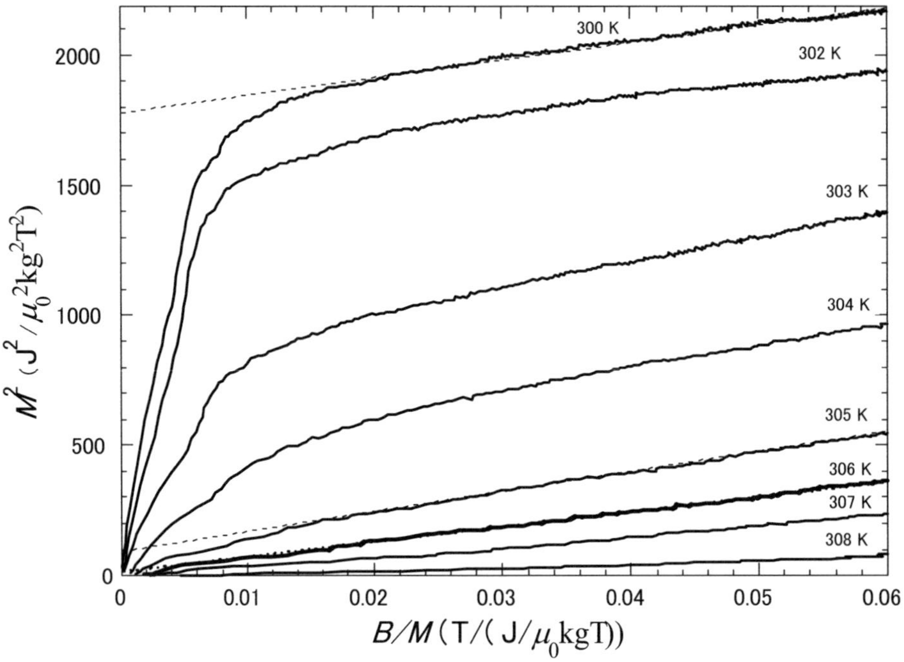

(c)

T. Sakon

Figure 7. 7 (a) Magnetization of $Ni_2Mn_{0.75}Cu_{0.25}Ga$ in a pulsed magnetic field up to 3 T. (b) High field magnetization of $Ni_2Mn_{0.75}Cu_{0.25}Ga$ using a pulsed magnet. (c) Arott plot of the magnetization of $Ni_2Mn_{0.75}Cu_{0.25}Ga$. Dotted lines at 305 K and 306 K are extrapolated linear lines.

Figure 7 (c) shows the Arrott plot of $Ni_2Mn_{0.75}Cu_{0.25}Ga$. The spontaneous magnetization at 300 K in a Ferro-M state is 42.4 J/$\mu_0$kgT. The obtained T_C of the martensite phase is 307 K, which is almost the same as T_M = 308 K and this is consistent with the *x*-*T* phase diagram of $Ni_2Mn_{1-x}Cu_xGa$, which is obtained experimental and theoretical calculations [23].

3.2. $Ni_2MnGa_{0.88}Cu_{0.12}$

Figure 8 shows the temperature dependence of magnetic permeability. When heating from 310 K, the signal gradually increased. A slightly peak was observed at 338 K and a sudden decrease occurred around 342 K. When cooling from a high temperature, the permeability shows a sharp peak at about 337 K. A dip was observed around 324 K. Figure 9 shows the linear thermal expansion. When heating from 305 K, slight expansion was observed at zero magnetic fields. Around 343 K, a sharp expansion was observed. Considering the results of a previous study [19], this is due to the reverse martensitic transition and T_R = 343 K, which is defined as the midpoint temperature of the transition. When cooling from 360 K, a sudden shrinking was observed at 336 K. Considering the lattice structure, the martensitic transition temperature T_M is 336 K. When cooling below 336 K, the linear expansion shows a dip and below 320 K, the value of linear expansion is nearly constant. Mentioned above, the permeability measurement also shows a dip between 336 K and 320 K. As for the permeability of pure Fe, a large peak was observed just below T_C = 1040 K [26]. The half width of the peak ΔT is about 100 K and the ratio ΔT / T_C = 0.095. Meanwhile, the half width of the peak ΔT of the permeability of $Ni_2MnGa_{0.88}Cu_{0.12}$ is about 4 K and T_C = 345 K. Then the ratio ΔT / T_C = 0.012, which indicates the increase of the permeability of $Ni_2MnGa_{0.88}Cu_{0.12}$ occurs within a narrower temperature range than that of pure Fe. Concerning the permeability and linear expansion results, dips and drastic changes, the magnetism and the lattice affects one another. The permeability of the austenite phase is very low as compared with that at the martensite phase. The results of the permeability and the linear expansion measurements indicate that the region above T_M is a paramagnetic austenite phase (Para-A) and the region below T_M is a ferromagnetic martensite phase (Ferro-M).

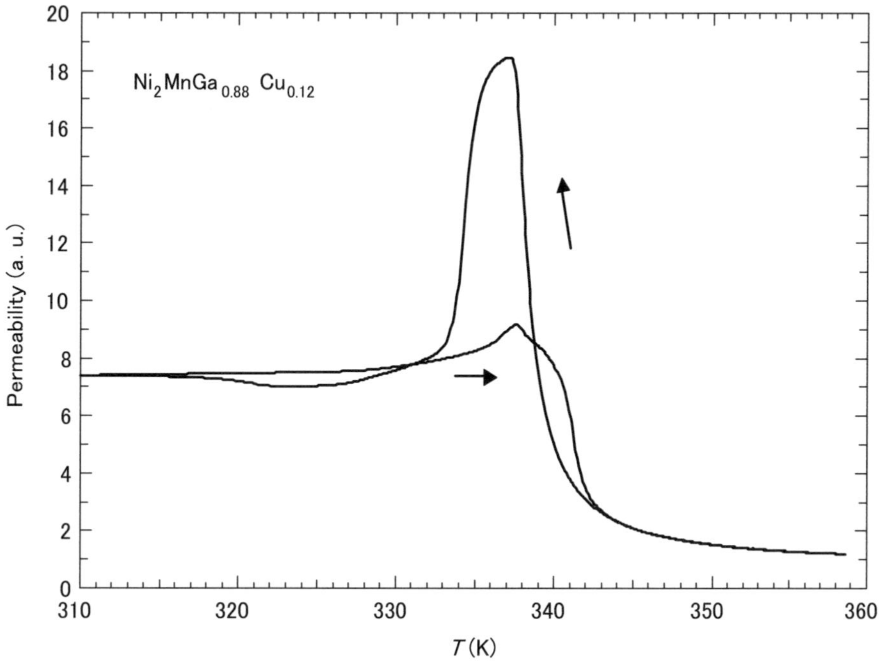

T. Sakon

Figure 8. Temperature dependence of the magnetic permeability μ of $Ni_2MnGa_{0.88}Cu_{0.12}$ in AC fields with f = 73 Hz and B_{max} = 0.0050 T. The origin of the vertical axis is the reference point when the sample is empty in the pick up coil of the magnetic permeability measurement system.

The contraction at T_M under zero fields is about 1.3 × 10^{-3} (0.13 %). As for $Ni_{52}Mn_{12.5}Fe_{12.5}Ga_{23}$ and $Ni_2Mn_{0.75}Cu_{0.25}Ga$, the contraction occurs at martensite temperature [27]. The strain at T_M of polycrystalline $Ni_{52}Mn_{12.5}Fe_{12.5}Ga_{23}$ was estimated as 0.14 % contraction. This value is almost the same as that of $Ni_2MnGa_{0.88}Cu_{0.12}$. After zero field measurements of the linear expansion, measurements in a magnetic field were performed from 3 T to 10 T. With increasing field, T_M and T_R are gradually increased. The shifts of T_M and T_R around zero magnetic fields were estimated as dT_M/dB =1.3 K/T and dT_R/dB = 1.5 K/T, as shown in Figure 3. This behavior is the same as that of the $Ni_{2+x}Mn_{1-x}Ga$

ferromagnetic alloys. The typical temperature T_L, which was defined as the kink point of the linear expansion for heating processes in Figure 9, also gradually increases with increasing fields.

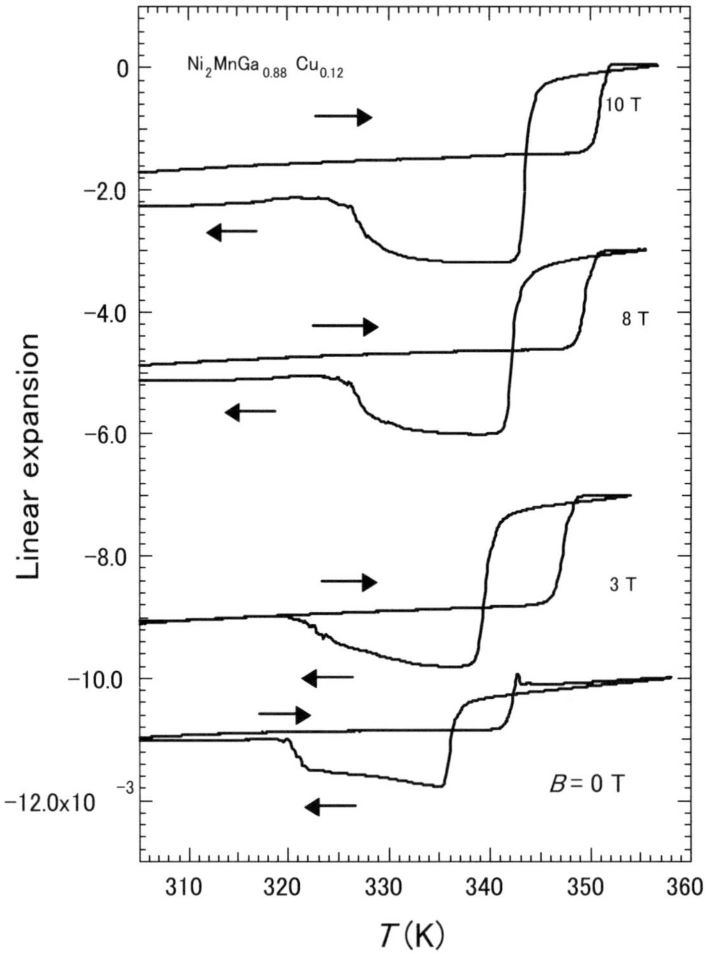

T. Sakon

Figure 9. Temperature dependences of the linear thermal expansion of $Ni_2MnGa_{0.88}Cu_{0.12}$, in static magnetic fields.

A noteworthy fact is that the dip of linear expansion measurements in magnetic fields is larger than that in zero fields. The variation of the strain between zero fields and non-zero field was observed for $Ni_{2.19}Mn_{0.81}Ga$ [28]. The contraction in magnetic fields was larger than that in zero fields. The reason is considered that the magnetic moments of Mn and Ni atoms are aligned parallel to the magnetic field just below T_M and the *14M* and/or $D0_{22}$ tetragonal lattices are rearranged by the magnetic moments. Therefore the rearrangement of these lattices due to magnetic fields occurred in high magnetic fields.

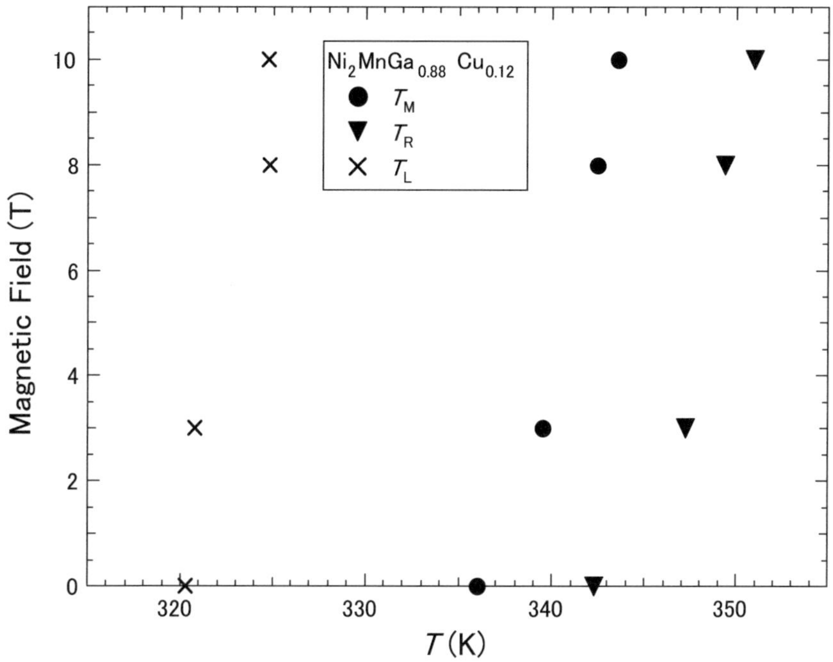

T. Sakon

Figure 10. Magnetic phase diagram of $Ni_2MnGa_{0.88}Cu_{0.12}$. Filled circles indicate the martensitic transition temperature T_M. Filled triangles indicate reverse martensitic temperature T_R. Crosses indicate the typical temperature T_L.

Figure 11 (a) shows the magnetization of $Ni_2MnGa_{0.88}Cu_{0.12}$ in a pulsed magnetic field up to 10 T. The unit of the magnetization *M*, J/$\mu_0$kgT in SI unit system is equal to emu/g in CGS unit system. The hysteresis of the *M-B* curve is considerably small. In other magnetic material, for example $Gd_3Ga_5O_{12}$, the magnetocaloric effect was reported [21]. They performed the magnetization measurements at initial temperature 4.2 K, then the magnetic contribution to heat capacity is comparable to the lattice heat capacity. In our experiment, the temperature change of the sample due to the magnetocaloric effect is considered within 1 K. This is due that these experiments were performed around room temperature, then the lattice heat capacity is much larger than the heating or cooling power by the magnetocaloric effect. Figure 11 (b) shows the magnetization of $Ni_2MnGa_{0.88}Cu_{0.12}$ in a pulsed magnetic field up to 2.2 T. The *M-B* curves with increasing field processes are shown. The *M-B* curves show ferromagnetic behavior below 333 K. The prominent decrease of magnetization occurred between 333 K and 336 K. Figure 12 shows the temperature dependence of the magnetization *M-T* at 0.5 T and 1 T, which were obtained by magnetization measurements in pulsed magnetic fields. A sudden decrease is apparent between 333 K and 336 K for each field. This temperature region corresponds to the sharp increase of the permeability when heating from low temperature in Figure 8, and just below T_M, which was obtained by the linear expansion measurement in Figure 9. The *M-T* curve shows a shallow depression between 310 K and 330 K, which corresponds to the dip of the permeability and the linear expansion results.

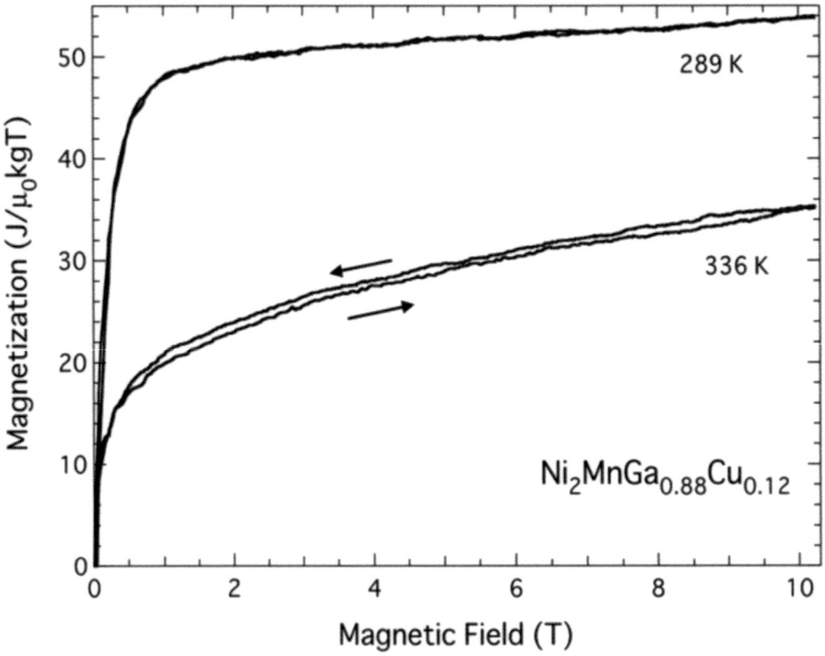

(a)

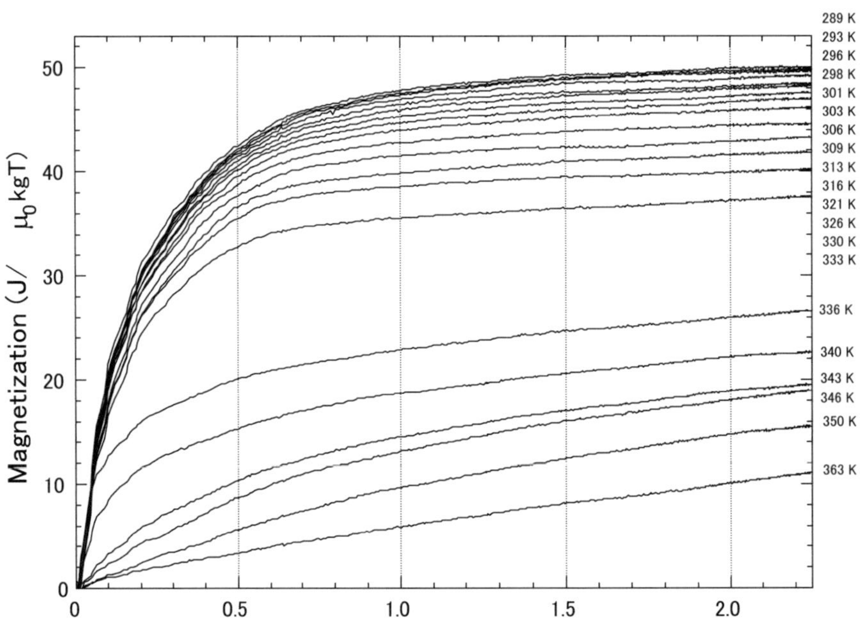

(b)

T. Sakon

Figure 11. (a). Magnetization of $Ni_2MnGa_{0.88}Cu_{0.12}$ in a pulsed magnetic field up to 10 T. (b). Magnetization of $Ni_2MnGa_{0.88}Cu_{0.12}$ in a pulsed magnetic field up to 2.2 T.

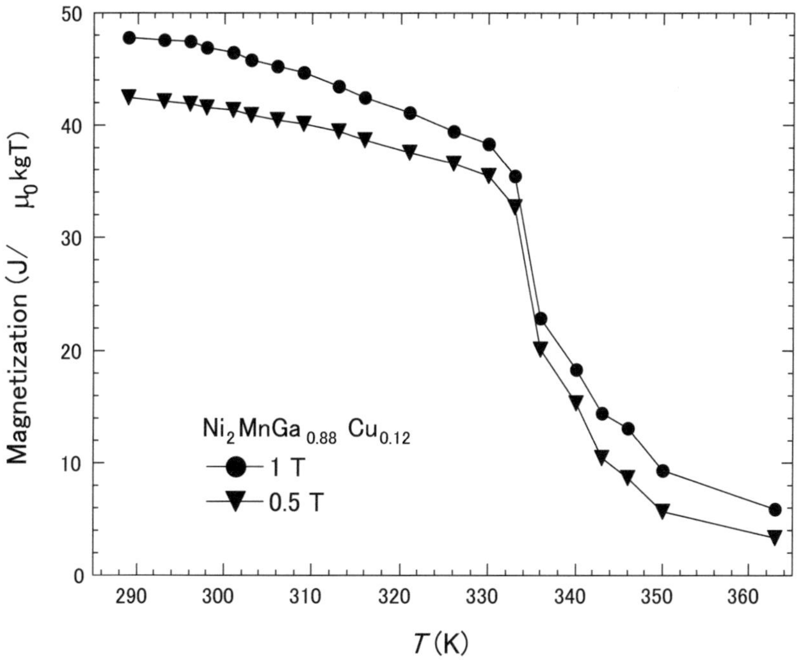

T. Sakon

Figure 12. *M-B* curves of $Ni_2MnGa_{0.88}Cu_{0.12}$.

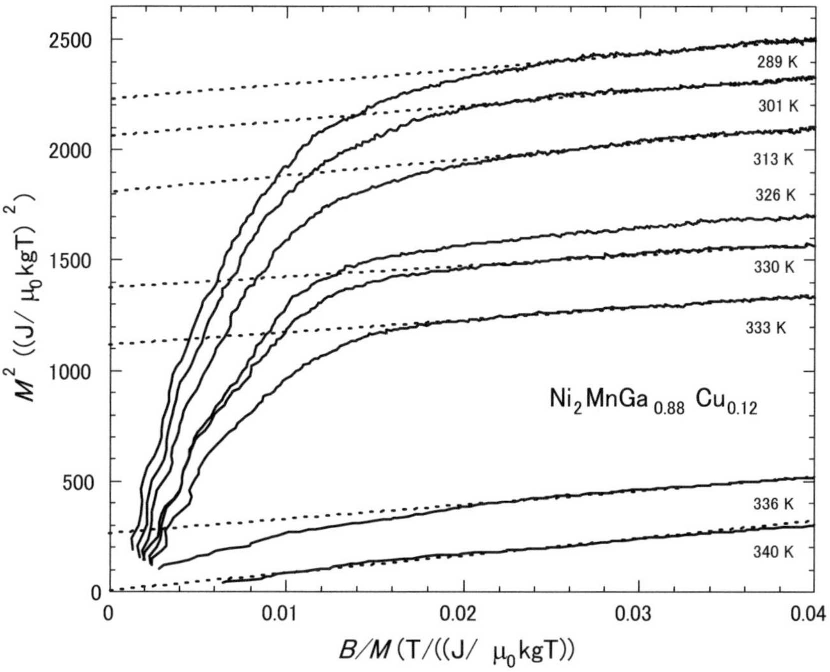

T. Sakon

Figure 13. Arrott plot of the magnetization of $Ni_2MnGa_{0.88}Cu_{0.12}$. Dotted straight lines are extrapolated lines.

Figure 13 shows the Arrott plot of $Ni_2MnGa_{0.88}Cu_{0.12}$. The spontaneous magnetization at 289 K in a Ferro-M state is 47.1 J/$\mu_0$kgT. The obtained T_C of the martensite phase by Arrott plots in Figure 13 is 340 K, which is almost the same as T_M = 337 K. This is consistent with the x-T phase diagram of $Ni_2MnGa_{1-x}Cu_x$, which is obtained in reference 19.

Figure 14 shows the magnetization of $Ni_2MnGa_{0.88}Cu_{0.12}$ in a pulsed high magnetic field up to 18.6 T. The difference of the magnetization between 333 K and 336 K is clearly seen. In high magnetic fields, an almost linear increase can be seen for each M-B curve. $Ni_2Mn_{0.75}Cu_{0.25}Ga$ also shows the difference of the magnetization between 302 K and 305 K, which is little lower than T_C = 307 K or T_M = 308 K [27]. It is noticeable that the Arrott plots of $Ni_2MnGa_{0.88}Cu_{0.12}$ left a space between 333 K and 336 K, and $Ni_2Mn_{0.75}Cu_{0.25}Ga$ also left a space between 302 K and 303 K. The spontaneous magnetizations of $Ni_2MnGa_{0.88}Cu_{0.12}$ are 33.4 J/$\mu_0$kgT at 333 K and 16.7 J/$\mu_0$kgT at 336K, which was obtained by the Arrott plot shown in Figure 13. As for $Ni_2Mn_{0.75}Cu_{0.25}Ga$, the spontaneous magnetizations are 40.0 J/$\mu_0$kgT at 302 K and 28.3 J/$\mu_0$kgT at 303 K.

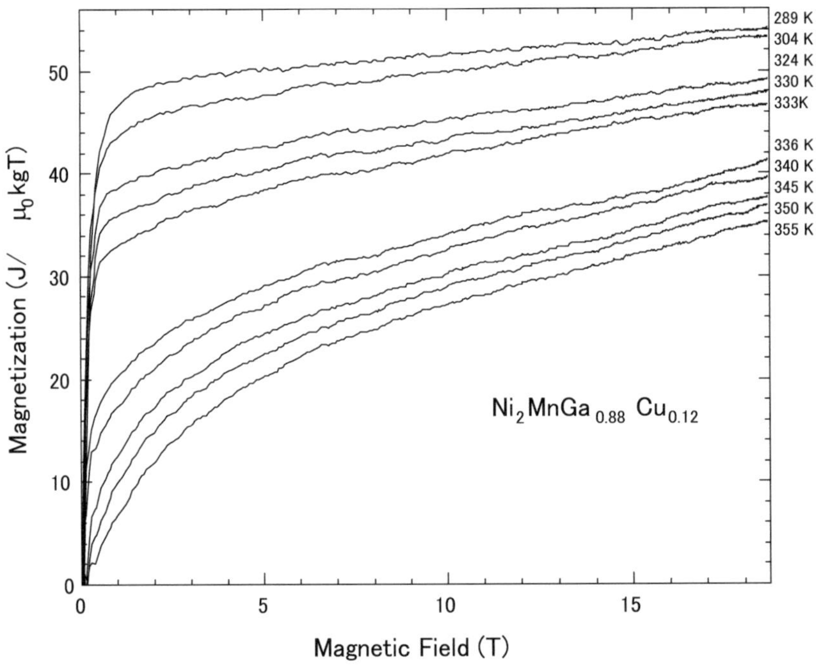

T. Sakon

Figure 14. Magnetization of $Ni_2MnGa_{0.88}Cu_{0.12}$ in a pulsed high magnetic field.

3.3. $Ni_{52}Mn_{25}Ga_{23}$

Figure 15 shows the temperature dependence of permeability. When heating from 300 K, permeability increases gradually. As shown in Figure 15, permeability increases above 330 K and suddenly decreases around 360 K. When cooling from a high temperature, permeability shows a sudden increase at about 356 K and decreases at 325 K. The sudden changes in permeability indicate that the ferrromagnetic transition occurs around 358 K. The temperature dependence of permeability for $Ni_{52}Mn_{25}Ga_{23}$ is similar to that for $Ni_{52}Mn_{12.5}Fe_{12.5}Ga_{23}$, which shows a transition of a ferromagnetic–martensite (Ferro–M) phase to a ferromagnetic–austenite (Ferro–A) phase [29]. The step around 330 K (heating process) and 325 K (cooling process) reflects stronger magnetic anisotropy in the tetragonal martensite phase [8,18]. Polycrystalline $Ni_{49.5}Mn_{28.5}Ga_{22}$, $Ni_{50}Mn_{28}Ga_{22}$ and $Ni_{52}Mn_{12.5}Fe_{12.5}Ga_{23}$ alloys also indicate the magnetization (or permeability) step at T_M [9,18,27] below the field of 10 mT.

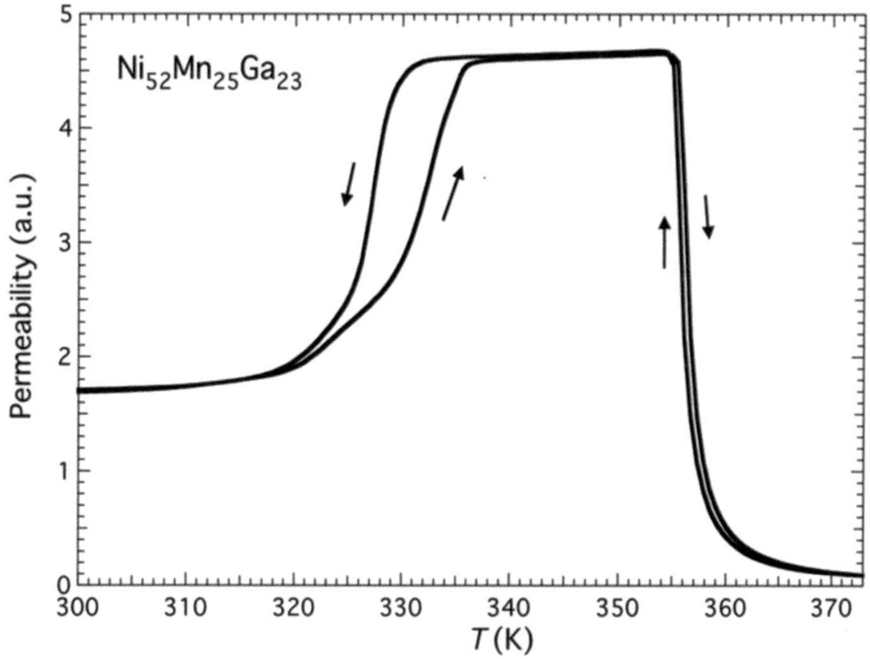

T. Sakon

Figure 15. Temperature dependence of the magnetic permeability μ of $Ni_{52}Mn_{25}Ga_{23}$ in AC fields with f = 73 Hz and B_{max} = 0.0050 T. The origin of the vertical axis is the reference point when the sample is empty in the pickup coil of the magnetic permeability measurement system.

Figure 16 (a) shows the linear thermal strain of $Ni_{52}Mn_{25}Ga_{23}$. Solid lines are the experimental data and dotted lines are the extrapolated lines. At zero magnetic fields, the memory strain was observed, as polycrystalline $Ni_{53.6}Mn_{27.1}Ga_{19.3}$ [10]. When heating from 300 K, slight strain is observed first at zero magnetic fields. Around 334 K, a sharp strain is observed. The results of previous studies [6,7] suggest that this is because of the reverse martensite transition T_R = 334 K, which is defined as the midpoint temperature of the transition. When cooling from 370 K, a sudden decrease is observed at 328 K. Given the lattice structure, the martensite transition temperature T_M is 328 K. The permeability at the Ferro–M phase is very low compared with that at the Ferro–A phase. The results of permeability and linear strain measurements indicate that the region above T_M is a Ferro–A phase and that below T_M is a Ferro–M phase. The permeability measurement results indicate that the ferromagnetic transition from the paramagnetic–austenite (Para–A) phase to the

Ferro–A phase occurs around 358 K (see Figure 15). On the other hand, the linear strain does not show noticeable anomaly at the ferromagnetic transition around 358 K.

When cooling from 370 K, the thermal strain shows a peak at 329 K. This may be attributed to the intermingling of the *L2₁* austenite lattices and the *M14* martensite lattices at the martensite transition. The sequential phenomenon is observed in single crystalline $Ni_{2.19}Mn_{0.81}Ga$ [31]. Zhu *et al.* suggests that the small satellite peaks in heat flow plot, which flanks the central peak indicates the structural transition takes place in multiple steps [11]. The contraction at T_M under zero fields is about 0.5×10^{-3} (0.05%). As for other Heusler alloys, $Ni_{52}Mn_{12.5}Fe_{12.5}Ga_{23}$ and $Ni_2Mn_{0.75}Cu_{0.25}Ga$, the contraction occurs at martensite temperature [27]. The strain at T_M of polycrystalline $Ni_{52}Mn_{12.5}Fe_{12.5}Ga_{23}$ was estimated as 0.14% contraction. This value is larger than that of $Ni_{52}Mn_{25}Ga_{23}$. After zero field measurements of the linear strain, measurements in

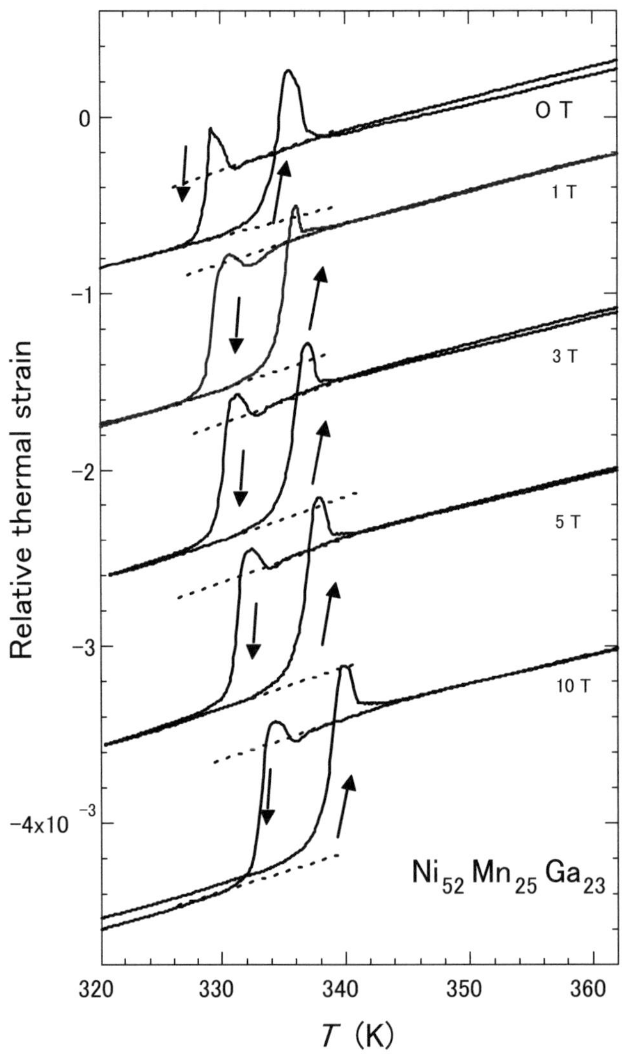

(a)

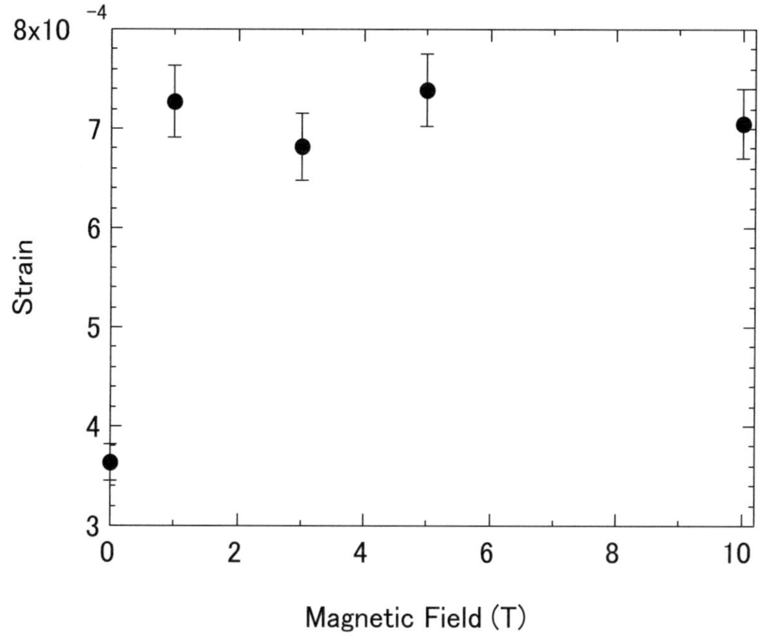

(b)

T. Sakon

Figure 16. (a). Temperature dependence of the linear thermal strain of $Ni_{52}Mn_{25}Ga_{23}$ in static magnetic fields. The dotted lines are the extrapolated lines of the thermal strain. (b). Magnetic field dependence of the strain at the martensite transition temperature obtained from the thermal strain in Figure. 16 (a).

magnetic fields from 1 T to 10 T were performed. The strain at T_M under the magnetic field of 1 T was estimated as 0.10% contraction, which is twice that under zero magnetic field (0.05%). These results indicate that the magnetic fields influence the structural phase transition. After these thermal cycles in magnetic fields, the thermal strain in zero fields was 0.05 % contraction, which is as same as the first cycle in zero fields. Around 358 K, which is the ferromagnetic transition temperature, no anomaly was observed in the magnetic fields. Figure 16 (b) shows the magnetic field dependence of the strain at T_M. At zero field, the strain is 3.6×10^{-4}. On the other hand, the strain in a magnetic field is about 7.1×10^{-4}, which is almost twice that in zero field. Ullakko *et al.* measured the magnetic-field-induced strain of a Ni_2MnGa single crystal [1]. The strain at T_M in zero field was 2×10^{-4}. This is only a small fraction compared with the lattice constant change for *c*-axis from the austenite to martensite phases, which was $\Delta c / c = 6.56\%$. It is proposed that the strain accommodation is occurred by different twin variant orientations. As shown in Figure. 16 (b), the thermal strain under the magnetic field of 1 T was 7.2×10^{-4}, indicating the field aligned some of the twin variants.

In the martensite phase, the magnetic moment in the magnetic easy direction was coupled with the strain along the short *c*-axis of the martensite variant structure. As a result, under the applied magnetic field, the variant rearrangement occurs with the assistance of twin boundary motion, such that the magnetic easy axis is parallel to the applied field. Therefore, the total magnetic free energy is minimal. The variant rearrangement results in field influence on the thermal expansion as shown in Figure 16(b).

Variation in the strain between zero field and non-zero field was observed for $Ni_{2.19}Mn_{0.81}Ga$ and $Ni_{2.20}Mn_{0.80}Ga$ polycrystalline samples [30]. The change in the sample length by means of the thermal strain measurements at the martensite phase transition was 0.04 % for $Ni_{2.19}Mn_{0.81}Ga$ and 0.12 % for $Ni_{2.20}Mn_{0.80}Ga$. The thermal strain for $Ni_{2.19}Mn_{0.81}Ga$ in the presence of 1.4 T magnetic field, the change was increased to 0.13 %, which means 3.2 times increase of the strain. The increase of the strain was 2.6 times (0.31 % strain) for $Ni_{2.20}Mn_{0.80}Ga$. The variation in the strain between zero fields and non-zero field was also observed for $Ni_{49.6}Mn_{27.3}Ga_{23.1}$ polycrystalline samples [31]. With increasing measuring magnetic fields, the difference in the strain increased. Aksoy *et al.* proposed that the strain increase is due to increase of the preferred alignment of the short *c* axis along the applied field, and, high twin boundary mobility in Ni-Mn-Ga is expected to be the main case of the alignment, although the martensite variant nucleation with preferred *c* axis orientation in the external field already just at the martensite transition temperature is also the influence of the shrinkage [31]. Further they mentioned that, when a sample was cooled from the austenite down to the martensite phase in zero fields, no preferred orientation is given to the variant growth during nucleation, whether the easy axis is a long axis or a short axis. When a magnetic field is applied in the austenite phase and the sample is cooled down through T_M in the constant field, a preferred growth direction is provided to the variants. Consequently, the variants with easy axis along the applied field direction nucleate more and more. If the easy axis is short axis, the sample length decreases. Then the contraction at T_M is observed in thermal strain measurements.

As for Ni_2MnGa single crystal, in zero-field cooling process, strains of nearly 0.02 % have been observed at T_M = 276 K [1]. The strain at transformation in 1.0 T is 0.145 %, indicating that the field has aligned some of the twin variants. Now we compare the strain and the magnetization results of $Ni_{2+x}Mn_{1-x}Ga$ alloys [28]. For the alloys which showed increase of the strain for x = 0.18 and 0.20, the T_M and T_c are almost same temperature. Consequently, the magnetization change is large. For these composition alloys, clear hysteresis in the magnetization was observed, which indicates first order magnetic transition. From these results, it is supposed that the magnetic field influences the orientation of the easy *c* axis along the magnetic field. As for $Ni_{52}Mn_{25}Ga_{23}$, The magnetization change is large at T_M, as shown in Figure 22. The permeability in Figure 15 shows clear change and hysteresis, which indicates the first order transition. It is also supposed that the magnetic field influences the orientation of the easy *c* axis along the magnetic field, and then the variant rearrangement was occurred. Consequently, the variation in the strain between zero fields and non-zero field was observed.

Figure 17 shows the magnetic-field-induced strain at 300 K (Ferro–M phase) in a static magnetic field. When increasing the magnetic field from zero fields, a sudden contraction occurs up to 1 T. Above 1 T, a gradual contraction is observed. When decreasing the magnetic field from 10 T, a modicum of strain occurs. Below 1 T, a sudden strain is observed. The magnetic-field-induced strain at 10 T is −100 ppm or −0.010%, which is considerably smaller than the contraction value at T_M. The sudden contraction between 0 and 1 T when increasing the field is supposed to be related to the temperature dependences of the linear strains between zero fields and above 1 T and below T_M. The variant rearrangement results in a magnetic-field-induced strain, which is the origin of the magnetostriction shown in Figure 17. The reason of smallness of the magnetic field induced strain is supposed that; when the sample is cooled down from the austenite phase to the martensite phase in a constant field, variant arrangement is occurred and the contraction is occurred, as mentioned above. In zero fields, cooling from the austenite phase to the martensite phase, the variant arrangement is fixed. When the magnetic field is applied with constant temperature, the variant rearrangement is considered to be difficult. Therefore the magnetic field induced strain is smaller than the strain at T_M. in the linear strain measurements.

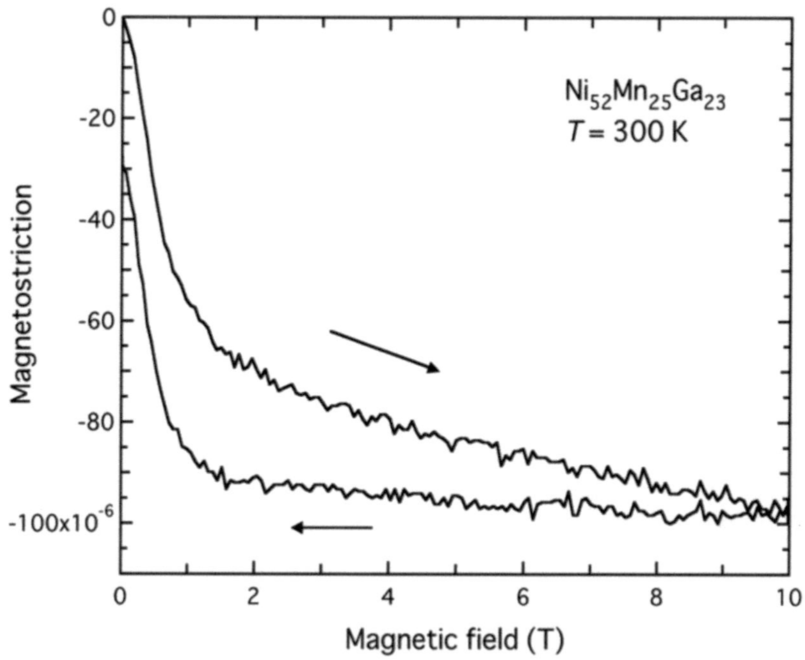

T. Sakon

Figure 17. Magnetostriction of $Ni_{52}Mn_{25}Ga_{23}$ at 300 K in a static magnetic field up to 10 T.

The magnetic-field-induced strain of the polycrystalline $Ni_{50}Mn_{28}Ga_{22}$ alloy was reported by Murray *et al.* [18]. They mentioned that the strain in the martensite phase below T_M is an order of magnitude smaller than that of a single crystal of the stoichiometric compounds [1]. They attributed this to the polycrystalline nature of the material or to the presence of impurities that impede twin boundary motion. The field-induced strain of $Ni_{50}Mn_{28}Ga_{22}$ increases on cooling from the austenite phase, leading to an abrupt increase with the appearance of the twin variant below T_M. On heating from the martensite phase, an abrupt increase occurs in the field-induced strain around T_M. They suggest that this is caused by lattice softening near T_M. As for the thermal strain of $Ni_{52}Mn_{25}Ga_{23}$, shown in Figure 16 (a), peaks appear for both T_M and T_R in zero field and all values of the magnetic field. The peak at T_R, associated with heating, is larger than that at T_M, associated with cooling. These peaks indicate that the lattice expands abruptly. Dai *et al.* studied the elastic constants of a $Ni_{0.50}Mn_{0.284}Ga_{0.216}$ single crystal using the ultrasonic continuous-wave method [31]. C_{11}, C_{33}, C_{66}, and C_{44} modes were investigated; every mode indicated abrupt softening around T_M. This lattice softening appears to be affected by the abrupt expansion just above T_M when cooling from the austenite phase.

Figure 18 shows the magnetic phase diagram of $Ni_{52}Mn_{25}Ga_{23}$. With increasing field, T_M and T_R gradually increase. The shifts in T_M and T_R around zero magnetic field are estimated as $dT_M/dB = 0.46$ K/T and $dT_R/dB = 0.43$ K/T, which are similar to those of the $Ni_{52}Mn_{12.5}Fe_{12.5}Ga_{23}$ alloy ($dT_M/dB = 0.5$ K/T) [27].

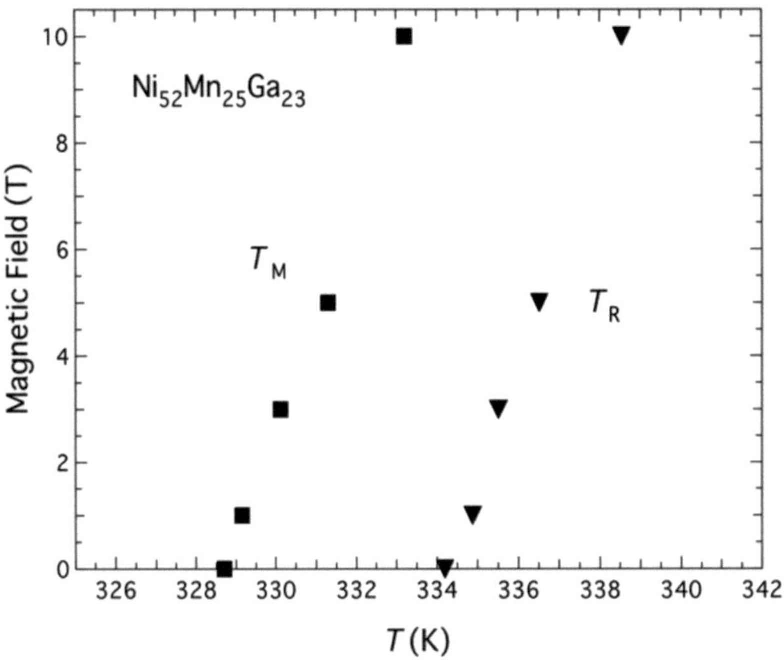

T. Sakon

Figure 18. Magnetic phase diagram of $Ni_{52}Mn_{25}Ga_{23}$. Filled squares indicate the martensite transition temperature T_M. Filled triangles indicate reverse martensite temperature T_R.

Figure 19 shows the magnetization curves of $Ni_{52}Mn_{25}Ga_{23}$ in a pulsed magnetic field up to 2.2 T. The unit of magnetization *M* is J/μ_0kgT in the SI unit system or emu/g in the CGS unit system (both having identical numerical values). The *M–B* curves were measured from low temperature. The hysteresis of the *M–B* curve is considerably small. The magneto caloric effects in other magnetic materials were also reported; for example, Levitin *et al.* reported for $Gd_3Ga_5O_{12}$ [32]. They performed magnetization measurements at an initial temperature of 4.2 K, where the magnetic contribution to heat capacity is comparable to the lattice heat capacity. In our experiment, the temperature change of the sample due to the magneto caloric effect is considered to be within 1 K. This is because these experiments were performed around room temperature, where the lattice heat capacity is much larger than the heating or cooling power by the magneto caloric effect.

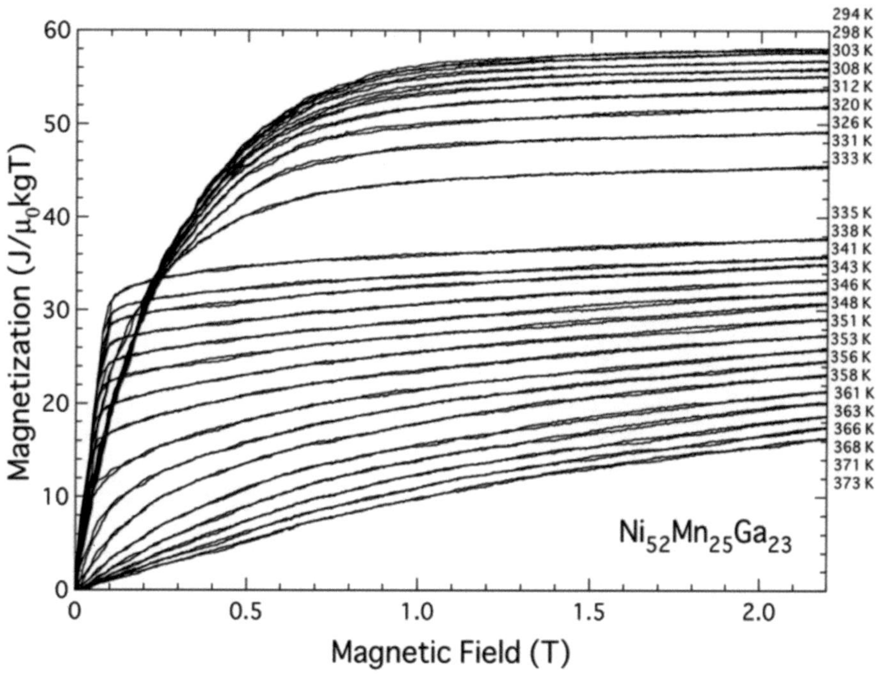

T. Sakon

Figure 19. Magnetization of $Ni_{52}Mn_{25}Ga_{23}$ in a pulsed magnetic field up to 2.2 T.

The *M–B* curves show ferromagnetic behavior below 356 K. It is clear that the field dependence of the magnetization at the Ferro–A phase above T_R = 334 K is different from that at the Ferro–M phase below T_R. At the Ferro–M phase, magnetization increases with magnetic fields. On the other hand, at the Ferro–A phase between 334 and 356 K, a sudden increase in magnetization occurs between 0 and 0.1 T.

Figure 20 shows magnetization in a magnetic field up to 15 T. In high magnetic fields, an almost linear increase can be seen for each *M–B* curve. In particular, as for the *M–B* curve below 334 K, the high magnetic field susceptibility is quite small.

Figure 21 shows the Arrott plot of $Ni_{52}Mn_{25}Ga_{23}$. The spontaneous magnetization at 294 K in a Ferro–M phase is 55.0 J/$\mu_0$kgT. The Curie temperature of the austenite phase T_{CA} determined by Arrott plots in Figure 21 is 358 K. This is consistent with the *x–T* phase diagram of $Ni_{50+x}Mn_{25}Ga_{25-x}$ [6,7]. In high magnetic fields, an almost linear increase can be seen for each *M–B* curve. $Ni_2Mn_{0.75}Cu_{0.25}Ga$ also shows the difference in magnetization between 302 and 305 K, which is somewhat lower than T_C = 307 K or T_M = 308 K [27]. Note that the Arrott plots of $Ni_{52}Mn_{25}Ga_{23}$ left a space between 333 and 335 K, and $Ni_2Mn_{0.75}Cu_{0.25}Ga$ left a space between 302 and 303 K. The spontaneous magnetizations of $Ni_{52}Mn_{25}Ga_{23}$ are 42.2 J/$\mu_0$kgT at 333 K and 34.2 J/$\mu_0$kgT at 335 K, which were obtained by the Arrott plot shown in Figure 21. As for $Ni_2Mn_{0.75}Cu_{0.25}Ga$, the spontaneous magnetizations are 40.0 J/$\mu_0$kgT at 302 K and 28.3 J/$\mu_0$kgT at 303 K.

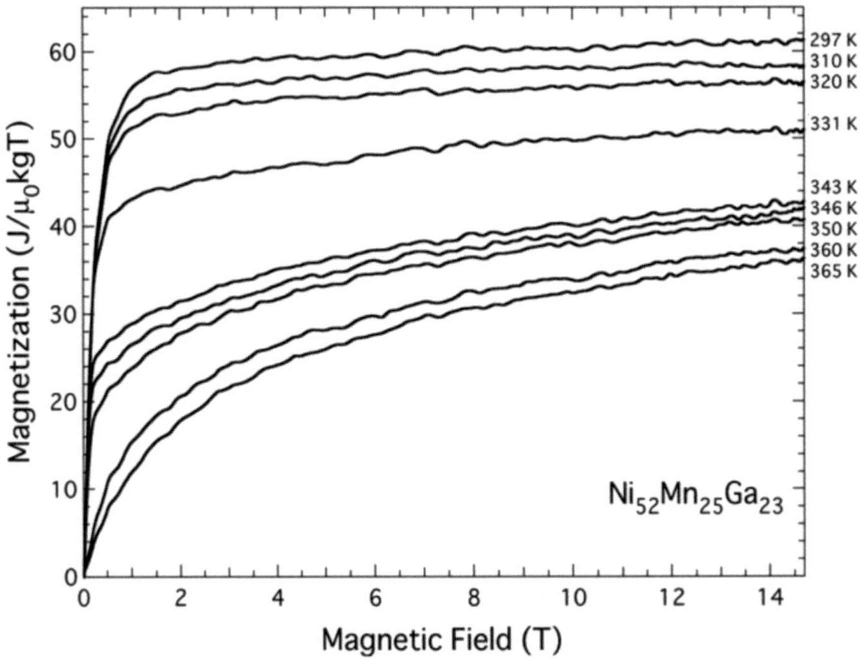

T. Sakon

Figure 20. Magnetization of $Ni_{52}Mn_{25}Ga_{23}$ in a pulsed magnetic field up to 15 T.

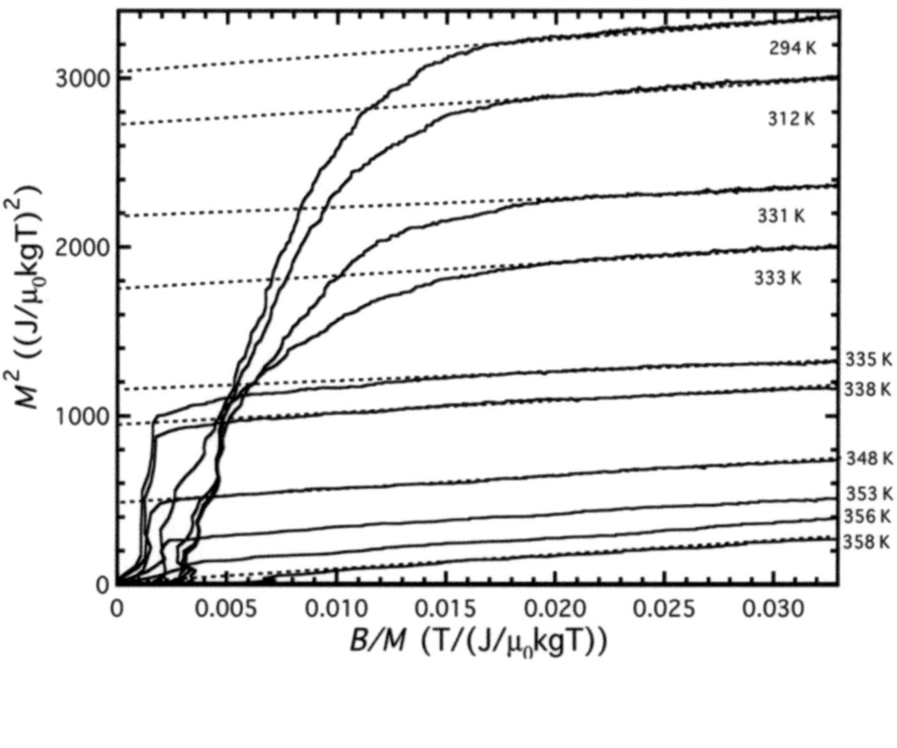

T. Sakon

Figure 21. Arrott plot of the magnetization of $Ni_{52}Mn_{25}Ga_{23}$. Dotted straight lines are extrapolated lines.

Figure 22 shows the temperature dependence of the magnetization *M–T* at 0.1, 0.5, and 1 T, which was obtained by magnetization measurements in pulsed magnetic fields. Open circles are the spontaneous magnetizations, which was obtained by the Arrott plot method. A sudden decrease is apparent between 333 and 336 K for each field, and also the spontaneous magnetization. This temperature region corresponds to the sharp increase in permeability when heating from low temperature in Figure 15, and just below T_R, which was obtained by the linear strain measurement in Figure 16 (a).

The *M-T* curve in Figure 22 can be seen as the combination of two single-phase *M-T* curves. One corresponds to the martensite phase, and the other corresponds to the austenite phase. The obtained Curie temperatures in the martensite phase and the austenite phase are T_{CM} = 333.5 ± 0.5 K and T_{CA} = 358.0 ± 0.5 K. This is due to the difference of the ferromagnetic interactions for both structural phases. These analyses of magnetic properties in $Ni_{51.9}Mn_{23.2}Ga_{24.9}$ were also reported in reference 11.

It is well known that the tetragonal martensite Ni–Mn–Ga has higher magnetocrystalline anisotropy in association with the multi-dominant structure of the martensite phase. Consequently, lower initial permeability and higher coercivity than the cubic austenite Ni–Mn–Ga alloys can occur [8,11–13,15]. The martensite transition occurs in the ferromagnetic phase, and the decrease in magnetization is observed at intermediate fields for $0 < B < 0.5$ T, as shown in Figure 22. This property is also shown by magnetization in many Ni–Mn–Ga alloys (e.g., $Ni_{49.5}Mn_{25.4}Ga_{25.1}$) and Ni–Mn–Sn alloys (e.g., $Ni_{50}Mn_{35}Sn_{15}$) [8,33,34]. Consequently, at low field, the austenitic Ni–Mn–Ga (with softer ferromagnetism) shows an abrupt increase in *M*, while the martensite Ni–Mn–Ga (with harder ferromagnetism) shows gradual increase in *M* with the field. On the other hand, the martensite Ni–Mn–Ga (in low-temperature phase) has higher saturation magnetization (typically, M_s increases with decreasing temperature) than the austenite Ni–Mn–Ga. As a result, at very high field or saturation field (>1 T), magnetization of the martensite is higher than that of the austenite, as shown in Figures 20 and 22. As for other Ni–Mn–Ga alloys, Kim *et al.* reported magnetization in a $Ni_{2.14}Mn_{0.84}Ga_{1.02}$ single crystal, which shows a transition from the Ferro–A phase to Ferro–M phases with *14M* structure [14]. The magnetization curve in $Ni_{2.14}Mn_{0.84}Ga_{1.02}$ at 290 K, just below the martensite transition temperature, sharply bend at the critical field, B_S = 0.6 T, and above 0.6 T, the magnetization slightly increases with increasing fields. On the other hand, a bend in the magnetization is not clear. We defined the critical field B_S in $Ni_{52}Mn_{25}Ga_{23}$ as the field where the magnetization Arrott plot was off from the extrapolated linear line, which is illustrated by the dotted line in Figure 21, and obtained B_S as 0.84 T, which is of the same order as that in $Ni_{2.14}Mn_{0.84}Ga_{1.02}$. The magnetization is the same as that in $Ni_{52}Mn_{25}Ga_{23}$. The magnetic anisotropy constant K_U in a Ni_2MnGa single crystal is 1.17×10^5 J/m^3 (11.7×10^5 erg/cm^3) in the martensite phase and 2.7×10^4 J/m^3 (2.7×10^5 erg/cm^3) in the austenite phase [1], indicating that the magnetic anisotropy is about four times larger in the martensite phase than that in the austenite phase. The Zeeman energy and/or magnetocrystalline anisotropy energy that is sufficient to induce motion of the twin boundary is denoted as $M_SB_S/2 = K_U$ [1]. Kim *et al.* also mentioned that the magnetocrystalline anisotropy energy is of the order of 10^5 J/m^3 [14]. The spontaneous magnetization in $Ni_{52}Mn_{25}Ga_{23}$ at 333 K, just below T_R is 42.2

J/$\mu_0$kgT, which was obtained by the Arrott plot in Figure 22. When using this value as M_S, the magnetocrystalline anisotropy energy in the martensite phase of $Ni_{52}Mn_{25}Ga_{23}$ is $M_SB_S/2 = K_U = 1.04 \times 10^5$ J/m^3, which is on the same order as that in the martensite phase of Ni_2MnGa. These magnetic properties were also shown for $Ni_{51.9}Mn_{23.2}Ga_{24.9}$ [11], $Ni_{49.5}Mn_{25.4}Ga_{25.1}$ [12], and $Ni_{54}Mn_{21}Ga_{25}$ [13].

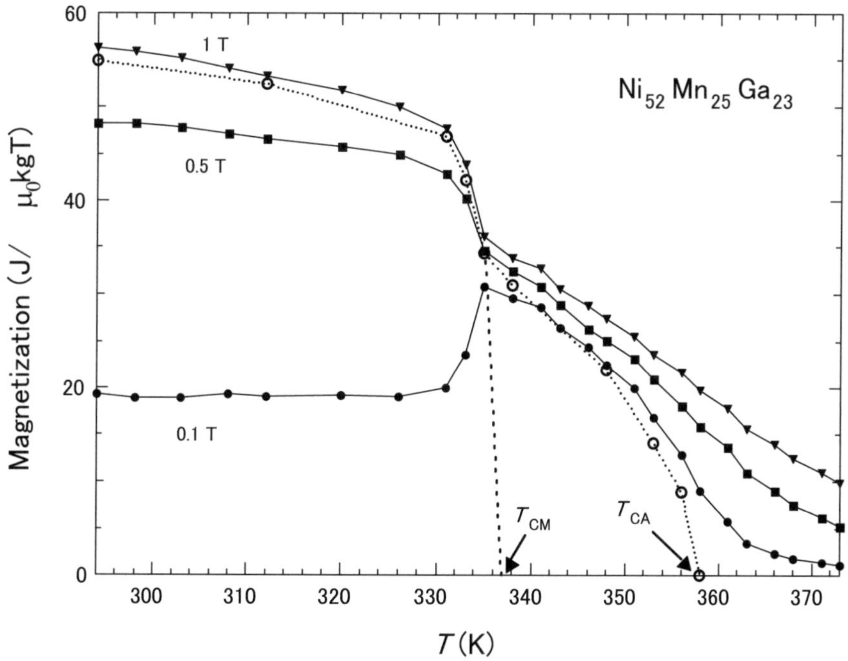

T. Sakon

Figure 22. Temperature dependence of the magnetization of $Ni_{52}Mn_{25}Ga_{23}$. Open circles are the spontaneous magnetizations, which was obtained by the Arrott plot method. Dotted lines are extrapolated lines of the spontaneous magnetization plots. T_{CM} and T_{CA} indicate the martensite Curie temperature and the austenite Curie temperature, respectively.

The relationship between magnetism and T_M in magnetic fields is discussed for Ni_2MnGa-type Heusler alloys. Table 1 shows the spontaneous magnetizations and dT_M/dB values of $Ni_{2+x}Mn_{1-x}Ga$, $Ni_{52}Mn_{12.5}Fe_{12.5}Ga_{23}$, $Ni_2Mn_{0.75}Cu_{0.25}Ga$, $Ni_2MnGa_{0.88}Cu_{0.12}$, and $Ni_{52}Mn_{25}Ga_{23}$. As for $Ni_{2+x}Mn_{1-x}Ga$ alloys, shifts in T_M in magnetic fields were observed by magnetization measurements [2,26–28]. T_M and T_C of Ni_2MnGa (x = 0) are 200 and 360 K, respectively. The region above T_M is the Ferro–A phase. The sample with x = 0 of $Ni_{2+x}Mn_{1-x}Ga$ shows phase transition from the Ferro–A to Ferro–M phases at T_M. The sample with x = 0.19 shows ferromagnetic transition and martensite transition at T_M. For x = 0, the shift in T_M is estimated as dT_M/dB = 0.2 K/T [35] and for x = 0.19, dT_M/dB = 1 .0 K/T [36]. The shift in T_M for x = 0.19 is higher than that for x = 0. These results indicate that the shift in T_M for the alloy that shows Para–A to Ferro–M phase transition is larger than that for the alloy that shows Ferro–A to Ferro–M phase transition. The values of dT_M/dB are roughly proportional to the change in spontaneous magnetization, (M_M – M_A)/M_M, as shown in Table 1. This indicates that the magnetic moments influence the martensite transition; in other words, the structural transition and the T_M increase in accordance with the magnetic fields are proportional to the difference between the magnetization of the austenite phase and that of the martensite phase. Therefore, it is considered that the alloys, in which T_M and T_C are close to each other, show a larger shift in T_M in magnetic fields.

Khovailo *et al.* discussed the correlation between the shifts in T_M for $Ni_{2+x}Mn_{1-x}Ga$ ($0 \leq x \leq 0.19$) using theoretical calculations [37,38]. The experimental values of this shift for $Ni_{2+x}Mn_{1-x}Ga$ ($0 \leq x \leq 0.19$) are in good agreement with the theoretical calculation results. In general, in a magnetic field, the Gibbs free energy is lowered by the Zeeman energy $-\Delta MB$ that enhances the motive force of the martensite phase transition. Thus, T_M of the ferromagnetic Heusler alloys $Ni_{52}Mn_{12.5}Fe_{12.5}Ga_{23}$, $Ni_2Mn_{0.75}Cu_{0.25}Ga$, and $Ni_2MnGa_{0.88}Cu_{0.12}$ in recent studies [27,39,40] and $Ni_{52}Mn_{25}Ga_{23}$ in this study are considered to have shifted in accordance with the magnetic fields because high magnetic fields are favorable for ferromagnetic martensite phases.

Chernenko *et al.* studied the temperature dependence of both the saturation magnetic field values and the x-ray powder diffraction patterns of Ni-Mn-Ga alloys and analyzed with the theoretical consideration [12]. The theory proposes that the free energy for ferromagnetic martensite phase, exposed to an external magnetic field, is expressed as three terms. First term is the magnetic anisotropy energy. The second and third terms describe the magnetostatic and the Zeeman energy, respectively. The *c/a* ratio was expressed as

$$c/a = 1 - \frac{\left[(H_S / M) + \left| D_1 - D_2 \right| \right]}{12\delta}, \tag{2}$$

where H_s indicates the saturation magnetic field. M denotes the absolute value of the magnetization. D_1 and D_2 denote the diagonal matrix elements, and δ is the dimensionless magnetoelastic parameter. The linear dependence of the magnetic anisotropy constant on the tetragonal distortion of the cubic crystal lattice arising in the course of the martensite transition.

In order to apply this theory to our present work, it is considered that further theoretical consideration is needed for apply this theory for analyzing the influence between the martensite variant structure and the magnetic field, which is reflected by the Zeeman term.

4. Conclusions

$Ni_{52}Mn_{12.5}Fe_{12.5}Ga_{23}$ and $Ni_2Mn_{0.75}Cu_{0.25}Ga$

Thermal expansion, magnetization, and permeability measurements were performed on the ferromagnetic Heusler alloys $Ni_{52}Mn_{12.5}Fe_{12.5}Ga_{23}$ and $Ni_2Mn_{0.75}Cu_{0.25}Ga$.

1. Thermal expansion

When cooling from austenite phase, steep decrease due to the martensitic transformation was obtained for both alloys. T_M and T_R increase gradually with increasing magnetic fields.

The shifts of T_M for $Ni_{52}Mn_{12.5}Fe_{12.5}Ga_{23}$ and $Ni_2Mn_{0.75}Cu_{0.25}Ga$ in magnetic fields were estimated as $dT_M/dB \approx 0.5$ K/T and 1.2 T/K, respectively.

2. Magnetization and permeability

$Ni_{52}Mn_{12.5}Fe_{12.5}Ga_{23}$ ---- The *M-B* curves indicate that the property of the Ferro-M phase is different from the Ferro-A phase. The Ferro-A phase is considered to be a more localized ferromagnetic phase as compared with Ferro-M phase.

$Ni_2Mn_{0.75}Cu_{0.25}Ga$ ---- The permeability abruptly changes around T_M. The permeability below T_M is about one-tenth times higher than that above T_M. The Arrott plot of magnetization indicates that T_C of the martensite phase is 307 K, which is almost the same as T_M = 308 K.

3. The values of dT_M/dB are roughly proportional to the change of the spontaneous magnetization $(M_M - M_A)/M_M$. T_M of the ferromagnetic Heusler alloys $Ni_{52}Mn_{12.5}Fe_{12.5}Ga_{23}$ and $Ni_2Mn_{0.75}Cu_{0.25}Ga$ in the magnetic field is considered to be shifted in accordance with the magnetic fields and proportional to the difference between the magnetization of austenite phase with that of martensite phase.

$Ni_2MnGa_{0.88}Cu_{0.12}$

Thermal expansion, permeability, magnetization measurements were performed on the Heusler alloy $Ni_2MnGa_{0.88}Cu_{0.12}$.

1. Thermal expansion

When cooling from austenite phase, a steep decrease due to the martensitic transition was obtained. T_M and T_R increase gradually with increasing magnetic fields. The shift of T_M was estimated as dT_M/dB = 1.3 K/T.

2. Magnetization and permeability

The permeability abruptly changes and shows the clear peak around T_M. The permeability below T_M is about one-tenth than that above T_M. The temperature dependence of the magnetization also shows a clear decrease around T_M. The Arrott plot of magnetization indicates that T_C of the martensite phase is 340 K, which is almost the same as T_M = 337 K, which was obtained by the linear expansion.

3. The values of dT_M/dB are roughly proportional to the change of the spontaneous magnetization $(M_M - M_A)/M_M$ in Ni_2MnGa type Heusler alloys. T_M of the ferromagnetic Heusler alloy $Ni_2MnGa_{0.88}Cu_{0.12}$ in the magnetic field is considered to be shifted in accordance with the magnetic fields and proportional to the difference between the magnetization of austenite and martensite phase.

$Ni_{52}Mn_{25}Ga_{23}$

Thermal strain, permeability, and magnetization measurements were performed on the Heusler alloy $Ni_{52}Mn_{25}Ga_{23}$.

1. Thermal strain: When cooling from the austenite phase, a steep decrease in the thermal strain is obtained because of the martensite transition. T_M and T_R increase gradually with increasing magnetic fields. The shifts in T_M and T_R in a magnetic field are estimated as $dT_M/dB = 0.46$ K/T and $dT_R/dB = 0.43$ K/T, respectively.
2. Magnetization and permeability: Permeability abruptly changes around T_M and T_R. Permeability below T_M is about one-third that above T_M. The temperature dependence of the magnetization also shows a clear discontinuity around T_M. The Arrott plot of magnetization indicates that T_C is 358 K. The sudden decrease in magnetization at the temperature of the martensite transition and the M–B curve indicate the magnetism of the hard Ferro–M phase and the soft Ferro–A phase.
3. The dT_M/dB values are roughly proportional to the change in spontaneous magnetization $[(M_M - M_A)/M_M]$ in Ni_2MnGa-type Heusler alloys. The T_M of the ferromagnetic Heusler alloy $Ni_{52}Mn_{25}Ga_{23}$ in the magnetic field is considered to be shifted in accordance with the magnetic fields and proportional to the difference in magnetization between the austenite and martensite phases.

Author details

T. Sakon*
Department of Mechanical System Engineering, Faculty of Science and Technology, Ryukoku University, Japan
Department of Mechanical Engineering, Graduate School of Engineering and Resource Science, Akita University, Japan

H. Nagashio, K. Sasaki, S. Susuga, D. Numakura and M. Abe
Department of Mechanical Engineering, Graduate School of Engineering and Resource Science, Akita University, Japan

K. Endo, S. Yamashita and T. Kanomata
Faculty of Engineering, Tohoku Gakuin University, Japan

H. Nojiri
Institute for Materials Research, Tohoku University, Japan

Acknowledgement

This study was supported by a Grant-in-Aid of the three universities cooperation project in North Tohoku area in Japan, and Japan Science and Technology project No. AS232Z02122B. This study was also partly supported by a Grant-in-Aid for Scientific Research (C) (Grant No. 21560693) from the Japan Society for the Promotion of Science (JSPS) of the Ministry of Education, Culture, Sports, Science and Technology, Japan.

---

* Corresponding Author

This study was technically supported by the Center for Integrated Nanotechnology Support, Tohoku University, and the High Field Laboratory for Superconducting Materials, Institute for Materials Research, Tohoku University. One of the authors (H. N.) acknowledges the support by GCOE-material integration.

5. References

[1] K. Ullakko, J.K. Huang, C. Kantner, R.C. O'Handley, V.V. Kokorin, *Appl. Phys. Lett.* 69 (1996) 1966.
[2] P.J. Webster, K.R.A. Ziebeck, S.L. Town, M.S. Peak, *Philos. Mag. B* 49 (1984) 295.
[3] P.J. Brown, J. Crangle, T. Kanomata, M. Matsumoto, K-.U. Neumann, B. Ouladdiaf, K.R. A. Ziebeck, *J. Phys: Condens. Matter* 14 (2002) 10159.
[4] J. Pons,R. Santamarta, V.A. Chernenko, E. Cesari, *J. Appl. Phys.* 97 (2005) 083516.
[5] R. Ranjan, S. Banik, S.R. Barman, U. Kumar, P.K. Mukhopadhyay, D. Pandey, *Phys. Rev. B* 74 (2006) 224443.
[6] C. Jiang, G. Feng, S. Gong, H. Xu, *Mater. Sci. Eng. A*, 342 (2003) 231.
[7] Y. Ma, C. Jiang, Y. Li, H. Xu, C. Wang, X. Liu, *Acta. Mater.* 55 (2007) 1533.
[8] L. Manosa, X. Moya, A. Planes, T. Krenke, M. Acet, E.F. Wassermann, *Mater. Sci. Eng. A* 481–482 (2008) 49.
[9] V. Sanchez-Ala'rcos J.I. Perez-Landaza'bal, C. Go'mez-Polo, V. Recarte, *J. Magn. Magn. Mater.* 320 (2008) e160.
[10] A. Rudajevov´a, *J. Alloys Compd.* 430 (2007) 153.
[11] F.Q. Zhu, F.Y. Yang, C.L. Chien, L. Ritchie, G. Xiao, G.H. Wu, *J. Magn. and Magn. Mater.* 288 (2005) 79.
[12] V.A. Chernenko, V.A. L'vov, V.V. Khovailo, T. Takagi, T. Kanomata, T. Suzuki, R. Kainuma, *J. Phys.: Condens. Matter* 16 (2004) 8345.
[13] V.O. Golub, A.Y. Vovk, C.J. O'Connor, V.V. Kotov, P.G. Yakovenko, K. Ullakko, *J. App. Phys.* 93 (2003) 8504.
[14] J. Kim, F. Inaba, T. Fukuda, T. Kakeshita, Acta Materialia 54 (2006) 493.
[15]A. Gonzalez-Comas, E. Obrado, L. Mafiosa, A. Planes, A. Labarta, *J. Magn. Magn. Mater.* 196-197 (1999) 637.
[16] N. Lanska, O. So¨derberg, A. Sozinov, Y. Ge, K. Ullakko, V.K. Lindroos, *J. App. Phys.* 95 (2004) 8074.
[17] A.A. Likhachev, K. Ullakko, *Phy. Lett. A* 275 (2000) 142.
[18] S.J. Murray M. Farinelli, C. Kantner, J.K. Huang, S.M. Allen, R.C. O'Handley, J. App. Phys. 83 (1998) 7297.
[19] K. Endo, T. Kanomata, A. Kimura, M. Kataoka, H. Nishihara, R. Y. Umetsu, K. Obara, T. Shishido, M. Nagasako, R. Kainuma, K. R. A. Ziebeck, edited by V.A. Chernenko, *Materials Science Forum* 684 (2011) 165.
[20] D. Kikuchi, T. Kanomata, Y. Yamauchi, H. Nishihara: *J. Alloys Compounds* 426 (2006) 223.
[21] A. N. Vasil'ev, E. I. Estrin, V. V. Khobailo, A. D. Bozhko, R. A. Ischuk, M. Matsumoto, T. Takagi, J. Tani: *Int. Appl. Electromagn. Mechan.* 12 (2000) 35.
[22] D. Kikuchi, Thesis, Tohoku-Gakuin Univ. 2010.

[23] M. Kataoka, K. Endo, N. Kudo, T. Kanomata, H. Nishihara, T. Shishido, R.Y. Umetsu, N. Nagasako, R. Kainuma: *Phys. Rev. B* 82 (2010) 214423.

[24] C. Kittel, *Introduction to Solid State Physics* (John Wiley & Sons, NJ, USA, 2005) 8th ed., p 75.

[25] P. Entel, V. D. Buchelnikov, V. V. Khovalio, A. T. Zayak, W. A. Adeagbo, M. E. Gruner, H. C. Herper, E. F. Wassermann: *J. Phys, D: Appl. Phys.* 39 (2006) 865-889.

[26] Bozorth R M 1951 *Ferromagnetism* p. 32 (D. van Nostrand)

[27] T. Sakon, H. Nagashio, K. Sasaki, S. Susuga, K. Endo, H. Nojiri, T. Kanomata, *Mater. Trans.* 52 (2011) 1142.

[28] A. N. Vasil'ev, E. I. Estrin, V. V. Khovailo, A. D. Bozhko, R. A. Ischuk, M. Matsumoto, T. Takagi, J. Tani, *Int. Appl. Electromagn. Mechan.* 12 (2000) 35.

[29] D. Kikuchi, T. Kanomata, Y. Yamaguchi, H. Nishihara, *J. Alloys Compd.* 426 (2006) 223.

[30] A.N. Vasil'ev, E.I. Estrin, V.V. Khovailo, A.D. Bozhko, R.A. Ischuk, M. Matsumoto, T. Takagi, J. Tani, *Int. J. Appl. Electromagn. Mechan.* 12 (2000) 35.

[31] L. Dai, J. Cullen, M. Witting, *J. Appl. Phys.* 95 (2004) 6957.

[32] R.Z. Levitin, V.V. Snegirev, A.V. Kopylov, A.S. Lagutin, A. Gerber, *J. Magn. Magn. Mater.* 170 (1997) 223.

[33] J. Markos, A. Planes, L. Manosa, F. Casanova, X. Battle, A. Labarta, B. Martinez, *Phys. Rev. B* 66 (2002) 224413.

[34] T. Krenke, M. Acet, E. F. Wassermann, X. Moya, L. Manosa, A. Planes, *Phys. Rev. B* 72 (2005) 014412.

[35] A.Gonzàlez-Comas, E. Obradó, L. Maños, A. Planes, V.A. Chernenko, B.J. Hattink, A. Labarta, *Phys. Rev. B* 60 (1999) 7085.

[36] D.A. Filippov, V.V. Khovailo, V.V. Koledov, E.P. Krasnoperov, R.Z. Levitin, V.G. Shavrov, T. Takagi, *J. Magn. Magn. Mater.* 258–259 (2003) 507.

[37] V.V. Khovailo, V. Novosad, T. Takagi, D.A. Filippov, R.Z. Levitin, A.N. Vasil'ev, *Phys. Rev. B* 70 (2004) 174413.

[38] V.V. Khovailo, T. Takagi, T.J. Tani, R.Z. Levitin, A.A. Cherechukin, M. Matsumoto, R. Note, *Phys. Rev. B* 65 (2002) 092410.

[39] T. Sakon, H. Nagashio, K. Sasaki, S. Susuga, D. Numakura, M. Abe, K. Endo, H. Nojiri, T. Kanomata, *Physica Scripta* 84 (2011) 045603.

[40] S. Aksoy, T. Krenke, M. Acet, E. F. Wassermann, *Appl. Phys. Lett.* 91 (2007) 251915.

Chapter 5

Iron Based Shape Memory Alloys: Mechanical and Structural Properties

Fabiana Cristina Nascimento Borges

Additional information is available at the end of the chapter

http://dx.doi.org/10.5772/51877

1. Introduction

The technological development is one of the reasons why there is a variety of new materials that can be applied to various situations. This situation enables many materials to be applied in different areas: engineering, medicine, agriculture, arts, space field, among others. Alloys with shape memory effect (SME) are materials that exhibit interesting characteristics and can be applied in various situations.

The SME in Fe-based alloys results from the reverse motion of Shockley partial dislocation during heating (Otubo, 2002) and (Bergeon et al. 1997). Figure 1 shows a schematic figure of

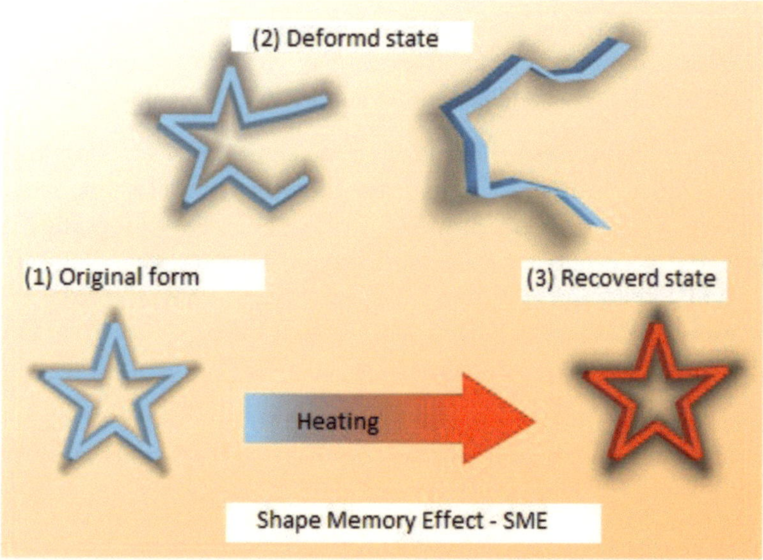

Figure 1. Shape Memory Effect (Nascimento, 2008).

SME. The original form of the material is a star, Fig. 1-(1). This star is deformed beyond its elastic limit, Fig. 1-(2) and the original crystal structure (f.c.c.) is transformed into h.c.p. structure. During the heating, reversion to the f.c.c. structure occurs and the original shape is recovered Fig. 1-(3). Reversion to the matrix phase (austenite) is not complete because a martensite residual amount exists which is not recovered during the heat treatment. Chemical composition and austenite grain size are important factors that affect the shape recovery in iron based shape memory alloys.

In this study, structural parameters of stress induced ε-martensite were analyzed for Fe-Mn-Si-Cr-Ni-(Co) different chemical compositions. The material was hot rolled at 1473 K followed by a heat-treatment at 1323 K for different times to obtain different austenite grain sizes samples. Two parameters were considered: austenitic grain and training cycles.

2. Iron shape memory alloys – history

Iron based shape memory alloys have been largely investigated during the last years. The Shape Memory Effect (SME) is a physical phenomenon which results in recovery of the original shape through temperature variation after the material has been deformed beyond its elastic limit. The alloys that exhibit this characteristic are known as Smart Materials - a group of materials that show reproducible and stable responses, through significant variations of at least one property, when subjected to external stimuli. Table 1 shows some of these materials and their properties.

In iron based alloys the SME, is related with the γ(f.c.c.) $\leftrightarrow$ ε(h.c.p.) nonthermoelastic martensitic transformation (Bergeon et al. 1997). This effect is the result of the reverse motion of Shockley partial dislocation during heating. In general, the technological development was largely responsible for, the emergence of new compositions with SME. The ferrous alloy was developed as an alternative to NiTi alloys and also the copper base compositions due to its low cost and properties similar to nitinol alloy.

Fe-Mn-Si alloys began to be studied in the 80s (Sato et al. 1982). The alloying elements Cr, Ni and Co were subsequently used to improve the properties of shape recovery. Fe-Mn-Si-Cr-Ni-Co alloys were developed, with several attractive properties and a more desirable shape recovery making them suitable for various technological applications (Shiming et al. 1991), (Bergeon et al. 1997), (Kajiwara et al. 1999), (Arruda, 1999). In Brazil the family of iron-based shape memory alloys has been extensively studied since 1995 (Otubo et al. 1995).

Tab. 2 presents a list of research groups registered in the CNPq (National Counsel of Technological and Scientific Development) investigating the ferrous alloys with EMF in Brazil.

Research groups are listed in Tab. 2 to present the several technological applications and basic studies. In this study we will focus on recovery as a function of the initial microstructure and training cycles.

Smart Materials	Properties
Shape memory alloys and shape memory polymers	Materials in which large deformation can be induced and recovered through temperature changes or stress changes.
Magnetic shape memory alloys	Materials that change their shape in response to significant change in the magnetic field.
Piezoelectric materials	Materials that produce a voltage when stress is applied.
Magnetostrictive materials	Materials that exhibit change in shape under the influence of magnetic field.
pH-sensitive polymers	Materials that change in volume when the pH of the surrounding medium changes.
Temperature-responsive polymers	Materials which undergo changes upon temperature.
Halochromic materials	Materials that change their color as a result of changing acidity.
Chromogenic systems	Materials that change color in response to electrical, optical or thermal changes.
Ferrofluid	
Photomechanical materials	Materials that change shape under exposure to light.
Self-healing materials	Materials that have the intrinsic ability to repair damage due to normal usage, thus expanding the material's lifetime
Dielectric elastomers	Smart material systems which produce large strains (up to 300%) under the influence of an external electric field.
Magnetocaloric materials	Compounds that undergo a reversible change in temperature upon exposure to a changing magnetic field.
Thermoelectric materials	Materials used to build devices that convert temperature differences into electricity and vice-versa.

Table 1. Smart Materials

Research groups -source: CNPq	Identification	Iron-based alloy
Development of metallic alloys	Fundação Centro Tecnológico de Minas Gerais – *CETEC*	Fe-Mn-Si-(Ni-Cr-Co)
Development of metallic alloys for industrial applications	Universidade Estadual de Campinas – *UNICAMP*	Fe-Mn-Si-(Ni-Cr-Co)
Shape memory alloys - characterization and application	Universidade Estadual de Ponta Grossa – *UEPG*	Fe-Mn-Si-Ni-Cr-(Co) and NiTi
Shape memory materials	Instituto Tecnológico da Aeronáutica – *ITA*	Stainless steel e NiTi

Table 2. Research groups of iron shape memory alloy in Brazil (source: CNPq)

3. Structural and mechanical properties

Technological applications of these alloys are directly related to the study of their mechanical and structural properties. There are several mechanical properties which may be mentioned. In this study the relationship between the effect of structural parameters on the mechanical properties and shape recovery will be presented through the analysis of samples subjected to cycles of training using compression test. Therefore, the results discussed refer to the reverse transformation of stress induced ε-martensite.

3.1. Structural characterizations

As it is known, the SME is directly related to processing and reversing the crystalline phases. In these materials the following transformations may occur:

$$g(f.c.c.) \leftrightarrow (h.c.p.), \gamma(f.c.c.) \leftrightarrow \alpha'(b.c.c.) \text{ and } \gamma(f.c.c.)\varepsilon(h.c.p.) \leftrightarrow \alpha'(b.c.c.)$$

The predominant type of transformation will depend on factors such as chemical composition and thermomechanical treatment cycles. The α' phase is bcc; Shockley partial dislocations are specific of compact structures f.c.c. and h.c.p. When the fraction of b.c.c. phase increases, there is a decrease in the fraction of compact structures, thus the recovery mechanism through partial dislocation Shockley is smaller.

The three types of crystal structure show interesting peculiarities which are discussed below.

a. Austenitic phase

The austenitic phase in iron based Fe alloys is known as a strong and stable phase. Crystallographically it presents characteristics similar to commercial stainless steels, AISI 304. It features a cubic crystal structure (f.c.c.) and space group Fm-3m.

b. Martensitic phase

In this study there are two important phases resulting from the martensitic transformation:

$$\gamma(f.c.c.) \rightarrow \varepsilon(h.c.p.) \rightarrow \alpha'(b.c.c.) or \gamma(f.c.c.) \rightarrow \alpha'(b.c.c.)$$

The ε-martensite-phase is of greater interest because the reversion to austenite results in SME. Literature data show that the hexagonal structure (h.c.p.) can be mechanically or thermally induced. In particular in this case priority is given for the stress induced ε-martensite.

The atomic stacking sequence for the f.c.c. phase is ABCABCABC ... and h.c.p. phase is ABABAB. According to studies on stacking faults, they are necessary in the f.c.c structure in order to generate the embryos which form the martensitic phase. The overlapping of stacking faults form an h.c.p. volume and a reversal movement of Shockley partial dislocation occurs. Figure 2 shows a diagram of the stacking sequence to cubic and hexagonal structures. The orientation relationship between these phases is shown in Figure 3.

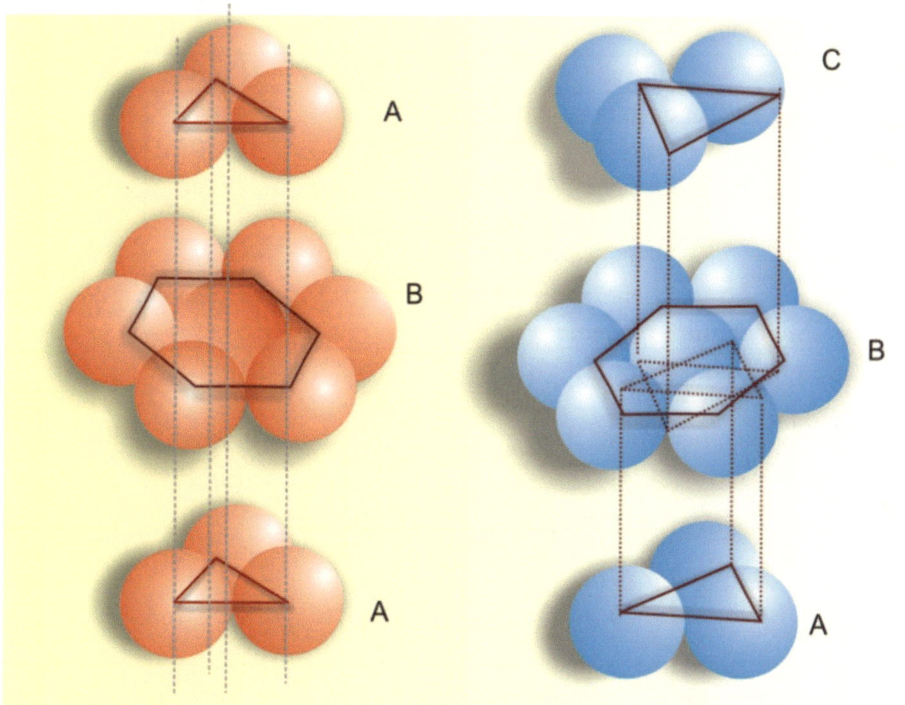

Figure 2. Atomic stacking sequence (ABCABCABC...) for f.c.c. structure with overlapping every third crystal plane (111) along [111]. Atomic stacking sequence (ABABAB...) for h.c.p. structure with overlapping crystal planes (0001) alternate along [0001] (Van Vlack, 1998).

The martensite and austenite phases can be identified using different techniques such as X-ray diffraction (XRD) and optical microscopy. The ferrous alloys, with SME, present a diffractogram similiar to AISI 304 commercial austenitic steels. Table 3 shows the position of 2θ reflections corresponding to the martensite and austenite phases. In this study, the XRD data were collected between 10 and 100°(2θ) at room temperature using a Philips diffractometer (PW1710) with Cu target and a graphite diffracted beam monochromator, step sizes of 0.02° and 2 seconds counting time.

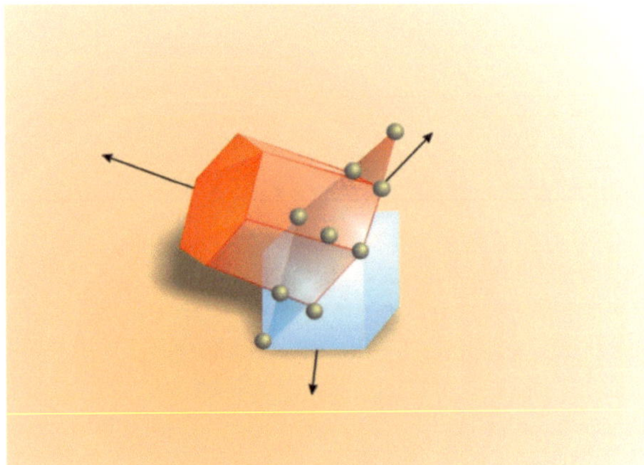

Figure 3. Orientation relationship between h.c.p. and f.c.c. phases (Huijun 1999).

Phase	(h.k.l.)	2θ
γ-austenite Structure: cubic Space group: Fm-3m	(111) (200) (220) (311)	43.7 50.7 74.8 90.8
ε-martensite Structure: hexagonal Space group: $P6_3/mm6$ (γ = 120°)	(10.0) (10.1) (10.2)	41.0 46.9 62.0

Table 3. Identification of austenite and martensite phases.

Figure 4 shows the identification of these phases and the effect of the training cycle on a sample with grain size 75 μm (4 - ASTM).

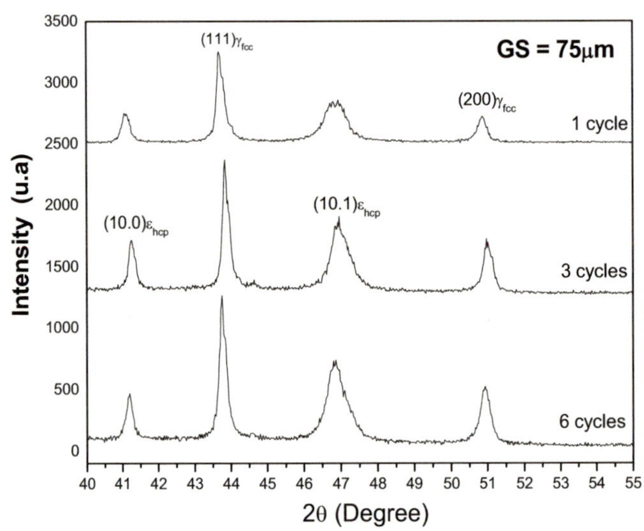

Figure 4. XRD patterns for 1st, 3rd and 6th thermo-mechanical cycles, deformed state, GS = 75 μm (Nascimento et al. 2008).

Figure 4 shows that with the increasing number of training cycles, the volumetric fraction of the martensitic phase increases. Using Rietveld refinement the quantitative analysis of phases was estimated considering the integrated intensity of the peaks $(10.1)_\varepsilon$ The small shift in the position 2θ of reflection (111)-austenitic phase shows variations in the lattice parameter of the unit cell this phase. These changes can be analyzed using Rietveld refinement.

In the Rietveld refinement the peak shape, width parameters and background parameters are considered. All these parameters were refined adopting the iterative least-squares method through minimization of residual parameter. Two structure types were considered: (a) cubic symmetry, space group **Fm-3m** for austenite phase, and (b) hexagonal symmetry, space group **P63/mmc** (with γ = 120°) for the martensite phase. Lattice parameters correspond to a similar composition alloy, AISI-304 steel. The thermal parameters (B's) initially used for both phases were $B_{overall}$ = 0.5 and the peak shape function used was the pseudo-Voigt. Figure 5 presents the experimental and refined X ray diffraction patterns as well as their difference.

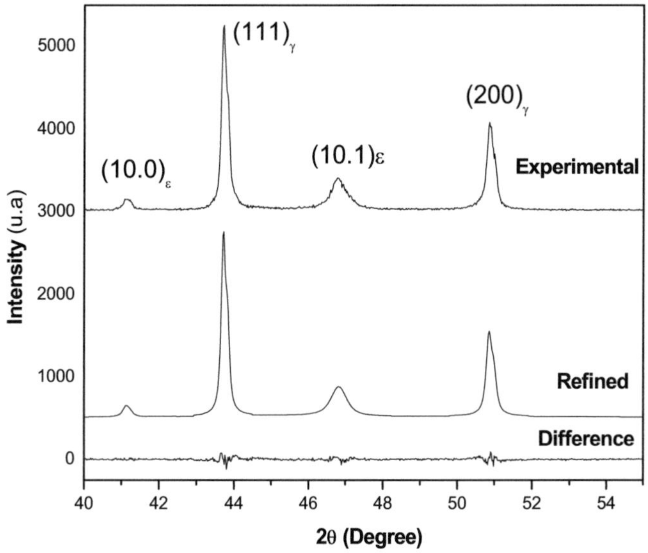

Figure 5. Rietveld refinement (GS = 75 μm), last thermo-mechanical cycle, deformed state (Nascimento et al. 2008).

The ε-martensite lattice parameters for the first cycle were: a_ε = 2.548(6) Å, c_ε = 4.162(2) Å. The ratio c/a found was c/a = 1.633(2). The standard deviations are shown in parenthesis. Austenitic phase indicated lattice parameters similar to those presented in the literature for stainless steel (Gauzzi et al. 1999): a_γ = 3.587(2) Å. Lattice parameters for the austenitic phase presented small variations (< 3%). The discrepancies between the experimental and refined profiles for all samples are small, indicating that the unit cell dimensions were accurately determined and that the chosen peak shape function pseudo-Voigt was a good choice for these samples. The thermal parameters (B's) presented a variation smaller than 0.5%. These structural variations are important because they affect the ratio c/a and also the reversion to the cubic austenitic phase (Nascimento et al. 2008).

Previous studies (Nascimento et al. 2008) show the effect of training cycles on the lattice parameter of the unit cell in stainless shape memory alloy, Figure 6. We note that for the sample with smaller grain size (75 μm) the a-parameters decreased with increased training cycles while the c-parameter increased. These changes affect the SME.

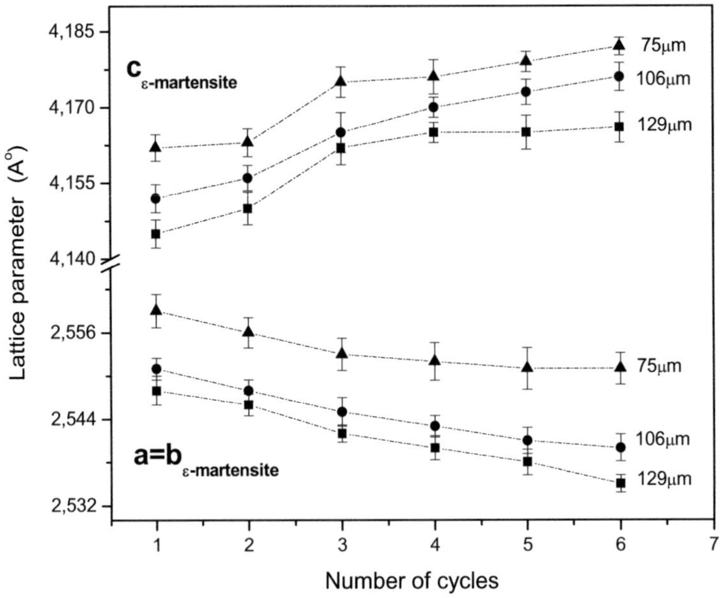

Figure 6. Structural parameters variation for austenite and martensite phases as a function of training cycles and grain size (Nascimento et al. 2008).

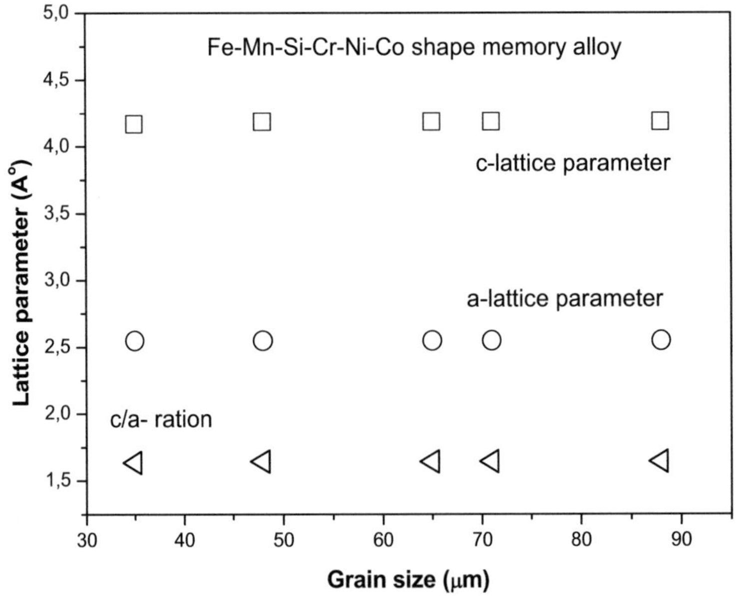

Figure 7. Structural parameters variation for austenite and martensite phases as a function of the grain size.

Figure 7 shows the variation of the structural parameters a, c, and the ratio a/c as a function of the initial microstructure. These samples showed a smaller variation of grain size and consequently lower variation of structural parameters.

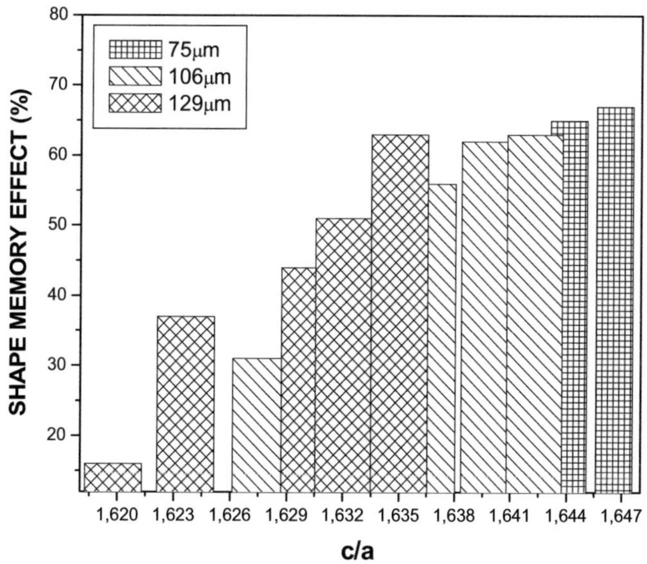

Figure 8. Shape memory effect as a function of ratio c/a.

Another way to identify the phases in alloys with SME is through optical microscopy using specific etching. The phases are differentiated through color (color etching method) that should be adapted to each sample (Nascimento et al. 2003). Figure 9 shows some images of the Fe-Mn-Si-Cr-Ni-Co alloy. In the first image, the austenitic grain boundaries are seen (Fig. 9a). Austenitic grain orientations are observed by different colors. Deformation twins can also be viewed (Fig. 9b). The coexistence of martensite and austenite phases can be observed in Fig. 9c. In this case, the darker regions have been identified as the martensitic phase. The color etching is also very important to verify the presence of the α′-martensite, considered as detrimental to the shape recovery process. This phase was not identified by X-ray diffraction because it has a low volumetric fraction (<2%). But, using optical microscopy, the α′-martensite was identified as spots throughout the sheets of ε-martensite, Fig. 9d.

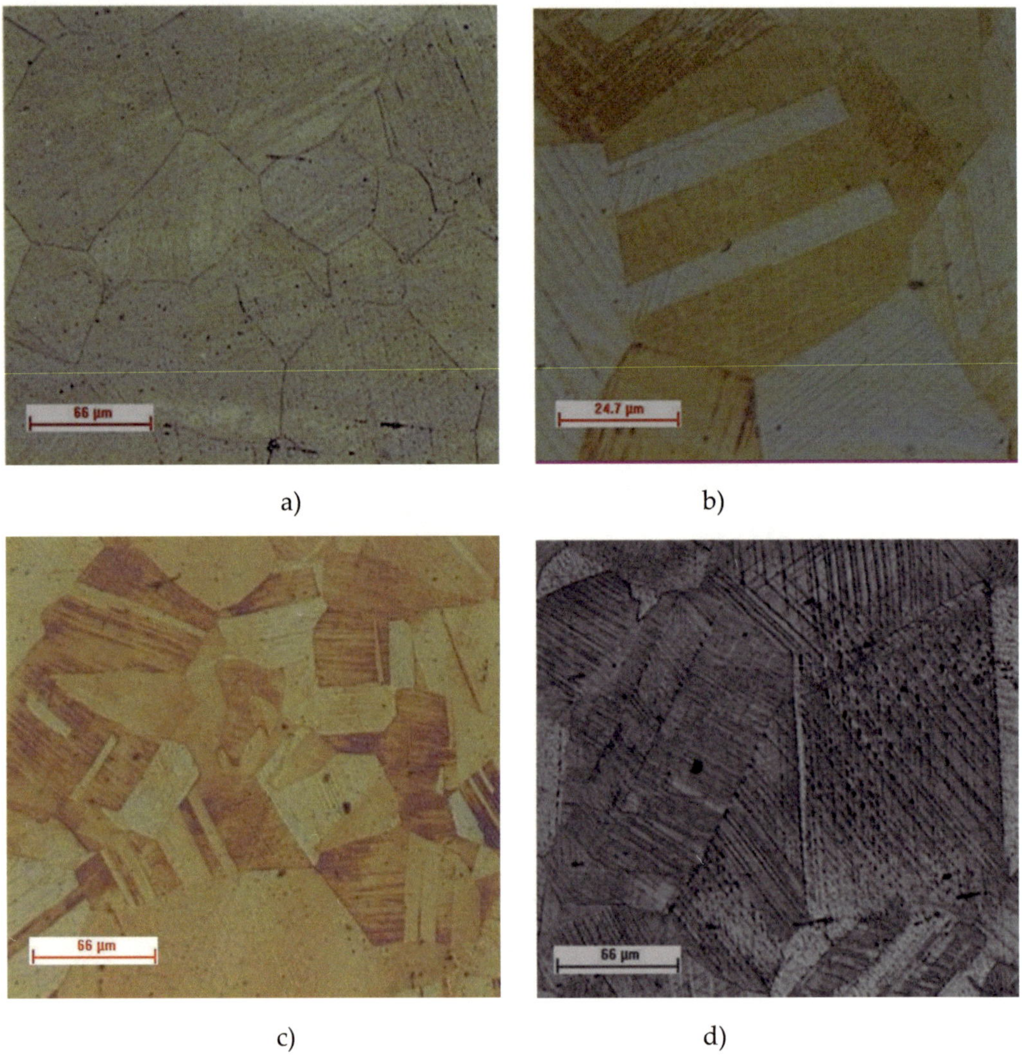

Figure 9. Identification of martensite and austenite phases using color etching: 2.0 g $K_2S_2O_5$ + 0.5g NH_4HF_2 + 50 ml H_2O (Bueno et al. 2003).

3.2. Mechanical properties

The mechanical properties such as hardness (Vickers hardness and nano hardness) were analyzed in samples subjected to compression cycles to study the stress induced ε-martensite (Nascimento, 2008). Figure 10 shows the influence of austenite grain size in Vickers hardness and the nano hardness of the Fe-Mn-Si-Cr-Ni-Co alloy.

The Vickers hardness (Fig-10a) shows the contribution of ε-martensite and austenite phases simultaneously. In this case, the behavior is similar to that of the commercial austenitic steel, the Vickers hardness decreases as a function of grain size (Nascimento, 2008). Literature data indicated a linear relationship between the yield stress (σ) and the inverse of the square root of grain diameter, according to Hall-Petch (Leslie, 1996), (Gladman, 1997). Using pyramidal indenter geometry it is possible to estimate the hardness (GPa) of these phases separately, Fig. 10(b) and Fig. 10(c).

The curve of hardness, Fig. 10(c) shows similar behavior to that of the austenite phase curve Fig. 10(a). But the martensitic phase, Fig. 10(b), shows an increased hardness due to increase in grain size. This result is explained by the fact that increased grain size makes shape

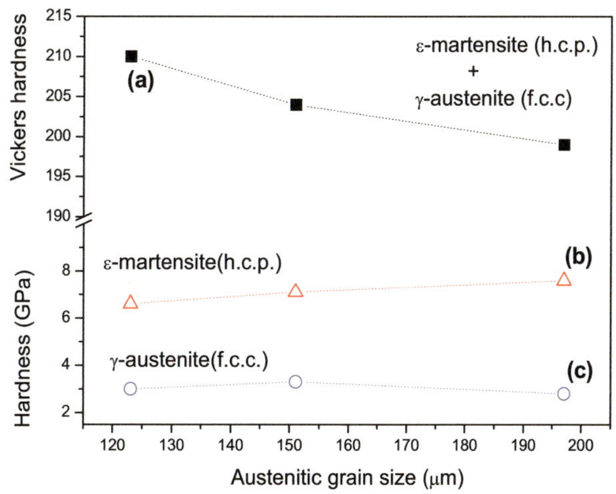

a)

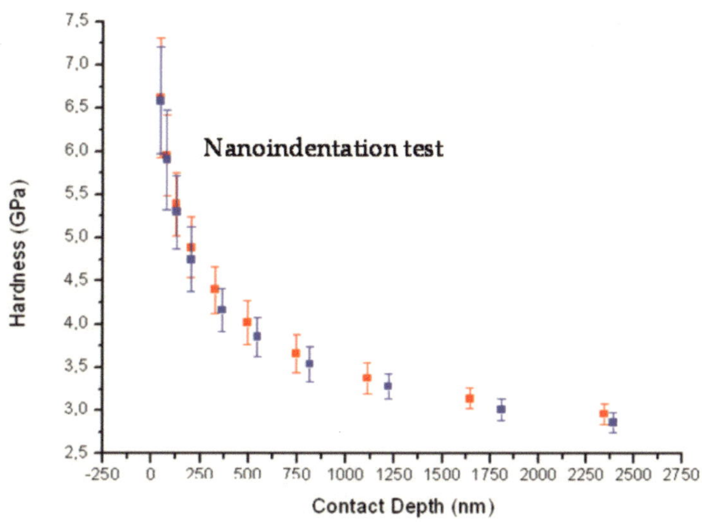

b)

Figure 10. a) Hardness (GPa) and Vickers hardness as a function of austenite grain size, recovery state, b) Hardness (GPa) curve obtained in nanoindentation tests in Fe-Mn-Si-Cr-Ni iron based shape memory alloy.

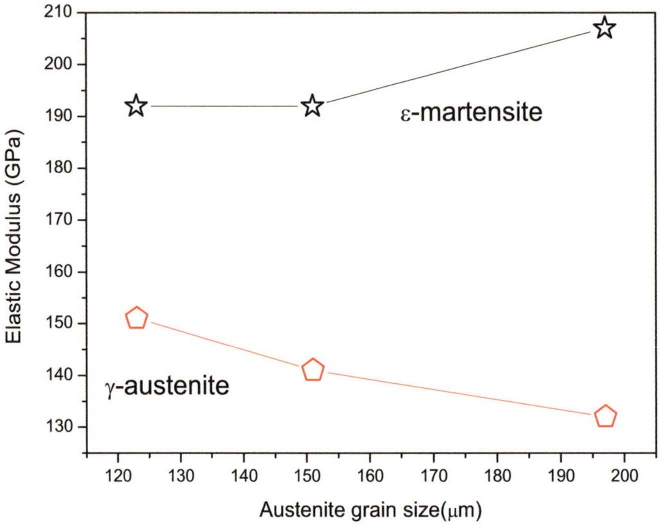

Figure 11. Effect of austenite grain size on the elastic modulus (GPa) to ε-martensite and austenite phases.

recovery more difficult. In this case, these samples have a higher volumetric fraction of ε-martensite residual, a phase which was not recovered at each cycle of thermomechanical treatment (Nascimento et al. 2003). Figure 10d shows the typical curve obtained in the nanoindentation test. The blue curve is the first training cycle and the red curve is the sixth cycle, or thermomechanical cycle. For small contact depths, hardness is analyzed on the surface of the material and for greater contact depths, the values of hardness are obtained in bulk, approaching conventional austenitic stainless steel.

The variations of the modulus of elasticity for the martensite and austenite phases are shown in Figure 11. For a commercial stainless steel the modulus of elasticity is around 210 GPa. When we analyze the phases separately, we observed a change in value. This variation is due to the difference in chemical composition and also alterations in the volumetric fraction of the phases.

4. Conclusion

The main conclusion of this study refers to the fact that the initial refinement of the microstructure in iron based alloys affects the performance of shape recovery of these materials. These changes occur in several aspects: morphology and microstructure of the phases, structural parameters, mechanical properties and shape memory effect. Changes in the ratio c/a of martensitic phase affect the reverse motion of partial dislocation that is also affected by grain size. Samples with larger grain size need to relax the strain by creating new guidelines facilitating the precipitation of the α′-martensite. Analysis using the Rietveld refinement are important because they allow better evaluation of the structural variations.

Author details

Fabiana Cristina Nascimento Borges
Universidade Estadual de Ponta Grossa, Departamento de Física – UEPG, Ponta Grossa, Paraná, Brazil

Acknowledgement

We would like to thank the Brazilian agency CNPq, Fapesp, AEB and Vilares Metals S.A. for its financial support, and the Phd. Jorge Otubo (ITA-SP-Brazil) and Phd. Paulo R. Mei (UNICAMP-SP-Brazil).

5. References

Araújo, L. A. ((Ed. Arte e Ciência). (1997)). *Manual de Siderurgia – Transformação*, Siemens.

Bergeon N, Guenin G, Esnouf C. (1997). Characterization of the stress-induced ε-martensite in a Fe-Mn-Si-Cr-Ni shape memory alloy: microstructural observation at different scales, mechanism of formation and growth. *Materials Science and Engineering A*. 1997; 238: 309-316.

Bueno J. C., Nascimento F. C., Otubo J., Lepienski C. M., Mei P. R. Development of phase identification by optical metallography with nanoindentation in shape memory alloys. In: *International Conference on Advanced in Materials and Processing Technologies*, 2006, Las Vegas. International Conference on Advanced in Materials and Processing Technologies, 2006. p. 01-04.

Bueno J. C., Nascimento F. C., Otubo J. (2003). Effect of training and the reduction of the austenitic grain size on the morphology of the stress-induced ε martensite in stainless SMAs. Cobem, 2003.

Gauzzi F, Montanari R. (1999). Martensite reversion in an Fe–21%Mn–0.1%Calloy. *Materials Science and Engineering A*. 273-275: 524-527.

Gladman, T. (1997). Microstructure property relationships. The physical metallurgy of microalloys steels, pp. 40.

Leslie, W.C. (1996). Physical metallurgy os steels. Physical Metallurgy, Chapter 17, Vol. II, pp. 1591.

Nascimento, F. C. (2008). *Ligas austeniticas com memória de forma- influência da microestrutura nas propriedades mecânicas e na recuperação de forma*, Edgar Blucher, ISBN 978-85-61209-38-4 São Paulo, Brazil.

Nascimento F. C., Mei P. R., Cardoso L. P., Otubo J. (2008). Grain size effect on the structural parameters of the stress induced epsilon hcp: martensite in iron-based shape memory alloy. *Materials Research*, Vol. 11, pp. 1516-1539, ISSN 1516-1439.

Nascimento F. C., Mei P. R., Otubo, J. (2003). Effects of grain size on the shape recovering properties of a stainless SMA. In: *Proceedings of Conference Advances in Materials and Processing technologies*, Las Vegas. 2003; 1436-1440.

Nascimento, F. C., Sorrila, F. V., Otubo, J., Mei P. R. (2003). Stainless shape memory alloys Microstructure by optical microsocpy using different etchants. Acta Microscopica, pp. 01-05.

Otubo J., Nascimento F. C., Mei P. R., Cardoso L. P., Kaufman M. (2002). Influence of austenite grain size on mechanical properties of stainless shape memory alloy. *Materials Transactions*, Vol. 43, pp. 916-919.

Vlack V., Vlack, Hall L. (1998). Princípios de Ciência dos Materiais. Edgar Blucher. pp.1-378.

Section 3

Mechanical Behavior

Chapter 6

Mechanics of Shape Memory Alloy Materials – Constitutive Modeling and Numerical Implications

V. P. Panoskaltsis

Additional information is available at the end of the chapter

http://dx.doi.org/10.5772/52228

1. Introduction

Shape Memory Alloys (SMAs) are a unique class of metal alloys which can be deformed severely and afterwards recover their original shape after a thermomechanical cycle (shape memory effect), or a stress cycle within some appropriate temperature regimes (pseudoelasticity, also sometimes called in the literature superelasticity, *not* to be confused with *hyperelasticity*). The mechanisms of this recovery are either a diffusionless transformation between the austenite phase (which is a highly ordered phase and is also called the parent phase) and the martensite phase (which is a less ordered one) or the reorientation (detwinning) of martensite variants. Detailed exposures to the physics of the subject may be found in Wayman (1964), Smallman and Bishop (2000) and Bhattacharya (2003). As is shown in these studies the thermomechanical response of SMAs is extremely complex, a fact that in conjunction with the continuously increasing use of SMAs in several innovating applications in many engineering fields results in a greater need for a better understanding of these materials. For the past decades several constitutive models have appeared within the literature (e.g., Raniecki et al., 1992; Abeyaratne and Knowles, 1993; Ivshin and Pence, 1994; Boyd and Lagoudas, 1996; Lubliner and Auricchio, 1996; Panoskaltsis et al., 2004), which within the context of a geometrical linear theory can capture several aspects of the experimentally observed response. Nevertheless, the physics of the problem (e.g., see Smallman and Bishop, 2000), together with some basic results of the crystallographic theory of martensitic phase transformations (e.g., Ball and James, 1987; James and Hane, 2000; Abeyaratne et al. 2001), suggest that a geometrically non – linear approach is more appropriate. Levitas and Preston, (2005) discuss the drawbacks of the infinitesimal models and they report that finite rotations of the crystal lattice can occur at small transformation strains (small strains and finite rotations) and can crucially affect the phase transformation conditions. Rather recently, several researchers have started to develop constitutive models for SMAs within the finite deformation regime. The approaches used for the description of the behavior of these materials are many and almost encompass all branches of mathematics, physics, material science and continuum mechanics. The existing models may be roughly classified in the following categories: (a) Constitutive models based on phase field or Landau – Ginzburg theory, (b) models based on irreversible thermodynamics and (c) models based on plastic flow theories.

The basic idea of the phase field theory is that out of all complexities of statistical mechanics one can reduce the behavior of a system undergoing a phase transformation to that of a few order parameters (i.e., parameters that give a measure of the transformation development), governed by a free energy function, which depends on stress (or deformation), temperature and those parameters. A characteristic example of modeling phase transformations by Landau – Ginsburg theory is provided by Levitas and Preston, 2005.

Also, in the realm of the so – called non equilibrium (or irreversible) thermodynamics several models have been proposed which are based on the use of a set of thermomechanical equations describing the kinetics of the martensitic transformations. The constitutive equations are developed in a non – linear manner on the basis of the laws of thermodynamics. Depending on whether they utilize the full microscopic deformation or the phenomenological one, the thermodynamical models may be classified further as microscopic (e.g., Levitas and Ozsoy, 2009) or macroscopic (e.g., Müller and Bruhns, 2006).

Another approach, which besides being thermodynamically consistent may also furnish a concrete micromechanical justification, is through the employment of plastic flow theories. Recall that the martensite transformation is a diffusionless one during which there is no interchange on the position of neighboring atoms but atom movements resulting in changes in the crystal structure (e.g., see Smallman and Bishop, 2000, pp. 278 – 279). Based on this observation the martensite formation has been explained by a shear mechanism or by a sequence of two shear mechanisms. The shear mechanism can take place either by twinning or by sliding, depending on the composition and on the thermodynamical conditions (Smallman and Bishop, 2000, p. 280). Although in the book of Smallman and Bishop mainly martensitic transformation in steel is described, the authors discuss efforts for the development of a general theory of the crystallography of martensitic transformations. The crystallographic mechanisms of martensite in nickel titanium (NiTi, also known as Nitinol) are similar, i.e., slip or twinning, as in the alloys described in the book of Smallman and Bishop. As a result it can be considered that the role played by the different transformation systems in the martensitic transformations may be suitably parallelized by the role played by the slip systems in crystal plasticity. Models based on this idea have been proposed among others by Diani and Parks (1998), Thamburaja and Anand (2000) and Anand and Gurtin (2003). It should be emphasized that these models are also computationally attractive because a lot of work has been put recently in the algorithms of crystal plasticity, both in their purely algorithmic as well as in their mathematical aspects, resulting in the development of robust algorithms well suited for finite element applications. Accordingly, complex constitutive representations may be considered, since their numerical implementation is no longer intractable, no matter how complex they may be.

An alternative approach, within the context of plastic flow theories, has been proposed by Lubliner and Auricchio (1996) and Panoskaltsis et al. (2004), who developed three – dimensional thermomechanical constitutive models based on generalized plasticity theory in the small deformation regime, and by Panoskaltsis et al. (2011a, 2011b) within finite strains and rotations.

Generalized plasticity is a general theory of rate – independent inelastic behavior which is physically motivated by loading – unloading irreversibility and it may be mathematically founded on set theory and topology (Lubliner 1974, 1984, 1987). Its particular structure provides the theory with the ability to address "non – standard" cases such as non – connected elastic domains.

The objective of this work is twofold: First, to extend the previous works of SMAs modeling based on generalized plasticity, providing a general geometrical framework. This general framework will in turn constitute a basis for the derivation of constitutive models for materials undergoing phase transformations and for *arbitrary* deformations. Second, as an application, to develop a finite strain model, which can simulate several patterns of the extremely complex response of SMAs under isothermal and *non – isothermal* loadings.

This chapter is organized as follows: In section 2, a general multi – surface formulation of non – isothermal generalized plasticity, capable of describing the multiple and interacting loading mechanisms which occur during phase transformations (see Panoskaltsis et al., 2011a, 2011b)) is presented within the context of tensor analysis in Euclidean spaces. The derivation of the thermomechanical state equations on the basis of *the invariance properties of the local form of the balance of energy equation under some groups of transformations,* is attained in section 3; this is a purely geometrical approach. In particular, the fundamental theorem of the covariant constitutive theory of non – linear elasticity (see Marsden and Hughes, 1994, pp. 202 – 203) is revisited and is used *in place of the second law of thermodynamics,* as a basic constitutive hypothesis for the subsequent derivation of the SMAs thermomechanical state equations. Rate constitutive equations are derived as well. Finally, as an application a specific model is derived within a fully thermomechanical framework in section 4. Computational aspects and numerical simulations are presented in section 5.

2. Generalized plasticity for phase transformations

2.1. Formulation of the governing equations in the reference configuration

Generalized plasticity is a local internal variable theory of rate – independent behavior, which is based primarily on loading – unloading irreversibility. As in all internal – variable type theories, it is assumed that the local thermomechanical state in a body is determined uniquely by the couple (**G**, **Q**) where **G** – belonging to a space G – stands for the vector of the controllable state variables and **Q** – belonging to a space Q – stands for the vector of the internal variables, which are related to *phase transformations.* Following the ideas presented in the review paper of Naghdi (1990) we follow a material (referential) approach within a strain – space formulation. Accordingly, **G** may be identified by (**E**, T) where **E** is the referential (Green – St. Venant) strain tensor and T is the (absolute) temperature. Depending on the nature of the (material) internal variable vector **Q**, the theory may, in principle, be formulated equivalently with respect to the *macro –, meso –, or micro – scale structure* of the material.

The central concept of generalized plasticity is that of the *elastic range,* which is defined at any material state as the region in the strain – temperature space comprising the strains which can be attained elastically (i.e., with no change in the internal variables) from the current strain – temperature point. It is assumed that the elastic range is a regular set in the sense that it is the closure of an open set. The boundary of this set is defined as a *loading surface* at **Q**, (see Eisenberg and Phillips, 1971; Lubliner, 1987). In turn, a *material state* may be defined as *elastic* if it is an interior point of its elastic range and *inelastic* if it is a boundary point of its elastic range; in the latter case the material state lies on a loading surface. It should be added that the notion of *process* is introduced implicitly here. By assuming that the loading surface is smooth at the current strain - temperature point and by invoking some basic axioms and results from set theory and topology, Lubliner (1987) showed that the rate equations for the evolution of the internal variable vector may be written in the form

$$\dot{\mathbf{Q}} = H\mathbf{L}(\mathbf{G},\mathbf{Q})\left\langle \mathbf{N}:\dot{\mathbf{G}} \right\rangle, \tag{1}$$

where <·> stands for the Macauley bracket which is defined as:

$$\langle x \rangle = \begin{cases} x & \text{if } x > 0 \\ 0 & \text{if } x \leq 0, \end{cases}$$

and H stands for a scalar function of the state variables. Accordingly, the value of H must be positive at any inelastic state and zero at any elastic one. Finally, **L** stands for a non - vanishing (tensorial) function of the state variables associated with the properties of the phase transformation, **N** is the outward normal to the loading surface at the current state, while the colon between two tensors denotes their double contraction operation. Furthermore, the set of the material states defined as $H = H(\mathbf{G},\mathbf{Q}) = 0,$ which comprises all the elastic states is called the *elastic domain* and its projection on the set defined by **Q** = const. is defined at the elastic domain at **Q**. In general, the elastic domain at **Q** is a subset of the elastic range (Lubliner, 1987). The particular case in which the two sets coincide corresponds to classical plasticity and the boundary of the elastic domain, that is the *initial loading surface,* constitutes the *yield surface* (see Eisenberg and Phillips, 1971; Lubliner, 1987; Panoskaltsis et al., 2008a, 2008b, 2011c).

It is emphasized that Eq. (1) has been derived under the assumption of a smooth loading surface at the current strain – temperature point, which implies that only one loading mechanism can be considered. On the other hand, phase transformations include multiple and sometimes interacting loading mechanisms, which may result in the appearance of a vertex or a corner at the current strain – temperature point. This fact calls for an appropriate modification of the rate Equation (1).

In order to accomplish this goal we assume that the loading surfaces are defined in the state space by a number – say n – of smooth surfaces, which are defined by expressions of the form

$$\Phi_i(\mathbf{G},\mathbf{Q}) = 0, \quad i=1, 2,..., n. \tag{2}$$

These surfaces can be either disjoint, or intersect in a possibly non – smooth fashion. Each of these surfaces is associated with a particular transformation mechanism which may be active at the current strain – temperature point. Then, by assuming that each equation $\Phi_i(\mathbf{G},\mathbf{Q}) = 0$ defines an independent (non – redundant) active surface at the current stress temperature point, and in view of Eq. (1), we can state the rate equations for the evolution of the internal variables in the following general form

$$\dot{\mathbf{Q}} = \sum_{i=1}^{n} H_i \mathbf{L}_i(\mathbf{G},\mathbf{Q}) \left\langle \mathbf{N}_i : \dot{\mathbf{G}} \right\rangle, \tag{3}$$

where H_i, $\mathbf{L}_i$ and $\mathbf{N}_i$ are functions of the state variables defined as in Eq. (1) and each set of them – defined by the index i – refers to the specific transformation associated with the part of the loading surface defined by $\Phi_i(\mathbf{G},\mathbf{Q}) = 0$. From Eq. (3) one can deduce directly the *loading – unloading criteria* for the proposed formulation as follows: Let us denote by $n_{adm} \le n$ the number of loading surfaces that may be active at an inelastic state i.e. $H_i > 0$, and let us denote by J_{adm} the set of n_{adm} indices associated with those surfaces, i.e.

$$J_{adm} = \{\alpha \in \{1,2,...,n\} / H_\alpha > 0\}.$$

Then Eq. (3) implies the following loading – unloading conditions:

If $J_{adm} = \varnothing$, then $\dot{\mathbf{Q}} = \mathbf{0}$.
If $J_{adm} \neq \varnothing$, then:
i If $\mathbf{N}_\alpha : \dot{\mathbf{G}} \le 0$ for all $\alpha \in J_{adm}$ then $\dot{\mathbf{Q}} = \mathbf{0}$,
ii If $\mathbf{N}_\alpha : \dot{\mathbf{G}} > 0$ for at least one $\alpha \in J_{adm}$ then $\dot{\mathbf{Q}} \neq \mathbf{0}$.

Hence, if we denote further by $n_{act} \le n_{adm}$ the number of parts for which (ii) holds, and we set:

$$J_{act} = \{\alpha \in J_{adm} / \mathbf{N}_\alpha : \dot{\mathbf{G}} > 0\},$$

the loading criteria in terms of the sets J_{adm} and J_{act} may be stated as:

$$\begin{cases} \text{If } J_{adm} = \varnothing: & \text{elastic state.} \\ \text{If } J_{adm} \neq \varnothing \text{ and } J_{act} = \varnothing: & \\ \quad \text{i. If } \mathbf{N}_\alpha : \dot{\mathbf{G}} < 0 \text{ for all } \alpha \in J_{adm}: & \text{elastic unloading,} \\ \quad \text{ii. If } \mathbf{N}_\alpha : \dot{\mathbf{G}} = 0 \text{ for at least one } \alpha \in J_{adm}: & \text{neutral loading,} \\ \text{If } J_{adm} \neq \varnothing \text{ and } J_{act} \neq \varnothing: & \text{inelastic loading.} \end{cases} \tag{4}$$

2.2. Equivalent spatial formulation

The equivalent assessment of the governing equations in the spatial configuration can be done on the basis of a push – forward operation (e.g., see Marsden and Hughes, 1994, pp. 67 – 68; Stumpf and Hoppe, 1997) to the basic equations. For instance, by performing a push – forward operation onto Eq. (3) the latter is written in the form

$$L_{\mathbf{V}}\mathbf{q} = \sum_{i=1}^{n} h_i \mathbf{l}_i(\mathbf{g},\mathbf{q},\mathbf{F})\langle r_i \rangle, \tag{5}$$

where **F** stands for the deformation gradient and **g** stands for the vector of the controllable variables in the spatial configuration and is composed by the Almansi strain tensor **e** – defined as the push – forward of the Green – St. Venant strain tensor – and the (scalar invariant) temperature T. Moreover in Eq. (5), **q** stands for the push forward of the internal variable vector, and $L_{\mathbf{V}}(\cdot)$ stands for the Lie derivative (e.g., see Marsden and Hughes, 1994, pp. 93 – 104; Schutz, 1999, pp. 73-79; Stumpf and Hoppe, 1997), defined as the convected derivative relative to the spatial configuration. Finally, h_i stands for the expression of the scalar invariant functions H_i in terms of the spatial variables (**e**, T, **q**) and the deformation gradient **F**, $\mathbf{l}_i$ stands for the push – forward of the tensorial functions $\mathbf{L}_i$ and r_i denotes the (scalar invariant) loading rates which are written in the form

$$r_i = \frac{\partial \varphi_i}{\partial \mathbf{e}} : L_{\mathbf{V}}\mathbf{e} + \frac{\partial \varphi_i}{\partial \mathrm{T}} \dot{\mathrm{T}}, \tag{6}$$

where φ_i is the expression for the loading surface associated with the index i, in terms of the spatial variables. The (spatial) loading – unloading criteria flow naturally from Eq. (5) as:

$$\begin{cases} \text{If } \mathrm{j_{adm}} = \varnothing : & \text{elastic state.} \\ \text{If } \mathrm{j_{adm}} \neq \varnothing \text{ and } \mathrm{j_{act}} = \varnothing : & \\ \quad \text{i. If } \mathrm{r}_\alpha < 0 \text{ for all } \alpha \in \mathrm{j_{adm}} : & \text{elastic unloading,} \\ \quad \text{ii. If } \mathrm{r}_\alpha = 0 \text{ for at least one } \alpha \in \mathrm{j_{adm}} : & \text{neutral loading,} \\ \text{If } \mathrm{j_{adm}} \neq \varnothing \text{ and } \mathrm{j_{act}} \neq \varnothing : & \text{inelastic loading,} \end{cases} \tag{7}$$

where the sets $\mathrm{j_{adm}}$ and $\mathrm{j_{act}}$ are now defined in terms of the spatial variables as, $\mathrm{j_{adm}} = \{\alpha \in \{1,2,\ldots,n\} / h_\alpha > 0\}$ and $\mathrm{j_{act}} = \{\alpha \in \mathrm{J_{adm}} / \mathrm{r}_\alpha > 0\}$.

2.3. Description of rate effects

Rather recent experimental results (see, Nemat – Nasser et al., 2005a, 2005b) on a NiTi shape memory alloy, show that some of the phase transformations depend on the rate of loading. Such a behavior can be accommodated by the (geometrical) framework developed here, by noting that generalized plasticity can be combined consistently with a rate – dependent (viscoplastic) theory. In this case the rate equations for the internal variables may be written in the form

$$\dot{\mathbf{Q}} = \sum_{i=1}^{n} H_i [\mathbf{L}_i(\mathbf{G},\mathbf{Q}) \langle \mathbf{N}_i : \dot{\mathbf{G}} \rangle + \mathbf{M}_i(\mathbf{G},\mathbf{Q})], \tag{8}$$

where the $\mathbf{M}_i$'s stand for additional functions of the state variables enforcing the rate – dependent properties of the transformation defined by the part of the loading surface associated with the index i. The crucial advantage of this approach lies on the compatibility of the two theories, in the sense that neither viscoplasticity, nor generalized plasticity employs the concept of the yield surface as its basic ingredient.

2.4. Transformation induced plasticity

From a further study of the experimental results of Nemat – Nasser et al. (2005a, 2005b) (see also Delville et al., 2011) it is observed that after a stress cycle within the appropriate limits for pseudo-elastic behavior permanent deformations appear, a fact which implies that a yielding behavior appears within the martensitic transformations.

Such a response can be described within our framework by introducing m additional (plastic) loading surfaces, which control the yielding characteristics of the material. These are assumed to be given by expressions of the form

$$G_i(\mathbf{G},\mathbf{Q},\mathbf{P}) = 0, \quad i=1, 2,..., m, \tag{9}$$

where $\mathbf{P}$ is *an additional internal variable vector,* which stands for the description of plastic phenomena within the material. In turn, the rate equations for the evolution of the plastic variables within the generalized plasticity context – which includes classical plasticity as a special case (see Lubliner, 1987; Panoskaltsis et al., 2008a, 2011c) – may be stated as

$$\dot{\mathbf{P}} = \sum_{i=1}^{m} K_i \mathbf{T}_i (\mathbf{G},\mathbf{Q},\mathbf{P}) \langle \mathbf{R}_i : \dot{\mathbf{G}} \rangle, \tag{10}$$

where the functions K_i, $\mathbf{T}_i$ and $\mathbf{R}_i$ have an identical meaning with the functions H_i, $\mathbf{L}_i$ and $\mathbf{N}_i$ which appear in Eq. (3). The constitutive modeling of plasticity phenomena within the martensitic transformations is nowadays a very active area of research. Recent contributions include the phenomenological models by Hallberg et al. (2007, 2010) and Christ and Reese (2009).

A further observation of Eqs. (9) and (10) and their comparison with the basic Eqs. (2) and (3) reveal that both sets of equations show exactly the same qualitative characteristics. Accordingly, it is concluded that from a geometrical standpoint the phase transformation loading surfaces are indistinguishable from the plastic loading surfaces, which means that the internal variable vector $\mathbf{P}$ may be absorbed in $\mathbf{Q}$ so that the basic equations can simulate both phase transformation and plasticity phenomena in a unified format. *This implies that plastic yielding can be understood within the proposed framework as a phase transition.*

3. The invariant energy balance equation and the thermomechanical state equations

The concept of invariance plays a fundamental role in several branches of mechanics and physics. In particular, within the context of continuum mechanics the invariance properties of the balance of energy equation, under some groups of transformations, may be systematically used in order to derive the conservation laws, the balance laws and/or to determine some restrictions imposed on the equations describing the material constitutive response (e.g., Ericksen, 1961; Green and Rivlin, 1964; Marsden and Hughes, 1994, pp. 163 – 167, 200 – 203; Yavari et al., 2006; Panoskaltsis et al., 2011c). For instance, Marsden and Hughes (1994, pp. 202 – 203) by studying the invariant properties of the local form of the material balance of energy equation, under the action of arbitrary spatial diffeomorphisms, determined the thermomechanical state equations for a non – linear elastic material. The basic objective of this section is to revisit the approach given in Marsden and Hughes (1994, pp. 202 – 203), within the context of the Euclidean space used herein and to show how this can used as a basic constitutive hypothesis *in place of the second law of thermodynamics* for the derivation of the constitutive response of the SMA material in question.

3.1. Revisiting Marsden and Hughes' theorem

Unlike the original approach of Marsden and Hughes where manifold spaces are used and the invariance of the local form of the material balance of energy equation is examined under the action of arbitrary spatial diffeomorphisms, which include also a temperature rescaling, we examine the invariance properties of the local form of the *spatial balance* of energy equation under the action of the same kind of transformations, within the context of a Euclidean space. Within this framework the basic axioms of Marsden and Hughes (1994, pp. 202 – 203) may be modified as follows:

Axiom 1 (Local energy balance or first law of thermodynamics): For a spatial point with coordinates x_1, x_2, x_3, and a given *elastic* thermomechanical process (**e**, T) the balance of energy holds

$$\rho\dot{e} + div\mathbf{h} = \boldsymbol{\sigma} : \mathrm{L}_{\mathbf{V}}\mathbf{e} + \rho r, \tag{11}$$

where e is the energy density, ρ is the spatial mass density, $\boldsymbol{\sigma}$ is the Cauchy stress tensor, **h** is the heat flux vector per unit of surface of the spatial configuration, r is the heat supply per unit mass and a superimposed dot indicates material time derivative. By introducing the Helmholtz free energy function ψ, obtained by the following Legendre transformation

$$\psi = e - \eta \mathrm{T},$$

where η is the specific entropy, the local form of the energy balance can be written in the form

$$\rho(\dot{\psi} + \dot{\eta}\mathrm{T} + \eta\dot{\mathrm{T}}) + div\mathbf{h} = \boldsymbol{\sigma} : \mathrm{L}_{\mathbf{V}}\mathbf{e} + \rho r. \tag{12}$$

Axiom 2 (Invariance of Helmholtz free energy): We denote by S the ambient space, by $\boldsymbol{\phi}$ the deformation mapping, by g the space of the control variables and by f the set of the C^S scalar fields all expressed in the spatial configuration. Next we assume the existence of a map $\hat{\hat{\psi}} : (S, g, R^+) \to f$ such that for any diffeomorphism which includes also a temperature rescaling, that is: (ξ, δ): $(S, R^+) \to (S, R^+)$, the following expression holds:

$$\hat{\psi}(\boldsymbol{\phi}, \mathbf{e}, \mathrm{T}) = \hat{\hat{\psi}}(\xi \circ \boldsymbol{\phi}, \xi_* \mathbf{e}, \delta \mathrm{T}, \delta),$$

where $(\cdot)_*$ denotes the push – forward operation.

Axiom 3 (Invariance of the energy balance): For curves $\xi_t : S \to S$ and $\delta_t(x) \in R^+$, $\boldsymbol{\phi}_t' = \xi_t \circ \boldsymbol{\phi}_t$, $\mathrm{T}_t' = \delta_t \mathrm{T}_t$ and by assuming that ρ, ψ and η are transformed as scalars, the heat flux vector is transformed as $\mathbf{h}_t' = \delta_t \xi_{t*} \mathbf{h}_t$ and the *"apparent heat supply" due to entropy production,* $r_t' - \mathrm{T}_t' \dot{\eta}_t'$, is transformed as $r_t' - \mathrm{T}_t' \dot{\eta}_t' = \delta_t (r_t - \mathrm{T}_t \dot{\eta}_t)$ the balance of energy holds, that is

$$\rho'(\dot{\psi}' + \dot{\eta}' \mathrm{T}' + \eta' \dot{\mathrm{T}}') + div \mathbf{h}' = \boldsymbol{\sigma}' : \mathrm{L}_{\mathbf{v}'} \mathbf{e}' + \rho' r'. \tag{13}$$

Then the basic theorem of Marsden and Hughes (Theorem 3.6 p. 203), takes in our case the form:

Theorem 1: Under axioms 1, 2 and 3 the thermomechanical state equations for the Cauchy stress tensor and the entropy density are given as:

$$\boldsymbol{\sigma} = \rho \frac{\partial \psi}{\partial \mathbf{e}}, \quad \eta = -\frac{\partial \psi}{\partial \mathrm{T}}. \tag{14}$$

Proof: The proof relies crucially, as in the case examined by Marsden and Hughes, on the evaluation of Eq. (13) at time t_0, when $\xi\big|_{t=t_0} = \mathbf{1}$ (identity), $\mathbf{w} = \left.\frac{\partial \xi}{\partial t}\right|_{t=t_0}$, and $\delta\big|_{t=t_0} = 1$, $\mathrm{u} = \left.\frac{\partial \delta}{\partial t}\right|_{t=t_0}$ where u is the velocity of δ at t_0. Then, since at $t = t_0\big| \to \mathbf{v}' = \mathbf{w} + \mathbf{v}$, $\dot{\mathrm{T}}' = \mathrm{uT} + \dot{\mathrm{T}}$, the time derivative of the Helmholtz free energy yields:

$$\dot{\psi}'\big|_{t=t_0} = \dot{\psi} + \frac{\partial \psi}{\partial \mathbf{e}} : L_{\mathbf{w}} \mathbf{e} + \frac{\partial \psi}{\partial \mathrm{T}} (\mathrm{uT}), \tag{15}$$

in which $L_{\mathbf{W}}(\cdot)$ stands for the *autonomous* Lie derivative (e.g., see Marsden and Hughes, 1994, pp. 96 – 98; Yavari et al., 2006). Furthermore it holds that

$$\boldsymbol{\sigma}' : \mathrm{L}_{\mathbf{v}'} \mathbf{e}'\big|_{t=t_0} = \boldsymbol{\sigma} : \mathrm{L}_{\mathbf{v}} \mathbf{e} + \boldsymbol{\sigma} : L_{\mathbf{w}} \mathbf{e}, \tag{16}$$

since (see Marsden and Hughes, 1994, p. 98)

$$\mathrm{L}_{\mathbf{v}+\mathbf{w}}\mathbf{e} = \dot{\mathbf{e}} + L_{\mathbf{v}+\mathbf{w}}\mathbf{e} = \dot{\mathbf{e}} + L_{\mathbf{v}}\mathbf{e} + L_{\mathbf{w}}\mathbf{e} = \mathrm{L}_{\mathbf{v}}\mathbf{e} + L_{\mathbf{w}}\mathbf{e}.$$

In light of Eqs. (15) and (16) and the transformation formulae for the heat flux vector and the "apparent heat supply" due to entropy production, Eq. (13) at $t = t_0$ can be written as:

$$\begin{aligned} &\rho\dot{\psi} + \rho\frac{\partial\psi}{\partial\mathbf{e}} : L_{\mathbf{w}}\mathbf{e} + \rho\frac{\partial\psi}{\partial \mathrm{T}}(\mathrm{uT}) + \\ &+\rho\eta\dot{\mathrm{T}} + \rho\eta(\mathrm{uT}) + \mathrm{div}\mathbf{h} = \boldsymbol{\sigma} : (\mathrm{L}_{\mathbf{V}}\mathbf{e} + L_{\mathbf{w}}\mathbf{e}) + \rho(r - \mathrm{T}\dot{\eta}). \end{aligned} \tag{17}$$

Subtracting Eq. (12) from Eq. (17) gives the identity

$$\rho\frac{\partial\psi}{\partial\mathbf{e}} : L_{\mathbf{w}}\mathbf{e} + \rho\frac{\partial\psi}{\partial \mathrm{T}}(\mathrm{uT}) + \rho\eta(\mathrm{uT}) - \boldsymbol{\sigma} : L_{\mathbf{w}}\mathbf{e} = 0, \tag{18}$$

or

$$(\rho\frac{\partial\psi}{\partial\mathbf{e}} - \boldsymbol{\sigma}) : L_{\mathbf{w}}\mathbf{e} + (\rho\frac{\partial\psi}{\partial \mathrm{T}} + \rho\eta)(\mathrm{uT}) = 0, \tag{19}$$

from which and by noting that $L_{\mathbf{w}}\mathbf{e}$ and u can be arbitrarily specified, the thermomechanical state Equations (14) follow. By performing a pull – back operation to Eqs. (14) the following relations are derived

$$\mathbf{S} = \rho_{\mathrm{ref}}\frac{\partial\Psi}{\partial\mathbf{E}}, \quad \eta = -\frac{\partial\Psi}{\partial\mathrm{T}}, \tag{20}$$

where $\mathbf{S}$ stands for the second Piola – Kirchhoff stress tensor, ρ_{ref} for the material mass density and Ψ for the expression of the Helmholtz free energy in the material configuration. It is concluded that Eqs. (20) are identical to the thermomechanical state equations of Marsden and Hughes (1994, p. 203). Thus, we can state the following proposition:

Proposition 1: The invariance of the local form of the balance of energy equation under the superposition of arbitrary spatial diffeomorphisms, which also include a temperature rescaling, gives identical results with respect to the thermomechanical state equations, irrespectively of whether the energy balance equation is considered in its material or its spatial form.

3.2. Thermomechanical state equations for a SMA material

Building on the previous developments we will derive the thermomechanical state equations for a *shape memory material with internal variables,* which obeys the rate Equations (5). The development relies crucially on establishing a new set of axioms which will incorporate the presence of the internal variables and their evolution in the course of the phase transformations. We proceed as follows:

Axiom 1: Since the internal variables are not involved explicitly in the balance laws, for a process which is either elastic or inelastic ($j_{adm} \neq \emptyset$ and $j_{act} \neq \emptyset$) axiom 1 remains unaltered, that is Eq. (12) holds.

Axiom 2 is modified as follows:

Axiom 2: In addition to the adopted notation, we denote by q the space of the internal variables in the spatial configuration and we assume the existence of a map $\hat{\hat{\psi}} : (S,g,q,R^+) \rightarrow f$ such that for any diffeomorphism which includes also a temperature rescaling, that is: (ξ,δ): $(S,R^+) \rightarrow (S,R^+)$, the following expression holds:

$$\hat{\psi}(\boldsymbol{\phi},\mathbf{e},\mathrm{T},\mathbf{q})=\hat{\hat{\psi}}(\xi \circ \boldsymbol{\phi},\delta\mathrm{T},\xi_*\mathbf{e},\mathbf{q}(\xi_*\mathbf{e},\delta\mathrm{T}),\delta).$$

Axiom 3: In addition to the energy invariance axiom it is assumed that under the application of the diffeomorphism (ξ,δ): $(S,R^+) \rightarrow (S,R^+)$, loading surfaces are transformed as scalars, that is $\varphi_i(\boldsymbol{\phi},\mathbf{e},\mathrm{T},\mathbf{q})=\varphi_i'(\xi \circ \boldsymbol{\phi},\delta\mathrm{T},\xi_*\mathbf{e},\mathbf{q}(\xi_*\mathbf{e},\delta\mathrm{T}),\delta)$.

In this case, as in the previous one, the derivation procedure is the following:

We evaluate Eq. (13) at time t_0, when $\xi\big|_{t=t_0} = \mathbf{1}$ (identity) and $\delta\big|_{t=t_0} = 1$ with $\mathbf{w}$ and u being the velocities of ξ and δ at t_0. Then

$$\dot{\psi}'\Big|_{t=t_0} = \dot{\psi} + \frac{\partial \psi}{\partial \mathbf{e}} : L_{\mathbf{w}}\mathbf{e} + \frac{\partial \psi}{\partial \mathrm{T}}(\mathrm{uT}) + \frac{\partial \psi}{\partial \mathbf{q}} : L_{\mathbf{w}}\mathbf{q}. \tag{21}$$

The critical step is the evaluation of the loading rates at $t = t_0$, which yields

$$\begin{aligned} r_i'\Big|_{t=t_0} &= \left(\frac{\partial \varphi_i'}{\partial \mathbf{e}'} : \mathrm{L_V}\mathbf{e}' + \frac{\partial \varphi_i'}{\partial \mathrm{T}'}\dot{\mathrm{T}}'\right)\Big|_{t=t_0} = \\ &= \frac{\partial \varphi_i}{\partial \mathbf{e}} : \mathrm{L_V}\mathbf{e} + \frac{\partial \varphi_i}{\partial \mathbf{e}} : L_{\mathbf{w}}\mathbf{e} + \frac{\partial \varphi_i}{\partial \mathrm{T}}\dot{\mathrm{T}} + \frac{\partial \varphi_i}{\partial \mathrm{T}}(\mathrm{uT}). \end{aligned} \tag{22}$$

Accordingly, the rate equation for the internal variables evaluated at time $t = t_0$, yields

$$\begin{aligned} \mathrm{L_{V'}}\mathbf{q}'\Big|_{t=t_0} &= \sum_{i=1}^{n} h_i'\mathbf{l}_i'(\mathbf{e}',\mathrm{T}',\mathbf{q}',\mathbf{F}')r_i'\Bigg|_{t=t_0} = \\ &= \sum_{i=1}^{n}[h_i\mathbf{l}_i(\mathbf{e},\mathrm{T},\mathbf{q},\mathbf{F})(\frac{\partial \varphi_i}{\partial \mathbf{e}} : \mathrm{L_V}\mathbf{e} + \frac{\partial \varphi_i}{\partial \mathrm{T}}\dot{\mathrm{T}})] + \sum_{i=1}^{n}[h_i\mathbf{l}_i(\mathbf{e},\mathrm{T},\mathbf{q},\mathbf{F})[\frac{\partial \varphi_i}{\partial \mathbf{e}} : L_{\mathbf{w}}\mathbf{e} + \frac{\partial \varphi_i}{\partial \mathrm{T}}(\mathrm{uT})], \end{aligned} \tag{23}$$

which in view of the rate Equations (5), reads

$$L_{\mathbf{w}}\mathbf{q} = \sum_{i=1}^{n} h_i\mathbf{l}_i(\mathbf{e},\mathrm{T},\mathbf{q},\mathbf{F})[\frac{\partial \varphi_i}{\partial \mathbf{e}} : L_{\mathbf{w}}\mathbf{e} + \frac{\partial \varphi_i}{\partial \mathrm{T}}(\mathrm{uT})]. \tag{24}$$

Upon substitution of Eq. (24), Eq. (21) takes the form

$$\dot{\psi}'\big|_{t=t_0} = \dot{\psi} + \frac{\partial\psi}{\partial \mathbf{e}} : L_{\mathbf{w}}\mathbf{e} + \frac{\partial\psi}{\partial T}(uT) + \frac{\partial\psi}{\partial \mathbf{q}} : \sum_{i=1}^{n} h_i \mathbf{l}_i(\mathbf{e},T,\mathbf{q},\mathbf{F})[\frac{\partial\varphi_i}{\partial \mathbf{e}} : L_{\mathbf{w}}\mathbf{e} + \frac{\partial\varphi_i}{\partial T}(uT)]. \qquad (25)$$

Now, by working in a similar manner as in the previous (i.e., the elastic) case, in light of Eq. (25) the basic Eq. (13) evaluated at $t = t_0$, yields

$$\begin{aligned}&\rho\dot{\psi} + \rho\frac{\partial\psi}{\partial \mathbf{e}} : L_{\mathbf{w}}\mathbf{e} + \rho\frac{\partial\psi}{\partial T}(uT) + \rho\frac{\partial\psi}{\partial \mathbf{q}} : \sum_{i=1}^{n} h_i \mathbf{l}_i(\mathbf{e},T,\mathbf{q},\mathbf{F})[\frac{\partial\varphi_i}{\partial \mathbf{e}} : L_{\mathbf{w}}\mathbf{e} + \frac{\partial\varphi_i}{\partial T}(uT)] + \\ &+ \rho\eta\dot{T} + \rho\eta(uT) + \mathrm{div}\mathbf{h} = \boldsymbol{\sigma} : (L_{\mathbf{V}}\mathbf{e} + L_{\mathbf{w}}\mathbf{e}) + \rho(r - T\dot{\eta});\end{aligned} \qquad (26)$$

from which by subtracting the balance of energy Eq. (12) we can derive the identity

$$\begin{aligned}&\rho\frac{\partial\psi}{\partial \mathbf{e}} : L_{\mathbf{w}}\mathbf{e} + \rho\frac{\partial\psi}{\partial T}(uT) + \rho[\frac{\partial\psi}{\partial \mathbf{q}} : \sum_{i=1}^{n} h_i \mathbf{l}_i(\mathbf{e},T,\mathbf{q},\mathbf{F})\frac{\partial\varphi_i}{\partial \mathbf{e}}] : L_{\mathbf{w}}\mathbf{e} + \\ &+ [\rho\frac{\partial\psi}{\partial \mathbf{q}} : \sum_{i=1}^{n} h_i \mathbf{l}_i(\mathbf{e},T,\mathbf{q},\mathbf{F})\frac{\partial\varphi_i}{\partial T}](uT) + \rho\eta(uT) - \boldsymbol{\sigma} : L_{\mathbf{w}}\mathbf{e} = 0,\end{aligned} \qquad (27)$$

or equivalently

$$\begin{aligned}&\{\rho[\frac{\partial\psi}{\partial \mathbf{e}} + \frac{\partial\psi}{\partial \mathbf{q}} : \sum_{i=1}^{n} h_i \mathbf{l}_i(\mathbf{e},T,\mathbf{q},\mathbf{F})\frac{\partial\varphi_i}{\partial \mathbf{e}}] - \boldsymbol{\sigma}\} : L_{\mathbf{w}}\mathbf{e} + \\ &+ \rho[\frac{\partial\psi}{\partial T} + \frac{\partial\psi}{\partial \mathbf{q}} : \sum_{i=1}^{n} h_i \mathbf{l}_i(\mathbf{e},T,\mathbf{q},\mathbf{F})\frac{\partial\varphi_i}{\partial T} + \eta](uT) = 0,\end{aligned} \qquad (28)$$

from which and by noting that $L_{\mathbf{w}}\mathbf{e}$ and u can be specified arbitrarily, we arrive at the expressions:

$$\begin{aligned}\boldsymbol{\sigma} &= \rho[\frac{\partial\psi}{\partial \mathbf{e}} + \frac{\partial\psi}{\partial \mathbf{q}} : \sum_{i=1}^{n} h_i \mathbf{l}_i(\mathbf{e},T,\mathbf{q},\mathbf{F})\frac{\partial\varphi_i}{\partial \mathbf{e}}], \\ \eta &= -[\frac{\partial\psi}{\partial T} + \frac{\partial\psi}{\partial \mathbf{q}} : \sum_{i=1}^{n} h_i \mathbf{l}_i(\mathbf{e},T,\mathbf{q},\mathbf{F})\frac{\partial\varphi_i}{\partial T}].\end{aligned} \qquad (29)$$

Therefore, unlike the classical elastic case, for the SMA material considered, the invariance of the local form of the energy balance under superposed spatial diffeomorphisms does not yield the standard thermomechanical state equations *unless a further assumption is made,* namely that an unloading process from an inelastic state (i.e., a process with $j_{adm} \neq \varnothing$ and $j_{act} = \varnothing$) is *quasi-reversible,* which means that in such a process both the *mechanical* and the *thermal dissipations,* defined as

$$d_{\mathbf{mech}} = -\frac{\partial\psi}{\partial \mathbf{q}} : \sum_{i=1}^{n} h_i \mathbf{l}_i(\mathbf{e},T,\mathbf{q},\mathbf{F})(\frac{\partial\varphi_i}{\partial \mathbf{e}} : L_{\mathbf{V}}\mathbf{e}),\ d_{\mathbf{therm}} = -\frac{\partial\psi}{\partial \mathbf{q}} : \sum_{i=1}^{n} h_i \mathbf{l}_i(\mathbf{e},T,\mathbf{q},\mathbf{F})\frac{\partial\varphi_i}{\partial T}\dot{T}, \qquad (30)$$

vanish. If this is the case, the classical thermomechanical state equations (Eqs. (14)) can be derived, as in the classical elastic case, directly from Eqs. (29). Thus, we can state the following theorem:

Theorem 2: For the rate – independent SMA material with internal variables whose evolution in the course of martensitic transformations is described by the rate equations (5), (or equivalently by Eqs. (3)), the invariance of the spatial local balance of energy equation under superimposed diffeomorphisms, which also include a temperature rescaling, does not yield the standard thermomechanical state equations, unless further assumptions are made.

It is interesting to note that in the classical theory of thermodynamics with internal variables Lubliner (1974, 1987) has arrived at a similar result by working entirely in the reference configuration and *on the basis of the second law of thermodynamics expressed in the form of the Clausius – Plank inequality*, which is a stronger (i.e., less general) form of the Clausius - Duhem inequality since it ignores dissipation due to heat conduction. In order to obtain the standard thermomechanical state equations, Lubliner modifies further the Clausius – Planck inequality, by assuming that it holds as an *equality* for elastic unloading and neutral loading. It is remarkable to note that by working with the covariance axiom we do not have to ignore dissipation due to heat conduction. Also, in comparing the two approaches we note that while in the second law of thermodynamics *we focus on all processes, in the covariance axiom we focus on all transformations of a given process* (Marsden and Hughes, 1994, p. 201).

4. A constitutive model

Up to now, the proposed formulation was presented largely in an abstract manner by leaving the kinematics of the problem and the number and the nature of the internal variables unspecified. The basic objective of this section is the introduction of a material model that will help make the application of the generalized plasticity concept in modeling phase transformations clearer. The model is based on a geometrically linear model proposed earlier within a stress space formulation by Panoskaltsis and co-workers (Panoskaltsis et al., 2004, Ramanathan et al., 2002) and which has been extensively used in several applications of engineering interesting (e.g., see Freed et al., 2008; Videnic et al., 2008; Freed and Aboudi, 2008; Freed and Banks – Sills, 2007).

There are two fundamental assumptions underlying the new model which is developed here. The first consists of the additive decomposition of the material strain tensor $\mathbf{E}$ into elastic $\mathbf{E}^{\mathbf{e}}$ and inelastic (transformation induced) $\mathbf{E}^{\mathbf{Tr}}$ parts, i.e.,

$$\mathbf{E} = \mathbf{E}^{\mathbf{e}} + \mathbf{E}^{\mathbf{Tr}}. \tag{31}$$

Such a decomposition has its origins in the work of Green and Naghdi (1965). The second fundamental assumption is that the response of the material is isotropic. Accordingly, it is assumed that it can be described in terms of a *single* scalar internal variable Z, which, as it is common within the literature (e.g., Boyd and Lagoudas, 1996; Lubliner and Auricchio, 1996; Panoskaltsis et al., 2004; Müller and Bruhns, 2006; Thamburaja, 2010), is identified by the fraction of a single (favorably oriented) martensite variant. In turn, and in view of Eq. (31), the internal variable vector is assumed to be composed by the transformation strain tensor $\mathbf{E}^{\mathrm{Tr}}$ and the martensite fraction Z.

By noting that the martensitic transformations to be considered are accompanied by variations of the elastic properties of the SMA material and in view of the additive decomposition of strain (Eq. (31)), the Helmholtz free energy can be additively decomposed in elastic and inelastic (transformation) parts, as follows

$$\Psi = \Psi_{\mathbf{e}}(\mathbf{E}-\mathbf{E}^{\mathrm{Tr}}(\mathrm{Z}),\mathrm{T},\mathrm{Z}) + \Psi_{\mathbf{Tr}}(\mathrm{Z},\mathrm{T}). \tag{32}$$

It is emphasized that this is *not* the conventional decomposition of the free energy function performed within the classical inelastic theories (e.g., plasticity, viscoelasticity, viscoplasticity), since the elastic part $\Psi_{\mathbf{e}}$ depends on the internal variable Z. In this sense the decomposition (32) resembles the decompositions employed within the thermomechanical treatment of damage (see Panoskaltsis et al., 2004). The elastic part of the Helmholtz free energy is assumed to be given as

$$\Psi_{\mathbf{e}}(\mathbf{E}-\mathbf{E}^{\mathrm{Tr}}(\mathrm{Z}),\mathrm{T},\mathrm{Z}) = U\left(\mathbf{E}-\mathbf{E}^{\mathrm{Tr}}(\mathrm{Z}),\mathrm{Z}\right) + \Theta\left(\mathrm{T}\right) + M\left(\mathbf{E}-\mathbf{E}^{\mathrm{Tr}}(\mathrm{Z}),\mathrm{Z},\mathrm{T}\right), \tag{33}$$

where the terms U, Θ and M will be defined next. U is the mechanical part of $\Psi_{\mathbf{e}}$ and is assumed to be given by a similar expression to the stored energy function of a St. Venant – Kirchhoff material (e.g., see Holzapfel, 2000, pp. 250 – 251), that is

$$U\left(\mathbf{E}-\mathbf{E}^{\mathrm{Tr}}(\mathrm{Z}),\mathrm{Z}\right) = \frac{\lambda(\mathrm{Z})}{2}\{tr[\mathbf{E}-\mathbf{E}^{\mathrm{Tr}}(\mathrm{Z})\}^2 + \mu(\mathrm{Z})tr[(\mathbf{E}-\mathbf{E}^{\mathrm{Tr}}(\mathrm{Z})]^2, \tag{34}$$

where λ and μ are Lame′ type of parameters ($\lambda > 0$, $\mu > 0$) and tr denotes the trace operator. These parameters are assumed to be dependent on the martensite fraction of the SMA, according to the following (power) law

$$\lambda(\mathrm{Z}) = \lambda_{\mathbf{A}} + \mathrm{Z}^{\mathbf{n}}(\lambda_{\mathbf{M}} - \lambda_{\mathbf{A}}), \quad \mu(\mathrm{Z}) = \mu_{\mathbf{A}} + \mathrm{Z}^{\mathbf{m}}(\mu_{\mathbf{M}} - \mu_{\mathbf{A}}), \tag{35}$$

where $\lambda_{\mathbf{A}}$, $\mu_{\mathbf{A}}$ are the Lame′ type of parameters when the material is fully austenite, $\lambda_{\mathbf{M}}$, $\mu_{\mathbf{M}}$ are these when the material is fully martensite and n and m are two additional model parameters. For the particular case n = m = 1 the rule of mixtures, which has been used extensively within the literature (e.g., Anand and Gurtin, 2003; Hallberg et al., 2007) is derived.

For the thermal part of the stored energy function, that is for the functions $\Theta\left(\mathrm{T}\right)$ and $M\left(\mathbf{E}-\mathbf{E}^{\mathrm{Tr}}(\mathrm{Z})\right)$ we consider the following expressions:

$$\begin{aligned} &\Theta\left(\mathrm{T}\right) = c[(\mathrm{T}-\mathrm{T}_0) - \mathrm{T}\ln\left(\frac{\mathrm{T}}{\mathrm{T}_0}\right)], \\ &M\left(\mathbf{E}-\mathbf{E}^{\mathrm{Tr}}(\mathrm{Z}),\mathrm{Z},\mathrm{T}\right) = -[3\lambda(\mathrm{Z}) + 2\mu(\mathrm{Z})]\alpha_{\mathbf{t}}(\mathrm{T}-\mathrm{T}_0)tr[\mathbf{E}-\mathbf{E}^{\mathrm{Tr}}(\mathrm{Z})], \end{aligned} \tag{36}$$

where T_0 is the reference temperature, c is the specific heat and α_t the linear expansion coefficient, which may be assumed varying within the phase transformations according to expressions analogous to those given in Eq. (35).

Finally, the transformation part of the Helmholtz free energy is given as

$$\Psi_{\mathbf{Tr}} = -T\eta_{\mathbf{Tr}}(Z) + u_{\mathbf{Tr}}(Z), \tag{37}$$

where $\eta_{\mathbf{Tr}}(Z)$ and $u_{\mathbf{Tr}}(Z)$ stand for the configurational entropy and the configurational internal energy and for which we assume two expressions justified in the work of Müller and Bruhns (2006) (see also the thermomechanical theory of Raniecki et al., 1992; Raniecki and Lexcellent, 1998), namely

$$\begin{aligned} \eta_{\mathbf{Tr}}(Z) &= s_0^{*\mathbf{A}} + Z\Delta s^* - Z(1\text{-}Z)\overline{s}_0, \\ u_{\mathbf{Tr}}(Z) &= u_0^{*\mathbf{A}} + Z\Delta u^* - Z(1\text{-}Z)\overline{u}_0, \end{aligned} \tag{38}$$

where $s_0^{*\mathbf{A}}$, Δs^* , $\overline{s}_0$, $u_0^{*\mathbf{A}}$, Δu^* and $\overline{u}_0$ are the model thermal parameters.

Then in light of the first of Eqs. (20) the second Piola – Kirchhoff stress tensor, after extensive calculations, is found to be

$$\mathbf{S} = \lambda tr(\mathbf{E} - \mathbf{E}^{\mathbf{Tr}})\mathbf{1} + 2\mu(\mathbf{E} - \mathbf{E}^{\mathbf{Tr}}) - (3\lambda + 2\mu)\alpha_t(T - T_0)\mathbf{1}, \tag{39}$$

where the dependence of the involved quantities on Z has been dropped for convenience.

The loading surfaces are assumed to be given in the *stress – space* as a two parameter family of von - Mises type of surfaces, that is

$$F(\mathbf{S},T) = |DEV\mathbf{S}| - CT - R = 0, \tag{40}$$

where $|.|$ denotes the Euclidean norm, $DEV(.)$ stands for the deviatoric part of the stress tensor in the reference configuration and C and R are parameters. On substituting from Eq. (39) into Eq. (40) the equivalent expression for the loading surfaces in the *strain – space* may be derived as

$$\Phi(\mathbf{E}, \mathbf{E}^{\mathbf{Tr}}, T) = 2\mu\,|DEV(\mathbf{E} - \mathbf{E}^{\mathbf{Tr}})| - CT - R = 0. \tag{41}$$

For the evolution of the transformation strain we assume a normality rule in the strain – space which is given as

$$2\mu\dot{\mathbf{E}}^{\mathbf{Tr}} = \varepsilon_L \dot{Z} \frac{\partial \Phi}{\partial \mathbf{E}}, \tag{42}$$

where ε_L is a material constant, which is defined as the maximum inelastic strain (e.g., Boyd and Lagoudas, 1996; Lubliner and Auricchio, 1996; Panoskaltsis et al., 2004;

Ramanathan et al., 2002), which is attained in the case of one – dimensional unloading in simple tension when the material is fully martensite.

The rate equation for the evolution of the martensite fraction Z, is determined on the basis of the geometrical framework described in section 2 as follows:

For the austenite to martensite transformation $(A \rightarrow M)$ we consider the Φ_M – loading surfaces as:

$$\Phi_M(\mathbf{E}, \mathbf{E}^{Tr}, T) = 2\mu \left| DEV(\mathbf{E} - \mathbf{E}^{Tr}) \right| - C_M T - R_M = 0, \tag{43}$$

where C_M is a material parameter which can be determined by means of the well – known (e.g., see Lubliner and Auricchio, 1996; Panoskaltsis et al., 2004; Ramanathan et al., 2002; Christ and Reese, 2009) critical stress – temperature phase diagram for the SMAs transformation. Moreover we consider

$$\begin{aligned} \Phi_{Mf}(\mathbf{E}, \mathbf{E}^{Tr}, T) &= 2\mu \left| DEV(\mathbf{E} - \mathbf{E}^{Tr}) \right| + R_{Mf} = 0, \\ \Phi_{Ms}(\mathbf{E}, \mathbf{E}^{Tr}, T) &= 2\mu \left| DEV(\mathbf{E} - \mathbf{E}^{Tr}) \right| + R_{Ms} = 0, \end{aligned} \tag{44}$$

where

$$R_{Mf} = -\sigma_{Mf} - C_M(T - M_f), \quad R_{Ms} = -\sigma_{Ms} - C_M(T - M_s),$$

where the parameters M_f and M_s stand for the martensite finish and martensite start temperatures respectively, and σ_{Mf} and σ_{Ms} are two additional parameters which may be determined from experimental results. Since Φ_{Mf} is related to the finish values and Φ_{Ms} to the starting values of the $A \rightarrow M$ transformation, the loading surfaces $\Phi_{Mf} = 0$ and $\Phi_{Ms} = 0$ may be considered as the boundaries of the set of all states for which the $A \rightarrow M$ transformation can be active. Then the constant $H_1 = H_M$ (see Eq. (3)) may be defined as

$$H_M = \frac{\langle -\Phi_{Mf}\Phi_{Ms} \rangle}{\left| \Phi_{Mf}\Phi_{Ms} \right|}. \tag{45}$$

For the function $\mathbf{L}_1 = \mathbf{L}_M$ several choices are possible (see Panoskaltsis et al., 2004). In this work, we use a linear expression (see Lickachev and Koval, 1992), which within the present strain – space formulation may be written in the form

$$L_M(\mathbf{E}, \mathbf{E}^{Tr}, T, Z) = -\frac{1 - Z}{\Phi_{Mf} - 2\mu\varepsilon_L(1 - Z)}. \tag{46}$$

In view of Eqs. (45) and (46) the rate equation for the evolution of the martensite fraction of the material during the $A \rightarrow M$ transformation may be written in the form

$$\dot{Z} = -\frac{\langle -\Phi_{Mf}\Phi_{Ms}\rangle}{|\Phi_{Mf}\Phi_{Ms}|}\frac{1-Z}{\Phi_{Mf} - 2\mu\varepsilon_L(1-Z)}\langle L_M\rangle, \tag{47}$$

where $L_M = L_1$ stands for the loading rate in the material description, that is

$$L_M = \mathbf{N}_M : \dot{\mathbf{G}} = \frac{\partial\Phi_M}{\partial \mathbf{E}} : \dot{\mathbf{E}} + \frac{\partial\Phi_M}{\partial \mathrm{T}}\dot{\mathrm{T}}. \tag{48}$$

Similarly, for the inverse $\mathrm{M} \to \mathrm{A}$ transformation we define the Φ_A – loading surfaces as follows

$$\Phi_A(\mathbf{E}, \mathbf{E}^{Tr}, \mathrm{T}) = 2\mu\,\left|DEV(\mathbf{E} - \mathbf{E}^{Tr})\right| - C_A\mathrm{T} - R_A = 0, \tag{49}$$

$$\begin{aligned} \Phi_{Af}(\mathbf{E}, \mathbf{E}^{Tr}, \mathrm{T}) &= 2\mu\,\left|DEV(\mathbf{E} - \mathbf{E}^{Tr})\right| + R_{Af} = 0, \\ \Phi_{As}(\mathbf{E}, \mathbf{E}^{Tr}, \mathrm{T}) &= 2\mu\,\left|DEV(\mathbf{E} - \mathbf{E}^{Tr})\right| + R_{As} = 0, \end{aligned} \tag{50}$$

where

$$R_{Af} = -\sigma_{Af} - C_A(\mathrm{T} - \mathrm{A_f}), \quad R_{As} = -\sigma_{As} - C_A(\mathrm{T} - \mathrm{A_s}),$$

and the parameters $C_A, \mathrm{A_f}, \mathrm{A_s}, \sigma_{Af}$ and σ_{As} are material parameters, all related to the $\mathrm{M} \to \mathrm{A}$ transformation. By applying analogous to the $\mathrm{A} \to \mathrm{M}$ transformation case arguments, we derive the rate equation for the evolution of Z for the $\mathrm{M} \to \mathrm{A}$ transformation as

$$\dot{Z} = -\frac{\langle -\Phi_{Af}\Phi_{As}\rangle}{|\Phi_{Af}\Phi_{As}|}\frac{Z}{\Phi_{Af} + 2\mu\varepsilon_L Z}\langle L_A\rangle, \tag{51}$$

where

$$L_A = \mathbf{N}_A : \dot{\mathbf{G}} = -\left(\frac{\partial\Phi_A}{\partial \mathbf{E}} : \dot{\mathbf{E}} + \frac{\Phi_A}{\partial \mathrm{T}}\dot{\mathrm{T}}\right) = -L_M. \tag{52}$$

As a result, the final form for the rate equation for the evolution of the internal variable Z (see Eq. (3)) takes the form

$$\dot{Z} = -\frac{\langle -\Phi_{Mf}\Phi_{Ms}\rangle}{|\Phi_{Mf}\Phi_{Ms}|}\frac{1\text{-}Z}{\Phi_{Mf} - 2\mu\varepsilon_L(1-Z)}\langle L_M\rangle - \frac{\langle -\Phi_{Af}\Phi_{As}\rangle}{|\Phi_{Af}\Phi_{As}|}\frac{Z}{\Phi_{Af} + 2\mu\varepsilon_L Z}\langle L_A\rangle. \tag{53}$$

The thermomechanical coupling phenomena, which occur during the martensitic transformations may be studied on the basis of the energy balance equation. It should be mentioned here that with the aid of the fundamental concept of energy it is possible to relate different physical phenomena to one another, as well as to evaluate their relative significance in a given process in mechanics and more generally in physics (Lubliner, 2008, p. 44). This will be accomplished as follows:

The energy balance Eq. (12) can be written in a material setting as

$$\rho_{\text{ref}}(\dot{\Psi} + \dot{\eta}\text{T} + \eta\dot{\text{T}}) + DIV\mathbf{H} = \mathbf{S} : \dot{\mathbf{E}} + \rho_{\text{ref}}\text{R}, \tag{54}$$

where $DIV(.)$ is the divergence operator, **H** is the heat flux vector and R is the heat supply per unit mass, all expressed in the material description. By taking the time derivative of the Helmholtz free energy function and inserting it in Eq. (54) we obtain

$$\rho_{\text{ref}}(\frac{\partial\Psi}{\partial\mathbf{E}} : \dot{\mathbf{E}} + \frac{\partial\Psi}{\partial\mathbf{E}^{\text{Tr}}} : \dot{\mathbf{E}}^{\text{Tr}} + \frac{\partial\Psi}{\partial\text{Z}}\dot{\text{Z}} + \frac{\partial\Psi}{\partial\text{T}}\dot{\text{T}}) + \rho_{\text{ref}}\eta\dot{\text{T}} + \rho_{\text{ref}}\dot{\eta}\text{T} + DIV\mathbf{H} = \rho_{\text{ref}}\text{R} + \mathbf{S}:\dot{\mathbf{E}}. \tag{55}$$

This equation in turn, upon substitution of the thermomechanical state Eqs. (20), yields

$$\rho_{\text{ref}}(\frac{\partial\Psi}{\partial\mathbf{E}^{\text{Tr}}} : \dot{\mathbf{E}}^{\text{Tr}} + \frac{\partial\Psi}{\partial\text{Z}}\dot{\text{Z}}) + \rho_{\text{ref}}\dot{\eta}\text{T} + DIV\mathbf{H} = \rho_{\text{ref}}\text{R}. \tag{56}$$

The time derivative of the entropy density is determined by the second of Eqs. (20) as

$$\dot{\eta} = -\frac{\partial^2\Psi}{\partial\text{T}\partial\mathbf{E}} : \dot{\mathbf{E}} - \frac{\partial^2\Psi}{\partial\text{T}\partial\mathbf{E}^{\text{Tr}}} : \dot{\mathbf{E}}^{\text{Tr}} - \frac{\partial^2\Psi}{\partial\text{T}\partial\text{Z}}\dot{\text{Z}} - \frac{\partial^2\Psi}{\partial\text{T}^2}\dot{\text{T}}. \tag{57}$$

Upon definition of the specific heat c as

$$c = -\frac{\partial^2\Psi}{\partial\text{T}^2}\text{T}, \tag{58}$$

and upon substitution of Eqs. (57) and (58) into Eq. (56), the latter yields the temperature evolution equation as

$$c\dot{\text{T}} = -(\frac{\partial\Psi}{\partial\mathbf{E}^{\text{Tr}}} : \dot{\mathbf{E}}^{\text{Tr}} + \frac{\partial\Psi}{\partial\text{Z}}\dot{\text{Z}}) + (\frac{\partial^2\Psi}{\partial\text{T}\partial\mathbf{E}} : \dot{\mathbf{E}} + \frac{\partial^2\Psi}{\partial\text{T}\partial\mathbf{E}^{\text{Tr}}} : \dot{\mathbf{E}}^{\text{Tr}} + \frac{\partial^2\Psi}{\partial\text{T}\partial\text{Z}}\dot{\text{Z}})\text{T} + (\text{R} - \frac{1}{\rho_{\text{ref}}}DIV\mathbf{H}). \tag{59}$$

If we now define the *heating due to thermoelastic effects* as

$$\dot{Q}_{\mathbf{e}} = \text{T}(\frac{\partial^2\Psi}{\partial\text{T}\partial\mathbf{E}} : \dot{\mathbf{E}} + \frac{\partial^2\Psi}{\partial\text{T}\partial\mathbf{E}^{\text{Tr}}} : \dot{\mathbf{E}}^{\text{Tr}}) \tag{60}$$

and the *inelastic (transformation) contribution to heating* as

$$\dot{Q}_{\mathbf{Tr}} = -(\frac{\partial\Psi}{\partial\mathbf{E}^{\text{Tr}}} : \dot{\mathbf{E}}^{\text{Tr}} + \frac{\partial\Psi}{\partial\text{Z}}\dot{\text{Z}}) + \text{T}\frac{\partial^2\Psi}{\partial\text{T}\partial\text{Z}}\dot{\text{Z}} = D_{\mathbf{Tr}} + \text{T}\frac{\partial^2\Psi}{\partial\text{T}\partial\text{Z}}\dot{\text{Z}}, \tag{61}$$

where $D_{\mathbf{Tr}} = -(\frac{\partial \Psi}{\partial \mathbf{E}^{\mathbf{Tr}}} : \dot{\mathbf{E}}^{\mathbf{Tr}} + \frac{\partial \Psi}{\partial \mathrm{Z}} \dot{\mathrm{Z}})$ is the *inelastic dissipation due to phase transformations*, the temperature evolution equation takes the following, remarkably simple, form (see also Rosakis et al., 2000)

$$c\dot{\mathrm{T}} = \dot{Q}_{\mathbf{e}} + \dot{Q}_{\mathbf{Tr}} + (\mathrm{R} - \frac{1}{\rho_{\mathrm{ref}}} DIV\mathbf{H}). \tag{62}$$

This expression has the obvious advantage of decoupling the elastic and inelastic contributions to material heating and is well suited for computational use.

It is noted that in an adiabatic process, that is in a process with $\mathrm{R} - \frac{1}{\rho_{\mathrm{ref}}} DIV\mathbf{H} = 0$,

Eq. (62) takes the form

$$c\dot{\mathrm{T}} = \dot{Q}_{\mathbf{e}} + \dot{Q}_{\mathbf{Tr}}, \tag{63}$$

from which and by assuming that the temperature evolution due to *structural heating* $H_{\mathbf{str}}$, defined as $H_{\mathbf{str}} = -(\frac{\partial^2 \Psi}{\partial \mathrm{T} \partial \mathbf{E}} : \dot{\mathbf{E}} + \frac{\partial^2 \Psi}{\partial \mathrm{T} \partial \mathbf{E}^{\mathbf{Tr}}} : \dot{\mathbf{E}}^{\mathbf{Tr}} + \frac{\partial^2 \Psi}{\partial \mathrm{T} \partial \mathrm{Z}} \dot{\mathrm{Z}})\mathrm{T}$ (e.g., see Simo and Miehe, 1992), is negligible in comparison to that due to inelastic dissipation $D_{\mathbf{Tr}}$, the temperature evolution equation takes the following simple form

$$c\dot{\mathrm{T}} = D_{\mathbf{Tr}}. \tag{64}$$

Finally, as a constitutive law for the heat flux vector we assume the standard Fourier's law (e.g., Simo and Miehe, 1992; Müller and Bruhns, 2006):

$$\mathbf{H} = -kGRAD\mathrm{T}. \tag{65}$$

5. Computational aspects and numerical simulations

As a final step we examine the ability of our model in simulating qualitatively several patterns of the extremely complex behavior of SMAs under simple states of straining. Isothermal and non – isothermal problems are considered.

5.1. Isothermal problems

Focusing our attention first in the isothermal case we note that when the total strain tensor **E** is known, the rate equations for the evolution of the internal variables (Eqs. (42) and (53)) and the mechanical state (thermoelastic stress-strain law) equation (Eq. (39)) together with the appropriate initial and boundary conditions form a system of three equations in the three unknowns $\mathbf{E}^{\mathbf{Tr}}$, Z and **S**. The numerical solution of this system of equations and accordingly the numerical implementation of the proposed model relies crucially on the general loading – unloading criteria (see Eq. (4)), which can be expressed in a remarkably simple form, based on the following observation:

As it has been mentioned the $\mathrm{A} \rightarrow \mathrm{M}$ transformation is active when $L_{\mathbf{M}} > 0$, while the inverse transformation is active when $L_{\mathbf{A}} > 0$. Since we always have $L_{\mathbf{M}} = -L_{\mathbf{A}}$, it is clear that only one phase transformation can be active at a given time of interest. Then we can treat the two phase transformations as two different inelastic processes and replace the general loading – unloading criteria by the following decoupled ones:

$\mathrm{A} \rightarrow \mathrm{M}$ Transformation:

$$\begin{cases} \text{If } H_{\mathbf{M}} = \dfrac{\langle -\Phi_{\mathbf{Mf}} \Phi_{\mathbf{Ms}} \rangle}{|\Phi_{\mathbf{Mf}} \Phi_{\mathbf{Ms}}|} = 0: & \text{elastic state,} \\ \text{If } H_{\mathbf{M}} = \dfrac{\langle -\Phi_{\mathbf{Mf}} \Phi_{\mathbf{Ms}} \rangle}{|\Phi_{\mathbf{Mf}} \Phi_{\mathbf{Ms}}|} = 1 \text{ then} & \\ \quad \text{i. If } L_{\mathbf{M}} < 0: & \text{elastic unloading,} \\ \quad \text{ii. If } L_{\mathbf{M}} = 0: & \text{neutral loading,} \\ \quad \text{iii. If } L_{\mathbf{M}} > 0: & \text{inelastic loading.} \end{cases}$$

$\mathrm{M} \rightarrow \mathrm{A}$ Transformation:

$$\begin{cases} \text{If } H_{\mathbf{A}} = \dfrac{\langle -\Phi_{\mathbf{Af}} \Phi_{\mathbf{As}} \rangle}{|\Phi_{\mathbf{Af}} \Phi_{\mathbf{As}}|} = 0: & \text{elastic state,} \\ \text{If } H_{\mathbf{M}} = \dfrac{\langle -\Phi_{\mathbf{Af}} \Phi_{\mathbf{As}} \rangle}{|\Phi_{\mathbf{Af}} \Phi_{\mathbf{As}}|} = 1 \text{ then} & \\ \quad \text{i. If } L_{\mathbf{A}} < 0: & \text{elastic unloading,} \\ \quad \text{ii. If } L_{\mathbf{A}} = 0: & \text{neutral loading,} \\ \quad \text{iii. If } L_{\mathbf{A}} > 0: & \text{inelastic loading.} \end{cases}$$

Then the governing equations, along with the aforementioned loading – unloading criteria, can be solved by a time discretization scheme based on backward Euler. The resulting system of the discretized equations is solved by means of a *three step predictor - corrector algorithm,* the steps of which are dictated by the time discrete loading - unloading criteria. Algorithmic details regarding the enforcement of the time discrete loading – unloading criteria and the solution of the system, within the framework of large deformation generalized plasticity in the case of a single loading surface, can be found in Panoskaltsis et al. (2008a, b).

To this end it is emphasized that predictor – corrector algorithms work well in case of domains which are connected. The commonly used predictor – corrector algorithms for elastoplasticity employ an elastic predictor and an inelastic corrector. The most important assumption is that the solution is unique for a particular set of values of the state variables. The predictor step *freezes* the plastic flow and checks for an elastic solution. The yield criterion then is checked and if it is satisfied the elastic solution is acceptable, otherwise the inelastic corrector is activated. In the cases of elastic – plastic analysis there exists a set of consistency conditions the enforcement of which "returns" the (wrong) elastic solution onto the exact solution point on the evolving yield surface. However, in the case of disconnected elastic zones separated by inelastic zones the predictor – corrector algorithm is very sensitive on the strain step used, while going from an inelastic zone to an elastic one.

This is the case of SMAs, which have a transformation (inelastic) zone separating the fully martensite and fully austenite zones (being treated as elastic zones). During forward or reverse transformation, the predictor strain step is very important as we near the elastic – plastic (i.e. transformation) boundary. If the predicted solution lies within the transformation zone (i.e., outside the elastic range) the corrector step is activated and the resulting set of non –linear equations are solved. However, *as we approach the end of the transformation zone and therefore the boundary between the inelastic and the elastic zones, the predictor could predict an elastic solution,* which the algorithm accepts as a valid one, but which is within the next elastic zone, achieved *without the transformation being fully complete* (i.e., achieved while the state is still inelastic) and is therefore an unacceptable solution. This would cause errors in the minimization process and results in jumps in the solution and kinks in the stress strain curve. This problem is resolved here by making the strain step very small and by checking the limits of the transformation.

The first problem we study is a standard problem within the context of finite inelasticity and is that of *finite* shear, defined as

$$x_1 = X_1 + \gamma X_2, \;\; x_2 = X_2, \;\; x_3 = X_3,$$

where X_1, X_2, X_3 are the material coordinates and γ is the shearing parameter. For this problem the model parameters are set equal to those reported in the work of Boyd and Lagoudas (1994), that is:

$$\lambda_M = 9{,}486.95 \text{ MPa}, \; \mu_M = 4{,}887.22 \text{ MPa}, \; \lambda_A = 21{,}892.97 \text{ MPa}, \; \mu_A = 11{,}278.20 \text{ MPa},$$
$$M_f = 5^{o}\text{C}, \; M_s = 23^{o}\text{C}, \; A_s = 29^{o}\text{C}, \; A_f = 51^{o}\text{C}, \; C_M = 11.3 \text{ MPa}/^{o}\text{C}, \; C_A = 4.5 \text{ MPa}/^{o}\text{C},$$
$$\sigma_{Mf} = \sigma_{Ms} = \sigma_{Af} = \sigma_{As} = 0 \text{ MPa}, \; \varepsilon_L = 0.0635.$$

All numerical tests that performed start with the specimen in the parent (austenite) phase, (Z=0).

The first simulation demonstrates the pseudoelastic phenomena within the SMA material. In this case the temperature is held constant at some value above A_f. The purpose is to study a complete stress – induced transformation cycle. The results for this finite shear problem are shown for constant material stiffness ($\lambda = \lambda_A = 21{,}892.97$ MPa, $\mu = \mu_A = 11{,}278.20$ MPa), as well for a linear (n = m = 1) and a power (n = m = 5) type of stiffness variation in Figures 1, 2 and 3. On loading, the material initially remains austenite (elastic region and straight shear stress – strain curve). As loading is continuing and the shear strain attains the value at which the material point crosses the initial loading surface for the $A \rightarrow M$ transformation $(\Phi_{Ms} = 0)$, the transformation starts (inelasticity and curvilinear shear stress – strain curve; coexistence of the two phases). If the loading continues and the strain crosses the final loading surface for the $A \rightarrow M$ transformation $(\Phi_{Mf} = 0)$, the material is completely transformed into martensite and on further loading since the state of the material is elastic the shear stress – strain diagram is straight. Then, during unloading, the material is fully martensite (elastic region and straight shear stress – strain curve) until the strain crosses the initial loading surface $(\Phi_{As} = 0)$ of the $M \rightarrow A$ transformation, which is subsequently activated (phase coexistence, inelasticity and curvilinear shear stress – strain curve). On further unloading and when the strain meets the last boundary surface for the $M \rightarrow A$ transformation $(\Phi_{Af} = 0)$, the material becomes fully austenite and on further unloading the stress – strain curve is straight going back to zero, which means that no permanent deformation exists and the austenite is completely recovered. This is expected as the martensite phase is not stable at a temperature above A_f at zero stress level.

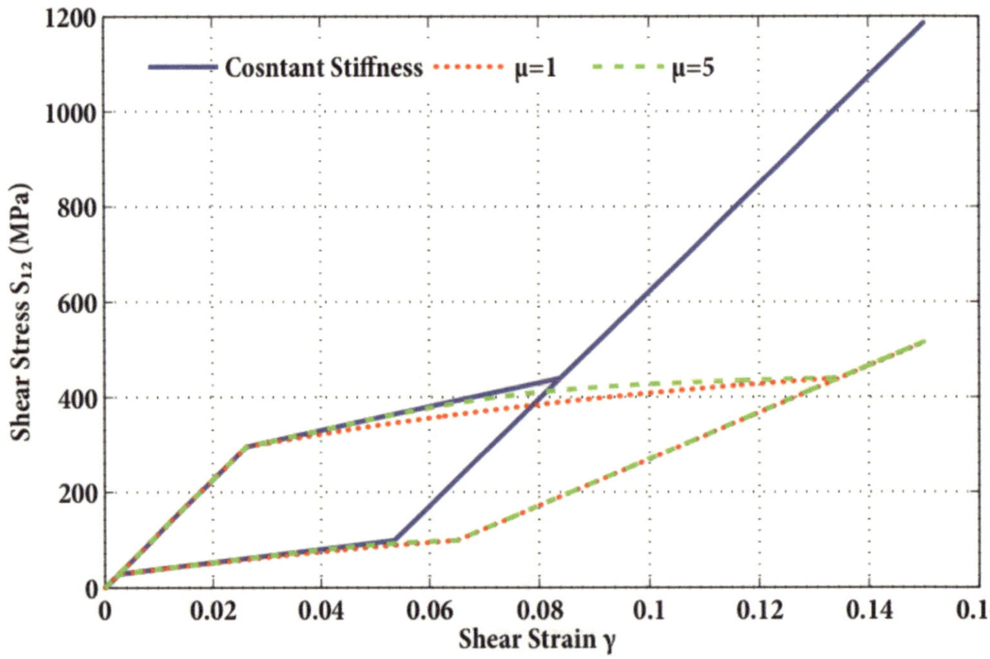

Figure 1. Finite shear. Isothermal one – dimensional behavior under monotonic loading. Shear stress S_{12} vs. shear strain γ.

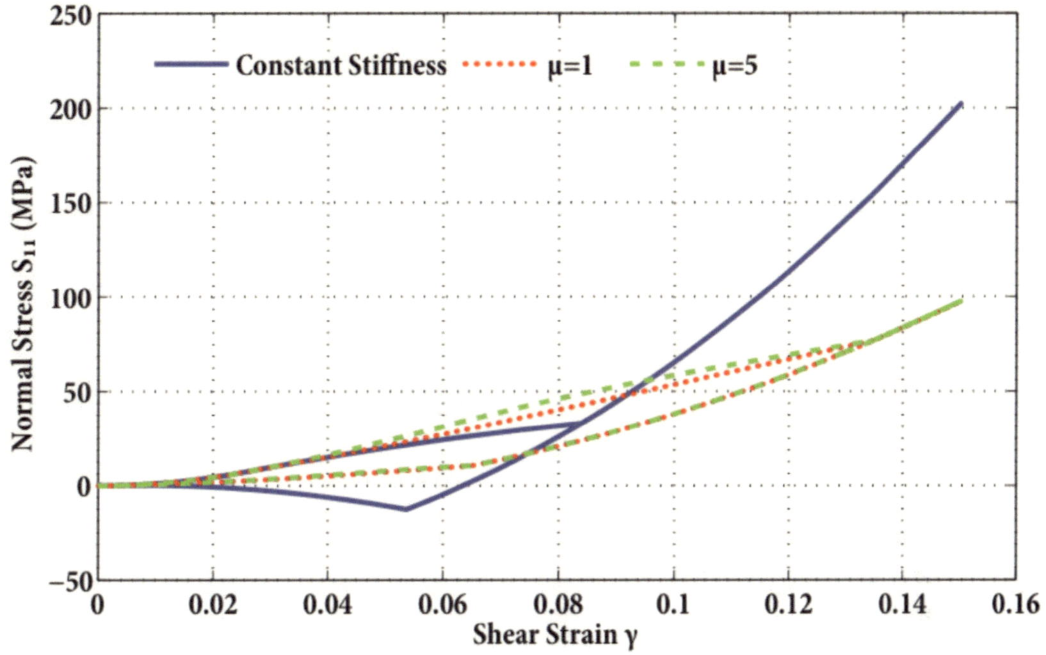

Figure 2. Finite Shear. Isothermal one – dimensional behavior under monotonic loading. Normal stress S_{11} vs. shear strain γ.

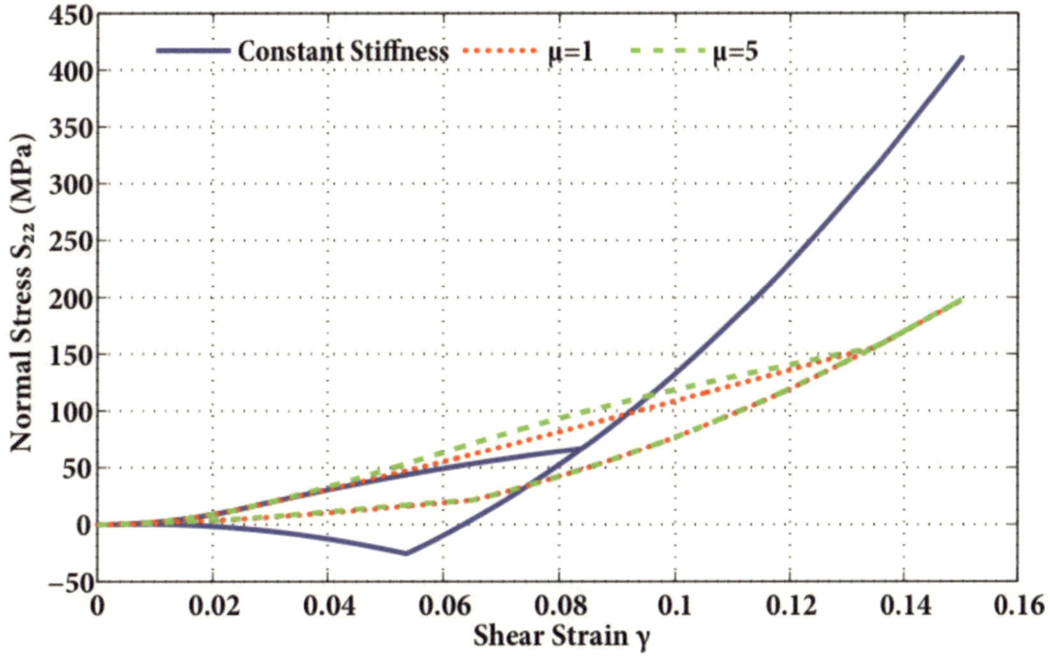

Figure 3. Finite Shear. Isothermal one –dimensional behavior under monotonic loading. Normal stress S_{22} vs. shear strain γ.

Next, the model is tested under *multiple shear stress cycles*, by subjecting it to partial unloading (incomplete $M \rightarrow A$ transformation) and partial reloading (incomplete $A \rightarrow M$ transformation). The results for linear stiffness variation are illustrated in Figure 4. A series of loops appears inside the complete loading – unloading cycle. These loops exhibit *slight ratcheting which stabilizes in a few cycles*. The response of the model is absolutely compatible with that described by other investigators (e.g., see Ivshin and Pence, 1994; Lubliner and Auricchio, 1996). In view of Figure 4 and since the dissipated energy can be estimated by the area of the $S_{12} - \gamma$ loop, the dissipated energy in the case of partial unloading and reloading is the area of the loop times the number of the loops. This explains the important property of the high internal damping of SMA materials. (For a discussion of the relation between areas of stress-strain diagrams and dissipated energy see Lubliner and Panoskaltsis, 1992.)

The ability of the model to simulate phase transformations and the corresponding stiffness variations under cyclic loading is demonstrated further by three additional tests. The first one illustrates the case of *partial loading with complete unloading*, the second the case of *partial unloading with complete loading* and the third the case of a series *of partial loading and partial unloading*. The results are shown in Figures 5, 6 and 7 respectively.

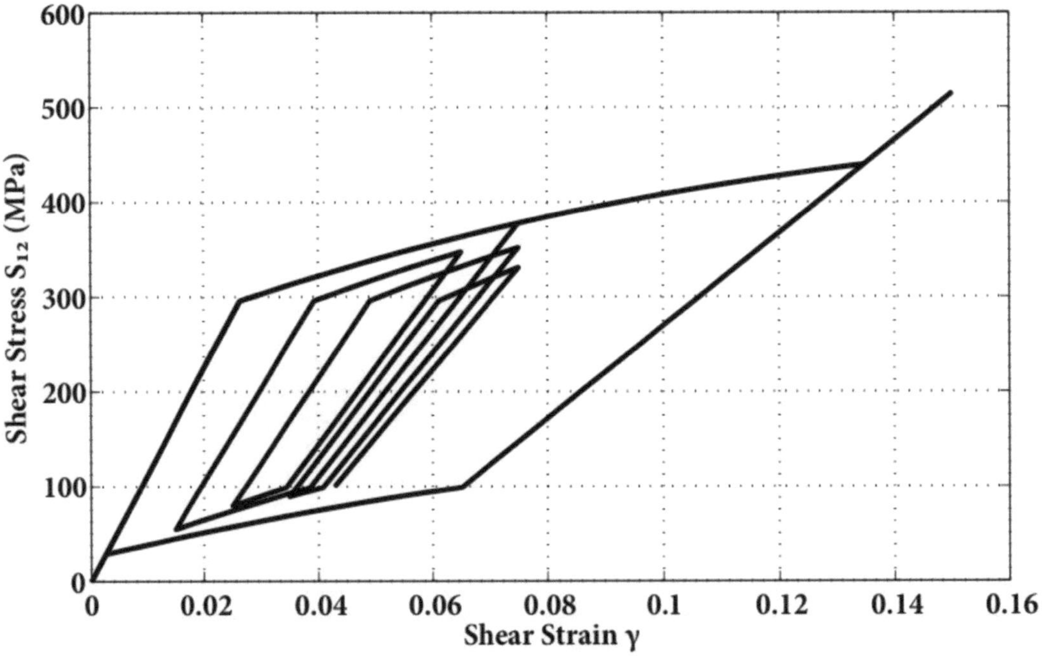

Figure 4. Finite Shear. Pseudoelasticity with partial loading and unloading. Shear stress S_{12} vs. shear strain γ.

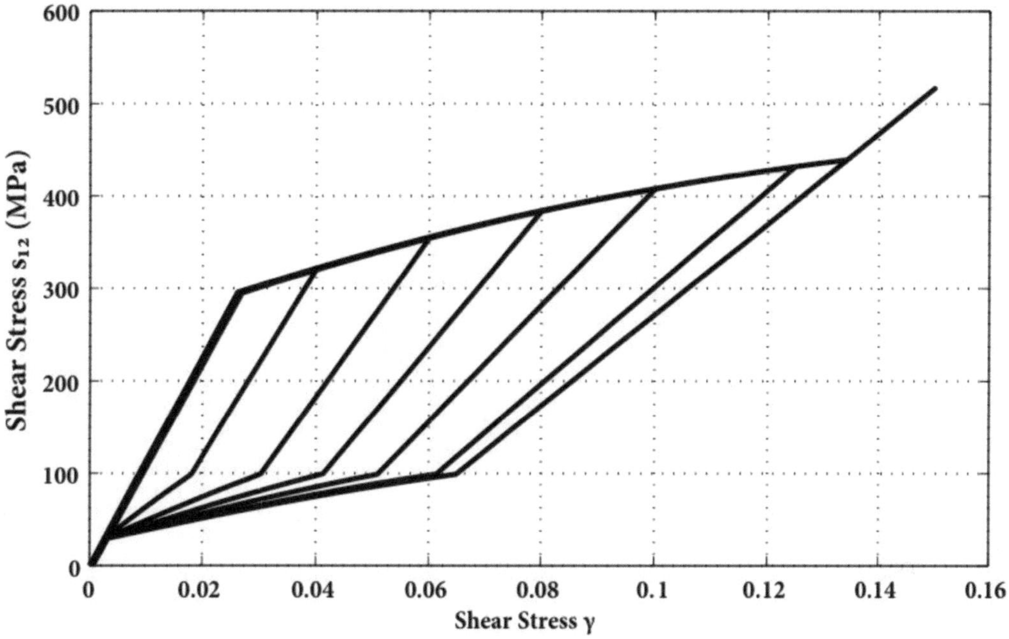

Figure 5. Finite Shear. Partial loading followed by complete unloading. Shear stress S_{12} vs. shear strain γ.

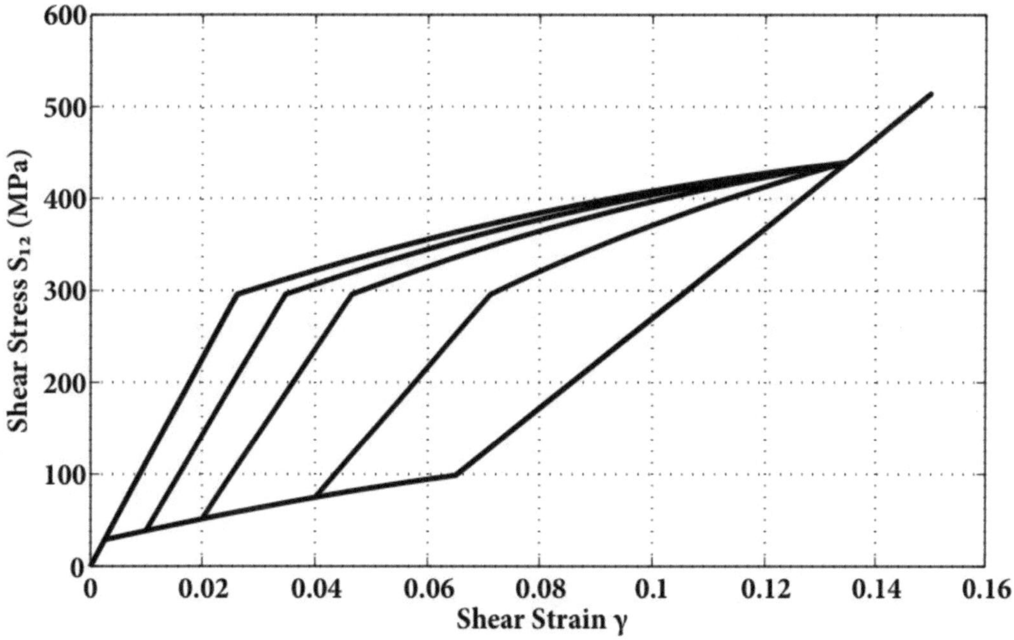

Figure 6. Finite Shear. Partial unloading followed by complete loading. Shear stress S_{12} vs. shear strain γ.

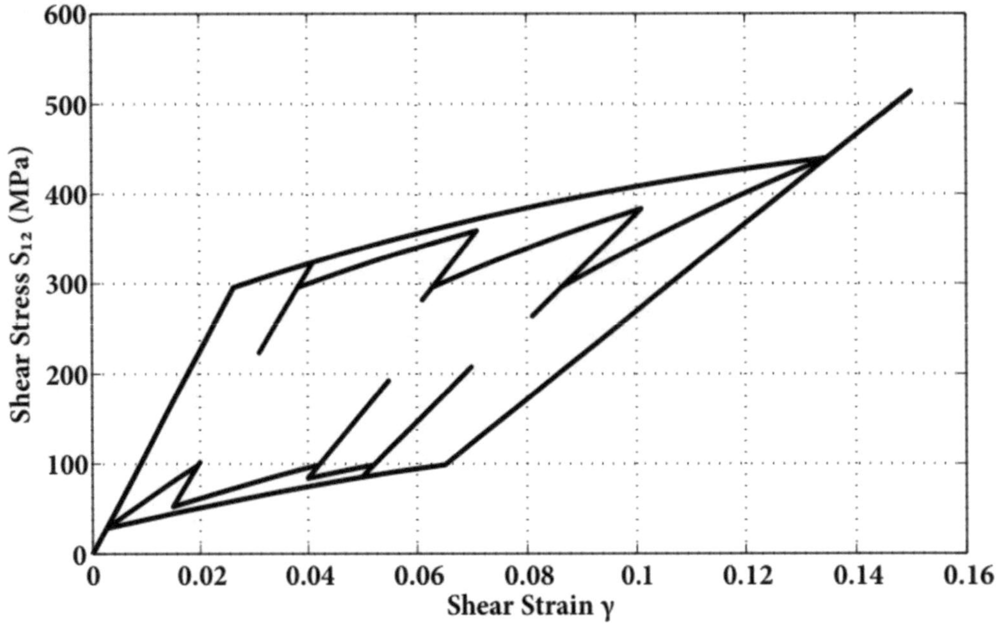

Figure 7. Finite Shear. Series of partial loading and partial unloading. Shear stress S_{12} vs. shear strain γ.

5.2. Non – Isothermal problems

In this section we examine the ability of the model in predicting pseudoelastic phenomena under *non – isothermal* conditions. In general, the numerical treatment of the coupled thermomechanical problems is performed on the basis of a split of the governing equations (equations of motion, constitutive equations, energy balance equation and the appropriate boundary conditions) into their mechanical and thermal parts. Most popular among the several computational schemes which have been proposed within the literature is the *isothermal split* proposed in the work of Simo and Miehe (1992). However, this rather straight forward numerical scheme has the disadvantage of not being unconditionally stable. An alternative time integration algorithm relies on the so-called adiabatic split (see Armero and Simo, 1993). In this approach, the problem is divided in a mechanical phase during which the entropy is held constant, followed by a thermal phase in which the configuration is held constant, leading to an unconditionally stable algorithm.

Nevertheless, since our objective is to discuss the proposed framework in its simplest setting, we consider two rather simple problems, namely a simple shear and a plane strain problem, where the equations of motion and the (mechanical) boundary conditions are trivially satisfied. Accordingly, within our simulations, *a simultaneous solution of the remaining governing equations, namely the constitutive equations and the balance of energy equation, is performed.*

First, an adiabatic test in finite simple shear is considered. We assume that due to the dynamic rates resulting in adiabatic response, heat exchanges due to conduction, convection and radiation can be neglected in comparison to the temperature changes induced by inelastic (transformation) dissipation, which leads to thermomechanical processes that can be considered as homogeneous. The elastic constants, the mass density and the thermal parameters used in this simulation are those considered in the work of Müller and Bruhns (2006), that is:

$$\lambda = \lambda_A = 25{,}541.80 \text{ MPa}, \ \mu = \mu_A = 13{,}157.90 \text{ MPa},$$
$$\rho_{ref} = 6.45 \times 10^{-3} \text{ k/m mm}^2, \ \alpha_t = 8.8 \times 10^{-6} 1/^{o}\text{ K, c=837.36 J/kg.K}$$
$$\Delta u^* = 16800.0 \text{ J/kg}, \ \Delta s^* = 64.50 \text{ J/kgK}, \ \overline{u}_0 = 4264.5 \text{ J/kg}, \ \overline{s}_0 = 11.5 \text{ J/kgK},$$

while the other parameters are set equal to those used in the isothermal problems studied before. The shear stress – strain curves predicted by the model, for both adiabatic and isothermal cases, are shown in Figure 8. It is observed that the stress – strain curves have similar qualitative characteristics with the adiabatic and the isothermal curves of a perfect gas in a pressure – volume diagram, with the adiabatic stress curve being above the corresponding isothermal one. This fact has to be attributed to material heating due to inelastic dissipation during the $A \rightarrow M$ transformation, which shifts the stress – strain curve upwards. Moreover, due to the higher stress attained during the $A \rightarrow M$ transformation, the initial loading surface for the inverse transformation $(\Phi_{As} = 0)$ is triggered at a higher stress level, a fact which results in a corresponding higher stress – strain unloading curve. The corresponding temperature – shear strain curve for the adiabatic specimen is shown in Figure 9 (for constant stiffness). Consistently with the experimentally observed adiabatic response of a SMA material, the model predicts heating of the material during the forward $A \rightarrow M$ transformation and cooling during the inverse $M \rightarrow A$ transformation.

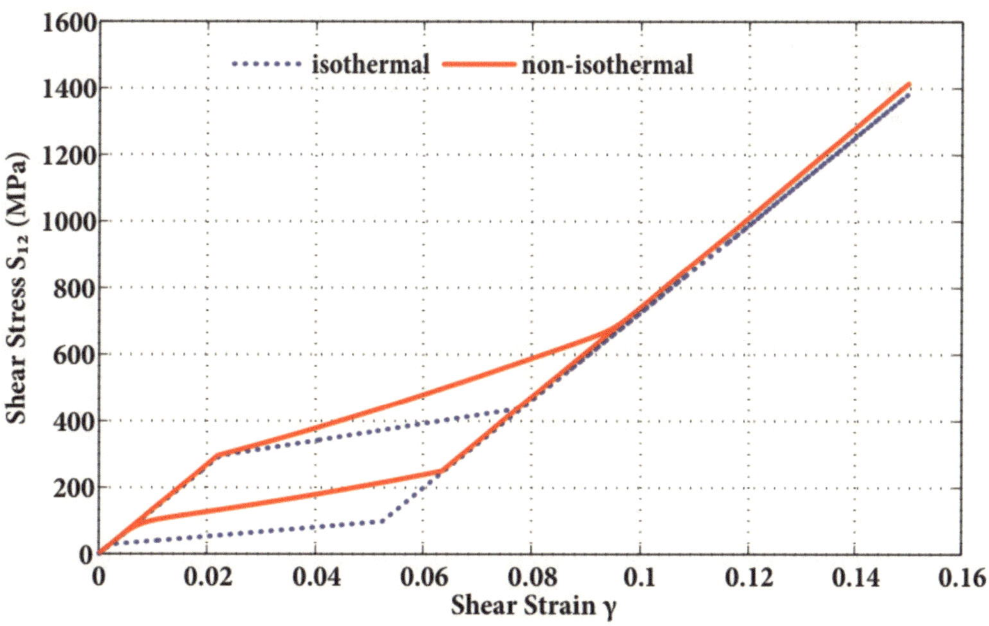

Figure 8. Finite Shear. Adiabatic and isothermal one – dimensional behavior under monotonic loading. Shear stress S_{12} vs. shear strain γ.

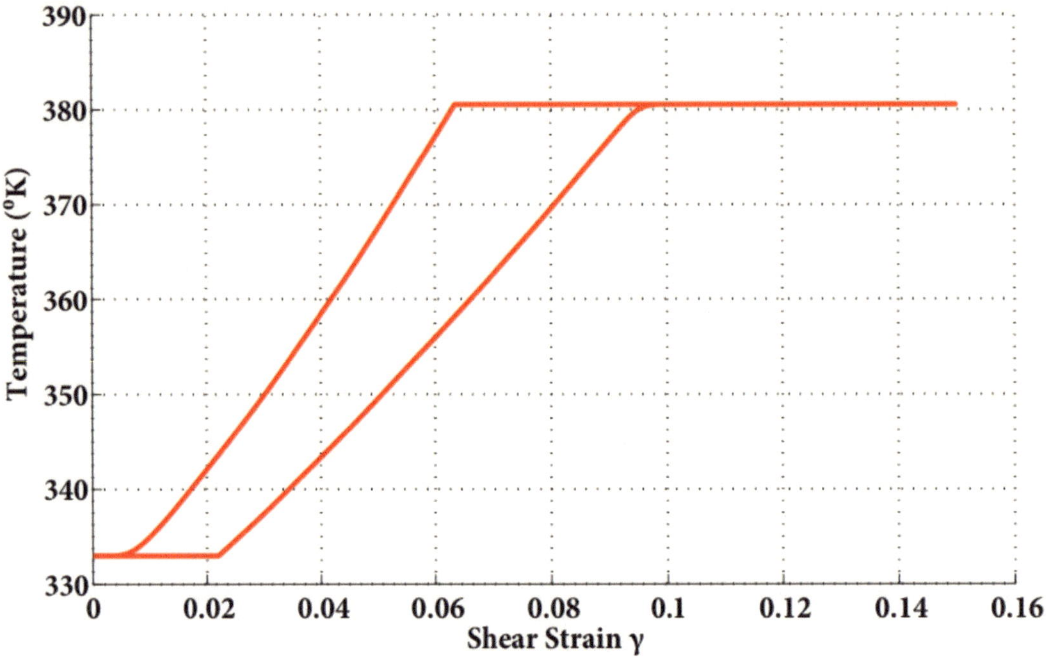

Figure 9. Finite Shear. Adiabatic one – dimensional behavior under monotonic loading. Temperature T vs. shear strain γ.

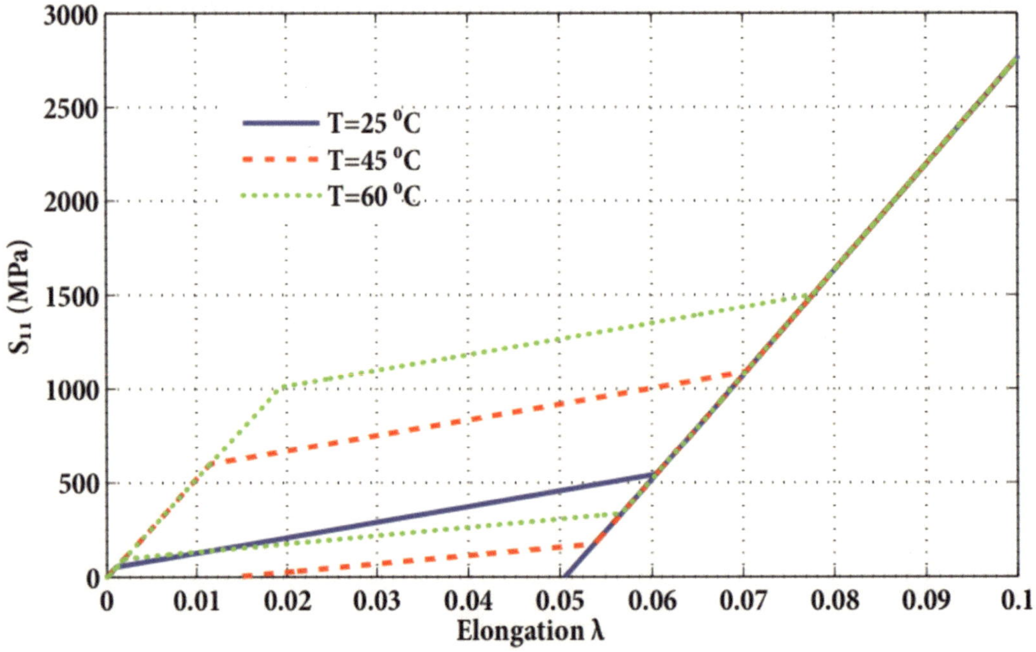

Figure 10. Plane strain (restrained tension). Monotonic loading at various temperatures. Normal stress S_{11} vs. axial displacement λ.

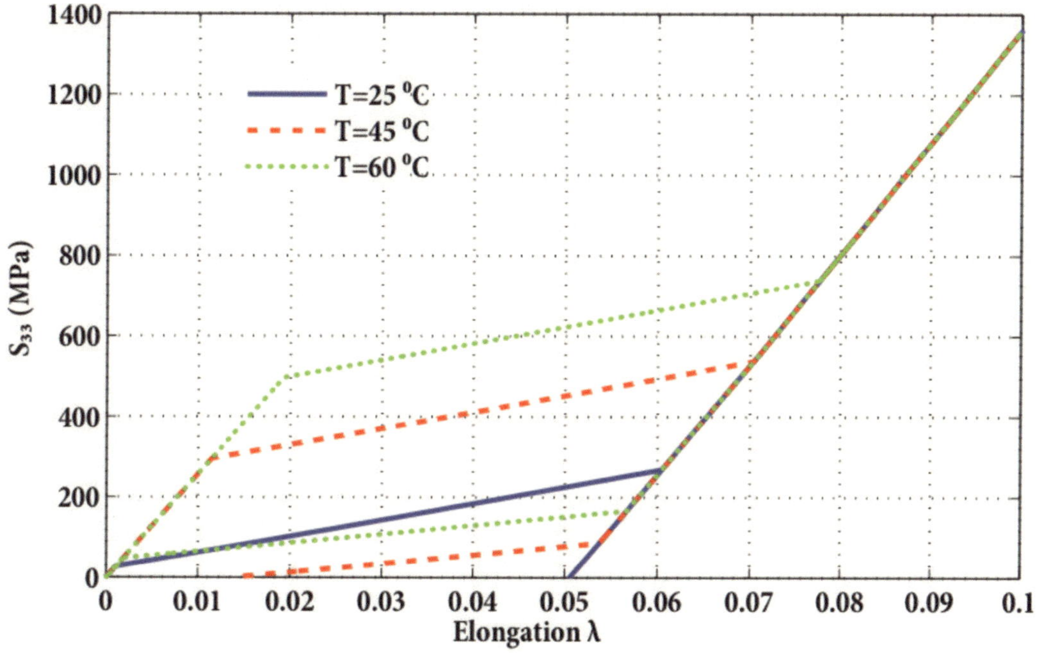

Figure 11. Plane strain (restrained tension). Monotonic loading at various temperatures. Normal stress S_{33} vs. axial displacement λ.

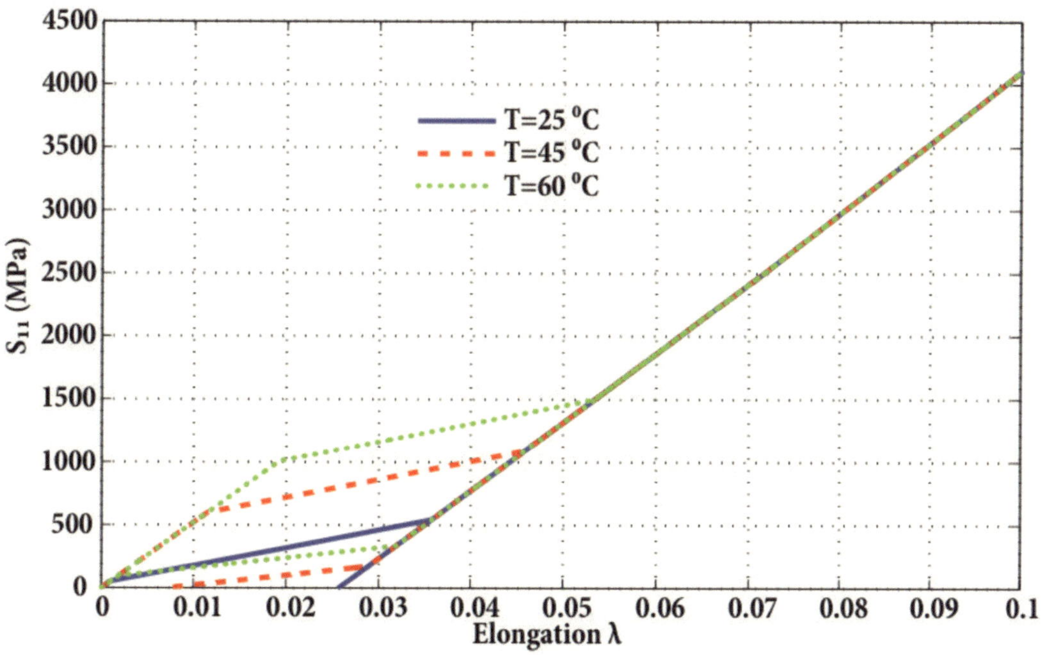

Figure 12. Plane Strain (biaxial extension). Monotonic loading at various temperatures. Normal stress S_{11} vs. axial displacement λ.

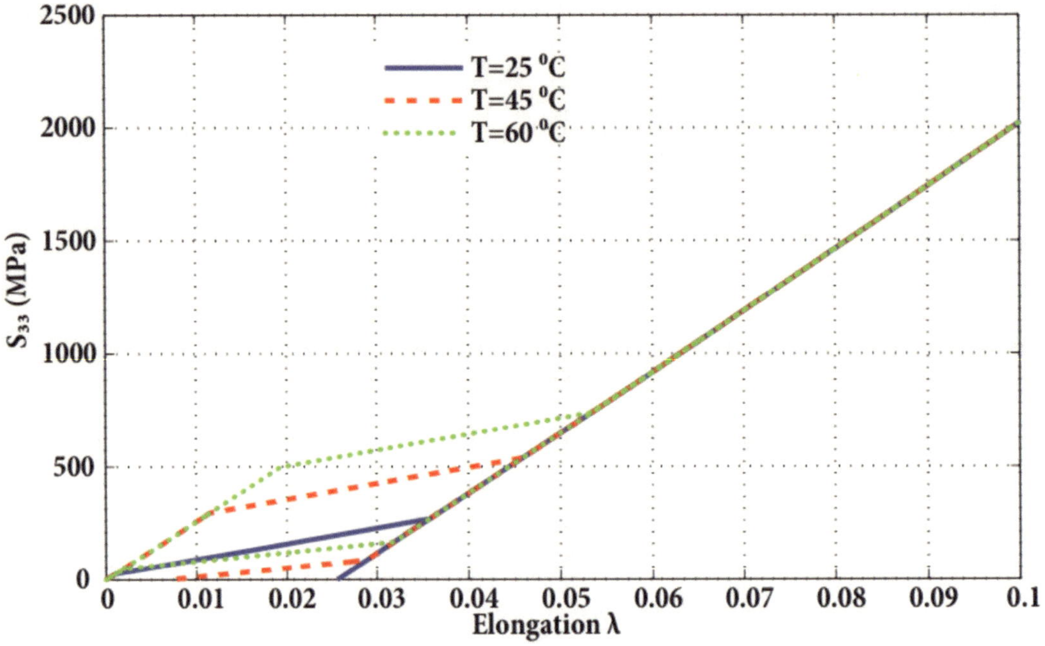

Figure 13. Plane strain (biaxial extension). Monotonic loading at various temperatures. Normal stress S_{33} vs. axial displacement λ.

Next, we study a plane strain model, that of the biaxial extension of a material block. The straining occurs along X_1 and X_2 axes while the block is assumed to be fixed along the X_3 direction. This problem is defined as

$$x_1 = (1+\lambda)X_1,\ x_2 = (1+\omega)X_2,\ x_3 = X_3,$$

where λ and ω are the straining parameters.

The *isothermal* stress – displacement curves for the limiting cases $\omega = 0$ (restrained tension) and $\omega = \lambda$ (biaxial tension) for three different material temperatures ($T_0 = 60^{o}C > A_f$, $A_f > T_0 = 45^{o}C > A_s$, $A_s > T_0 = 25^{o}C$) are shown in Figures 10, 11, 12 and 13. By referring to these figures for $T_0 > A_f$, we easily verify the ability of the model in predicting pseudoelastic phenomena in two dimensions.

The isothermal tests for $A_f > T_0 > A_s$ and $A_s > T_0$ are conducted in order to show the ability of the model in predicting the shape memory effect. In the first of them, upon loading the $A \rightarrow M$ transformation is activated, but since the temperature is less than the temperature required for the complete reverse transformation at zero stress, upon unloading the two phases coexist and permanent deformations appear. However, these deformations are recovered after increasing the temperature. In the second test the temperature initially is kept constant at a value less than the austenite start temperature at zero stress. As a result, at the end of the stress cycle the material is completely in the martensite phase and large permanent deformation appears. Nevertheless, like in the previous test, this deformation may be eliminated upon heating. For these new non – isothermal (i.e. heating) problems we assume thermal boundary conditions corresponding to convective heat exchange between the specimen and the surrounding medium on the free faces (with area A) of the specimen. In this case the normal heat flux is given by Newton's law of cooling (e.g., see Simo and Miehe, 1992) as: $\mathrm{H_u} = hA(\mathrm{T}_\infty - \mathrm{T}_0)$, with h being the constant convection coefficient, which is chosen as $h = 17.510^{-3}$ N/mm°K , and T_∞ is the surrounding medium temperature. By assuming that the size of the tested material is small, the contribution to the material heating due to heat conduction can be neglected, so that the temperature evolution equation (see Eq. (62)) can be written in the form

$$c\dot{\mathrm{T}} = \dot{Q}_{\mathbf{e}} + \dot{Q}_{\mathbf{Tr}} - \frac{1}{\rho_{\mathrm{ref}}}\mathrm{H_u}.$$

The results of these tests are illustrated in Figures 14 and 15, where the elongation along X_1 axis is plotted versus the surrounding medium temperature. The slight increase of the elongation of the SMA material due to the (elastic) thermal expansion occurring prior to the activation of the $\mathrm{M} \rightarrow \mathrm{A}$ transformation, for initial temperature $\mathrm{T_0}<\mathrm{A_s}$, is noteworthy (Figure 15).

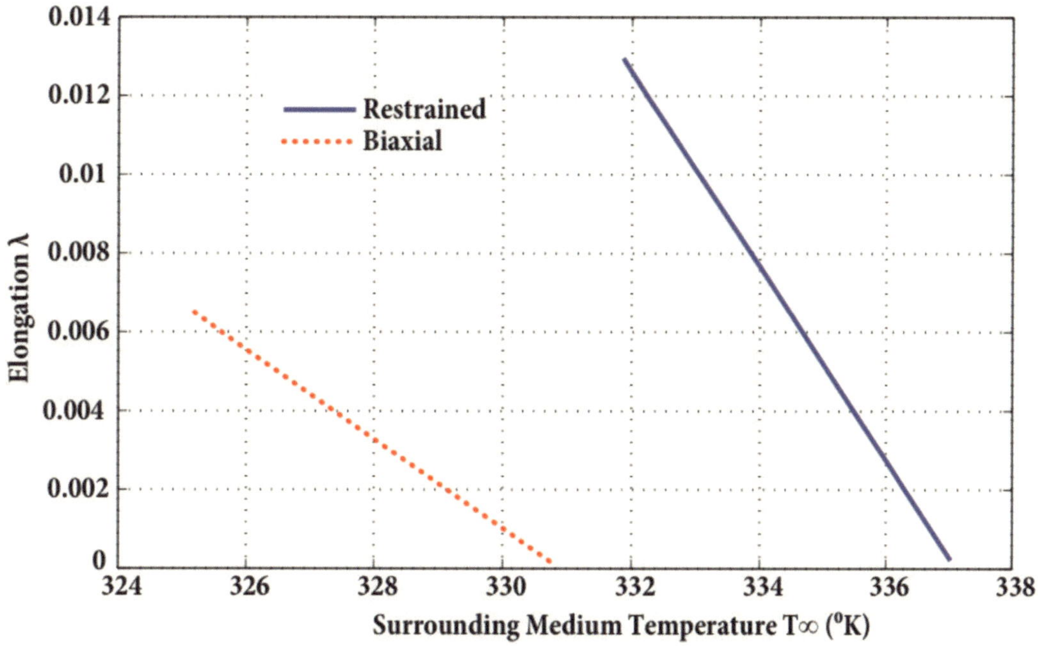

Figure 14. Plane strain. Shape memory effect $(\mathrm{A_s}<\mathrm{T_0}<\mathrm{A_f})$. Axial displacement λ vs. surrounding medium temperature T_∞.

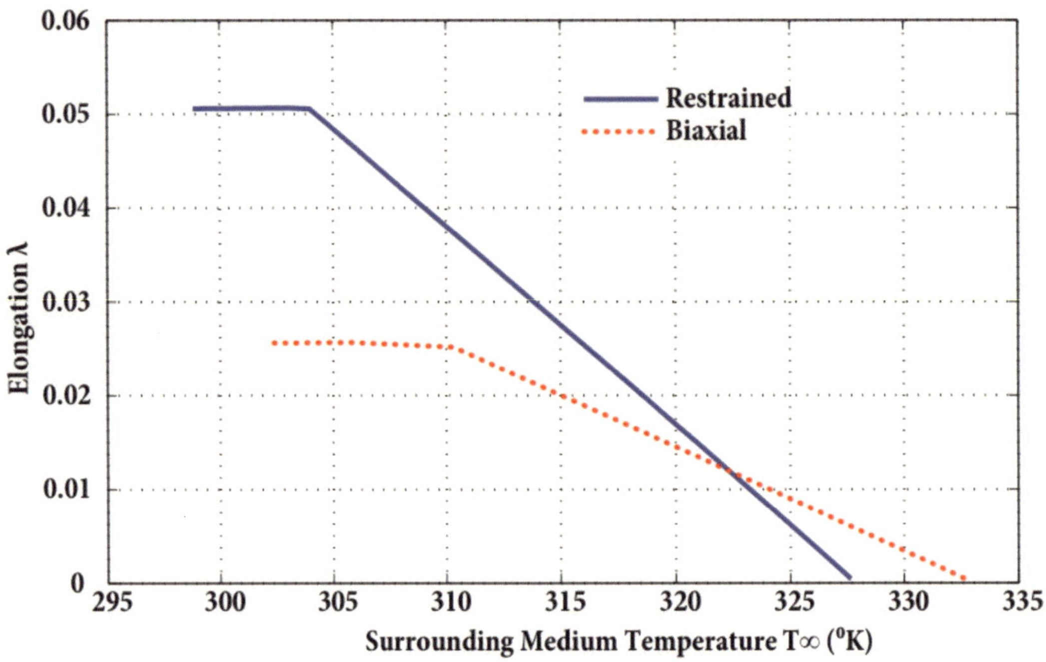

Figure 15. Plane strain. Shape memory effect $(T_0 < A_s)$. Axial displacement λ vs. surrounding medium temperature T_∞.

6. Concluding remarks

In this chapter we developed a geometrical framework for the establishment of constitutive models for materials undergoing phase transformations and in particular for shape memory alloys. The proposed framework has the following characteristics:

i. It is quite general for the derivation of the kinetic equations governing the transformation behavior and it can describe multiple and interacting loading mechanisms.
ii. It formulates general loading – unloading criteria, in both their material and spatial settings, that can be systematically employed for the numerical implementation of the derived constitutive models.
iii. It can describe rate effects.
iv. It can model non-isothermal conditions.
v. It can model transformation induced plasticity by considering it as an additional phase transformation.
vi. It employs the invariance of the spatial balance of energy equation under the superposition of arbitrary spatial diffeomorphisms – that is spatial transformations which can change the Euclidean metric – as a basic constitutive hypothesis, in place of the second law of thermodynamics.

As an application a specific three – dimensional thermomechanical constitutive model for SMA materials is derived. The model can simulate several patterns – under isothermal and *non-isothermal* conditions – of the extremely complex behavior of these materials such as:

a. The pseudoelastic behavior observed under monotonic loading.
b. The pseudoelastic behavior observed under several cyclic loadings.
c. The stiffness variations occurring during phase transformations.
d. The shape memory effect.

Additionally, the basic differences between the classical return mapping algorithms and the one used here for the case of not connected regions, have been outlined.

Author details

V. P. Panoskaltsis
Department of Civil Engineering, Demokritos University of Thrace, Xanthi, Greece

7. References

Abeyaratne, R., Bhattacharya, K., Knowles, J. K., 2001. Strain-energy functions with multiple local minima: modeling phase transformations using finite thermoelasticity. In: Y. Fu and R. W. Ogden (Eds.), Nonlinear Elasticity: Theory and applications, Cambridge Univ. Press, 433 – 490.

Abeyaratne, R., Knowles, J. K., 1993. A continuum model of a thermoelastic solid capable of undergoing phase transitions. J. Mech. Phys. Solids 41, 451 – 571.

Anand, L., Gurtin, M. E., 2003. Thermal effects in the superelasticity of crystalline shape – memory materials. J. Mech. Phys. Solids 51, 1015 – 1058.

Armero, F., Simo, J. C., 1993. A priori stability estimates and unconditionally stable product formula algorithms for nonlinear coupled thermoplasticity. Int. J. Plasticity 9, 749 – 782.

Ball, J. M., James, R. D., 1987. Fine phase mixtures and minimizers of energy. Arch. Rational Mech. Anal. 100, 13 – 52.

Bhattacharya, K., 2003. *Microstructure of Martensite. Why it forms and how it gives rise to the shape – memory effect?* Oxford Univ. Press, Oxford.

Boyd, J., Lagoudas, D., 1994. A thermodynamical model for shape memory materials. i. The monolithic shape memory alloy. Int. J. Plasticity 12, 805 – 842.

Christ, D., Reese, S., 2009. A finite element model for shape – memory alloys considering thermomechanical couplings at large strains. Int. J. Solids Struct. 46, 3694 – 3709.

Delville, R., Malard, B., Pilch, J., Sittner, P., Schryvers, D., 2011. Transmission electron microscopy investigation of dislocation slip during superelastic cycling of Ni – Ti wires. Int. J. Plasticity 27, 282 – 297.

Diani, J. M., Parks, D. M., 1998. Effects of strain state on the kinetics of strain – induced martensite in steels. J. Mech. Phys. Solids 46, 1613 – 1635.

Eisenberg, M.A., Phillips, A., 1971. A theory of plasticity with non-coincident yield and loading surfaces. Acta Mech. 11, 247 – 260.

Ericksen, J. L., 1961. Conservation laws for liquid crystals. Trans. Soc. Rheol. 5, 23 – 34.

Freed, Y., Banks – Sills, L., Aboudi, J., 2008. On the transformation toughening of a crack along an interface between a shape memory alloy and an isotropic medium. J. Mech. Phys. Solids 56, 3003 – 3020.

Freed, Y., Aboudi, J., 2008. Micromechanical investigation of plasticity – damage coupling of concrete reinforced by shape memory alloy fibers. Smart Matls Struct. 17 art. no. 015046.

Freed, Y., Banks – Sills, L., 2007. Crack growth resistance of shape memory alloys by means of a cohesive zone model. J. Mech. Phys. Solids 55, 2157 – 2180.

Green, A. E., Naghdi, P.M., 1965. A general theory of an elastic - plastic continuum. Arch. Rat. Mech. Anal. 18, 251 – 281.

Green, A. E., Rivlin, R. S., 1964. On Cauchy's equation of motion. Z. Angew. Math Phys 15, 290 – 293.

Hallberg, H., Hakansson, P. Ristinmaa, M., 2010. Thermo – mechanically coupled model of diffusionless phase transformation in austenitic steel. Int. J. Solids Struc. 47, 1580 – 1591.

Hallberg, H., Hakansson, P. Ristinmaa, M., 2007. A constitutive model for the formation of martensite in austenitic steels under large strain plasticity. Int. J. Plasticity 23, 1213 – 1239.

Holzapfel, G. A., 2000. *Nonlinear Solid Mechanics.* John Wiley and Sons, New York.

Ivshin, Y., Pence, T., 1994. A thermodynamical model for a one variant shape memory material. J. Intell. Mater. Systems Struct. 5, 455 – 473.

James, R. D., Hane, K. F., 2000. Martensitic transformations and shape – memory materials. Acta Mater. 48, 197 – 222.

Levitas, V. I., Ozsoy, I. B. 2009. Micromechanical modeling of stress – induced phase transformations. Part 1. Thermodynamics and kinetics of coupled interface propagation and reorientation. Int. J. Plasticity 25, 239 – 280.

Levitas, V. I., Preston, D. L., 2005. Thermomechanical lattice instability and phase field theory of martensitic phase transformations, twinning and dislocations at large strains. Physics Letters A 343, 32 – 39.

Likhachev, A. A., Koval, Y. N., 1992. On the differential equation describing the hysterisis behavior of shape memory alloys. Scrpt. Metal. Mater. 27, 223 – 227.

Lubliner, J., 1974. A simple theory of plasticity. Int. J. of Solids and Structures 10, 313 – 319.

Lubliner, J., 1984. A maximum – dissipation principle in generalized plasticity. Acta Mech. 52, 225 – 237.

Lubliner, J., 1987. Non-isothermal generalized plasticity. In: Thermomechanical Couplings in solids, eds. H. D. Bui and Q. S. Nyugen, 121 – 133.

Lubliner, J., 2008. *Plasticity Theory.* Dover Publications, New York.

Lubliner, J., Auricchio, F., 1996. Generalized plasticity and shape memory alloys. Int. J. of Solids and Structures 33, 991 – 1004.

Lubliner J., Panoskaltsis, V.P., 1992. The modified Kuhn model of linear viscoelasticity. Int. J. of Solids and Structures 29, 3099-3112.

Marsden, J. E., Hughes, T. J. R., 1994. *Mathematical Foundations of Elasticity,* Dover Publications, New York.

Müller, Ch., Bruhns O.T., 2006. A thermodynamic finite – strain model for pseudoelastic shape memory alloys. Int. J. Plasticity 22, 1658 – 1682.

Naghdi, P. M., 1990. A critical review of the state of finite plasticity. Z. Angew. Math. Phys. (Journal of Applied Mathematics and Physics, ZAMP), 41, 315 – 387.

Nemat – Nasser, S., Choi, J.Y., Wei – Guo, G., Isaacs, J. B., Taya, M., 2005a. High strain – rate, small strain response of a NiTi Shape – memory alloy. ASME J. Eng. Mater. Technol. 127, 83 – 89.

Nemat – Nasser, S., Choi, J.Y., Wei – Guo, G., Isaacs, J. B., 2005b. Very high strain – rate, response of a NiTi Shape – memory alloy. Mechanics of Materials 37, 287 – 298.

Panoskaltsis, V. P., Soldatos, D., Triantafyllou, S., P. 2011a. Generalized plasticity theory for phase transformations, Procedia Engineering 10 (2011), 3104 – 3108, (Science Direct). 11th International conference on the mechanical behavior of materials, M. Guagliano ed., ICM 11, 5-9 June 2011, Milano, Italy.

Panoskaltsis, V. P., Soldatos, D., Triantafyllou, S., P. 2011b. A new model for shape memory alloy materials under general states of deformation and temperature conditions. 7th GRACM International Congress on Computational Mechanics, A.G. Boudouvis and G.E. Stavroulakis eds., 30 June-2 July 2011, Athens, Greece.

Panoskaltsis, V. P., Soldatos, D., Triantafyllou, S., P. 2011c. The concept of physical metric in rate – independent generalized plasticity. Acta Mech. 221, 49 -64.

Panoskaltsis, V. P., Polymenakos L.C., Soldatos D., 2008a. Eulerian structure of generalized plasticity: Theoretical and computational aspects. ASCE, J. Eng. Mech., 134, 354 – 361.

Panoskaltsis, V.P., Polymenakos, L.C., Soldatos, D., 2008b. On large deformation generalized plasticity. J. of Mechanics of Materials and Structures, 3, 441 – 457.

Panoskaltsis, V. P., Bahuguna, S., Soldatos, D., 2004. On the thermomechanical modeling of shape memory alloys. Int. J. Non – Linear Mech. 39, 709 – 722.

Ramanathan, G., Panoskaltsis V.P., Mullen, R., Welsch, G., 2002. Experimental and Computational Methods for Shape Memory Alloys, Proceedings of the 15th ASCE Engineering Mechanics Conference, A. Smyth, Editor, June 2-5, 2002, Columbia University, N.Y., N.Y.

Raniecki, B., Lexcellent, C., 1998. Thermodynamics of isotropic pseudoelasticity in shape memory alloys. Eur. J. Mech. A/ Solids 17, 185 – 205.

Raniecki, B., Lexcellent, C., Tanaka, K. 1992. Thermodynamic models of pseudoelastic behaviour of shape memory alloys. Arch. Mech. 44, 261-284.

Rosakis, P., Rosakis, A. J., Ravichandran, G., Hodowany, J., 2000. A thermodynamic internal variable model for partition of plastic work into heat and stored energy in metals. J. Mech. Phys. Solids 48, 581 – 607.

Schutz, B., 1999. *Geometrical Methods of Mathematical Physics,* Cambridge University Press.

Simo, J. C., Miehe, C., 1992. Associative coupled thermoplasticity at finite strains: Formulation, numerical analysis and implementation. Comput. Methods Appl. Mech. Eng. 98, 41 – 104.

Smallman, R. E., Bishop, R. J., 2000. *Modern Physical Metallurgy and Materials Engineering*, 6th Edition, Butterworth – Heinemann, Stoneham, MA.

Stumpf, H., Hoppe, U. 1997. The application of tensor analysis on manifolds to nonlinear continuum mechanics – Invited survey article. Z. Agrew. Math. Mech. 77, 327 – 339.

Thamburaja, P., 2010. A finite-deformation-based theory for shape-memory alloys Int. J. Plasticity 26, 1195 – 1219.

Thamburaja, P., Anand, L., 2000. Polycrystalline shape – memory materials: effect of crystallographic texture. J. Mech. Phys. Solids 49, 709 – 737.

Yavari, A., Marsden, J. E., Ortiz, M., 2006. On spatial and material covariant balance laws in elasticity. J. Math. Phys. 47, 1 – 53.

Videnic, T., Kosel, F., Sajn, V., Brojan, M., 2008. Biaxial constrained recovery in shape memory alloy rings. J. Intell. Mater. Systems Struct. 19, 861 – 874.

Wayman, C. M., 1964. Introduction to the crystallography of martensitic transformation, Macmillan, New York.

Chapter 7

Determination of Elastic and Dissipative Energy Contributions to Martensitic Phase Transformation in Shape Memory Alloys

Dezso L. Beke, Lajos Daróczi and Tarek Y. Elrasasi

Additional information is available at the end of the chapter

http://dx.doi.org/10.5772/51511

1. Introduction

There is a long standing debate in the literature on shape memory alloys that while the contribution of the dissipative energy, D, to the austenite/martensite, A/M, (or reverse) transformation can be directly obtained from the experimental data (hysteresis loop, Differential Scanning Calorimeter, DSC, curves), the contributions from the elastic, E, and the chemical free energy, ΔG_c, can not be separated. The temperature dependence of $\Delta G_c=\Delta H-T\Delta S$ is described by $\Delta G_c=(T-T_o)\Delta S$, where $T_o=\Delta S/\Delta H$ is the equilibrium transformation temperature (at which the chemical free energies of the two phases equals, i.e. $\Delta G_c=0$) as well as ΔH and ΔS are the chemical enthalpy and entropy change of the phase transformation (they are negative for A to M transformation), respectively. The experimentally determined quantities (DSC or hysteresis curves) usually contain a combination of the chemical, elastic and dissipative terms in such a way [1] that always the sum of E and ΔG_c can be calculated and thus for their separation one would need the knowledge of ΔS and T_o (see also below in details). While the direct determination of ΔS is possible (e.g. from the measured DSC curves) the determination of T_o is rather difficult: it has been shown and experimentally illustrated that the simple expression proposed by Tong and Waynman [2]: $T_o=(M_s+A_f)/2$ (where M_s and A_f are the martensite start and austenite finish temperatures) can not be valid in general. Indeed Salzbrenner and Cohen [1] have been nicely illustrated that T_o can be calculated from the above relation only in those cases when the elastic energy contributions to M_s and A_f can be neglected.

In this review we will summarize our model [3-8] for the thermal hysteresis loops, $\xi=\xi(T)$ (at constant other driving fields such as uniaxial stress, σ, magnetic field, B, or pressure, p) in terms of T_o, and the derivatives $\partial\Delta S/\partial\xi=\Delta s$, $\partial E/\partial\xi=e$ and $\partial D/\partial\xi=d$, where ξ is the martensite transformed (volume) fraction. (In the following quantities given by small letters denote the quantity belonging to unit volume fraction.) Similar relations for example for the strain, $\varepsilon(\sim\xi)$, versus σ (or e.g. magnetization, $\Delta m(\sim\xi)$, versus B) hysteresis loops can be derived, where instead of Δs, ε^{tr} (or Δm^{tr}) appears. Here ε^{tr} is the transformation strain (and Δm^{tr} is the change of magnetization) of phase transformation. The results obtained from the application of this model to our experimental data measured in single and polycrystalline CuAlNi alloys will be summarized too.

2. Description of the model

Our model is in fact a local equilibrium formalism and based on the thermoelastic balance (see e.g. [9,10] and [11]) offering a simple form of the elastic and dissipative energy contributions to the start and finish parameters [3-8]. The *total* change of the Gibbs free energy versus the transformed martensite fraction (if the hydrostatic pressure and the magnetic field are zero), for the A/M transformation (denoted by $^{\downarrow}$), can be written in the form [3,8]:

$$\partial(\Delta G^{\downarrow}) / \partial\zeta = \partial(\Delta G_c^{\downarrow} + E + D) / \partial\zeta = \Delta g_c^{\downarrow} + e^{\downarrow}(\zeta) + d^{\downarrow}(\zeta) = 0. \qquad (1)$$

where

$$\Delta g^{\downarrow} = \Delta u^{\downarrow} - T\Delta s^{\downarrow} - \sigma V \varepsilon^{tr\downarrow}, \qquad (2)$$

with $\Delta s^{\downarrow} = s_M - s_A (= -\Delta s^{\uparrow} (<0))$, and V is the molar volume. Similar expression holds for the M/A transformation (with upper index $^{\uparrow}$):

$$\Delta g^{\uparrow} = \Delta u^{\uparrow} - T\Delta s^{\uparrow} - \sigma V \varepsilon^{tr\uparrow}. \qquad (3)$$

The elastic energy accumulates as well as releases during the processes down and up just because the formation of different variants of the martensite phase usually is accompanied by a development of an elastic energy field (due to the transformation strain). It is usually supposed that $E_{el}^{\downarrow} = - E_{el}^{\uparrow} > 0$. The dissipative energy is always positive in both directions.

In thermoelastic transformations the elastic term plays a determining role. For example at a given under-cooling, when the elastic term will be equal to the chemical one, for the further growth of the martensite an additional under-cooling is required. Thus if the sample is further cooled the M phase will grow further, while if the sample is heated it will become smaller. Indeed in *thermoelastic* materials it was observed that once a particle formed and reached a certain size its growth was stopped and increased or decreased as the temperature was decreased or raised. This is *the thermoelastic behaviour* (the thermal and elastic terms are balanced).

In principle, one more additional term, proportional to the entropy production, should be considered, but it can be supposed [12] that for thermoelastic transformations all the energy losses are mechanical works, which are dissipated without entropy production, i.e. the dissipation is mainly energy relaxation in the form of elastic waves. Indeed acoustic waves were detected as acoustic emissions during the transformation. Thus in the following the term proportional to the entropy production will be neglected. Furthermore, usually there is one more additional term in ΔG: this is the nucleation energy related to the formation of the interfaces between the nucleus of the new phase and the parent material. However, since this term, similarly to the dissipative energy, is positive in both directions and thus it is difficult to separate from *D*, it can be considered to be included in the dissipative term.

According to the definitions of the equilibrium transformation temperature and stress

$$T_o(0) = \Delta u^{\downarrow} / \Delta s^{\downarrow} = \Delta u^{\uparrow} / \Delta s^{\uparrow}, \tag{4}$$

$$\sigma_o(0) = \Delta u^{\downarrow} / V\varepsilon^{tr^{\downarrow}}(T=0) = \Delta u^{\uparrow} / V e^{tr^{\uparrow}}(T=0), \tag{5}$$

respectively.

$\Delta g_c{}^{\downarrow}$, if the external hydrostatic pressure, *p*, and magnetic field, *B*, are also not zero, can have the general form as:

$$\Delta g_c{}^{\downarrow} = \Delta u^{\downarrow} - T\Delta s^{\downarrow} - sVe^{tr\downarrow} + p\Delta v^{\downarrow} - B\Delta m^{tr\downarrow}, \tag{6}$$

where $\Delta v^{\downarrow}$ is the volume change of the phase transformation.

It is plausible to assume that $\Delta u^{\downarrow}$, $\Delta s^{\downarrow}$ and $\Delta v^{\downarrow}$ are independent of ξ, i.e. ΔU, ΔV and ΔS linearly depends on the transformed fraction. On the other hand the terms containing ε^{tr} and Δm^{tr} in general have tensor character and, as a consequence, even if one considers uniaxial loading condition, leading to scalar terms in (2), the field dependence of these quantities is related to the change of the variant/domain distribution in the martensite phase with increasing field parameters. Thus at zero σ (or *B*) values thermally oriented multi-variant martensite structure (or multi-variant magnetic domain structure) forms in thermal hysteresis, while at high enough values of σ (or *B*) a well oriented array i.e. a single variant (or single domain structure) develops. For the description of this, the volume fraction of the stress induced (single) variant martensite structure, η, can be introduced [8]: $\eta = V_{M\sigma}/V_M$, ($V_M = V_{MT} + V_{M\sigma}$ and $\xi = V_M/V$, with $V = V_M + V_A$, where V_M and V_A are the volume of the martensite and austenite phases, respectively and V_{MT} and $V_{M\sigma}$ denotes the volume of the thermally as well as the stress induced martensite variants, respectively). The concept of introduction of this parameter was based e.g. on works of [11, 13-15]. Accordingly, e.g. ε^{tr} is maximal for $\eta=1$, and $\varepsilon^{tr\uparrow}(\eta=1)=\varepsilon^{tr\uparrow}_{max}$ in single crystalline sample, while it can be close to zero for $\eta=0$. In the following only the case of simultaneous action of temperature and uniaxial stress will be treated (extension to more general cases is very plausible).

Thus, in (2) and (3) ε^{tr} depends on η. Since η depends on T and σ, $\varepsilon^{tr\uparrow}$ can also depend on T or σ at fixed σ or T, respectively.

From (1) with (2):

$$\Delta u^{\downarrow} - T\Delta s^{\downarrow} - \sigma V\varepsilon^{tr^{\downarrow}} + e^{\downarrow} + d^{\downarrow} = 0. \tag{7}$$

For fixed σ parameter(s) from (6) and using also (4) for *Δu* (for both up and down processes);

$$\begin{aligned} T^{\downarrow}(\zeta) &= T_o(0) - \sigma V\varepsilon^{tr\downarrow}(\sigma)/\Delta s^{\downarrow} + (e^{\downarrow}(\zeta) + d^{\downarrow}(\zeta))/\Delta s^{\downarrow} = T_o(\sigma) + (e^{\downarrow}(\zeta) + d^{\downarrow}(\zeta))/\Delta s^{\downarrow} && \text{(a)} \\ T^{\uparrow}(\zeta) &= T_o(0) - \sigma V\varepsilon^{tr\uparrow}(\sigma)/\Delta s^{\uparrow} + (e^{\uparrow}(\zeta) + d^{\uparrow}(\zeta))/\Delta s^{\uparrow} = T_o(\sigma) + (e^{\uparrow}(\zeta) + d^{\uparrow}(\zeta))/\Delta s^{\uparrow}. && \text{(b)} \end{aligned} \tag{8}$$

Here $T_o(\sigma)$ is the same for both directions, since $\varepsilon^{tr\downarrow}/\Delta s^{\downarrow} = \varepsilon^{tr\uparrow}/\Delta s^{\uparrow}$ ($\varepsilon^{tr\downarrow} = -\varepsilon^{tr\uparrow}$, as well as $\Delta s^{\downarrow} = -\Delta s^{\uparrow}$ and in our case $\varepsilon^{tr\downarrow} > 0$ and $\Delta s^{\downarrow} < 0$).

The inverses of (8a) and (8b), i.e. the $\xi(T^{\downarrow})$ and $\xi(T^{\uparrow})$ functions, are the down and up braches of the thermal hysteresis loops at fixed σ. Furthermore, the temperature at which (8a) is equal to zero at ξ=0 as well as ξ=1 is the martensite start (M_s) and finish (M_f) temperature, respectively. Similar definitions hold for the austenite start and finish temperatures, A_s and A_f, respectively (see eq. (8b)). Figure 1 illustrates the shape of the hysteresis curves for the following schematic cases: a) both d(ξ) and e(ξ) are zero; b) e(ξ)=0 and d(ξ)≠0, but d(ξ) is constant; c) d(ξ) is constant and e(ξ) linearly depends on ξ. It can be seen that in a) the transformation takes place at T_o, in both directions, in b) there is already a hysteresis, but the $\xi(T^{\downarrow})$ and $\xi(T^{\uparrow})$ branches are vertical. For the case of c) the hysteresis curve is tilted, reflecting the ξ dependence of e.

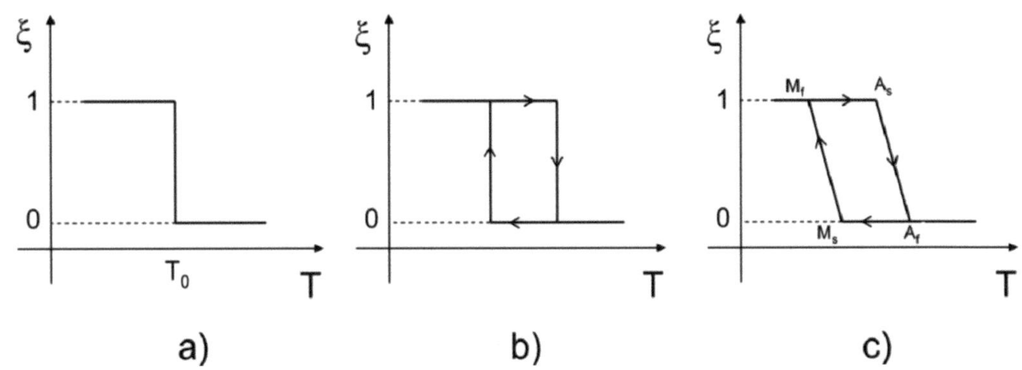

Figure 1. Thermal hysteresis loops schematically, see also the text.

Similarly as above but for fixed temperature(s), and now inserting *Δu* from (5),

$$\begin{aligned} \sigma^{\downarrow}(\zeta) &= \sigma_o(0)V\varepsilon^{tr\downarrow}(\sigma_o)/V\varepsilon^{tr\downarrow}(T) - T\Delta s^{\downarrow}/V\varepsilon^{tr\downarrow}(T) + (e^{\downarrow}(\zeta) + d^{\downarrow}(\zeta))/V\varepsilon^{tr\downarrow}(T) = && \text{(a)} \\ &\sigma_o(T) + (e^{\downarrow}(\zeta) + d^{\downarrow}(\zeta))/V\varepsilon^{tr\downarrow}(T) \\ \sigma^{\uparrow}(\zeta) &= \sigma_o(0)V\varepsilon^{tr\uparrow}(\sigma_o)/V\varepsilon^{tr\uparrow}(T) - T\Delta s^{\uparrow}/V\varepsilon^{tr\uparrow}(T) + (e^{\uparrow}(\zeta) + d^{\uparrow}(\zeta))/V\varepsilon^{tr\uparrow}(T) = && \text{(b)} \\ &\sigma_o(T) + (e^{\uparrow}(\zeta) + d^{\uparrow}\zeta))/V\varepsilon^{tr\uparrow}(T). \end{aligned} \tag{9}$$

Here again $\sigma_o(T)$ is the same taking also into account that $\varepsilon^{tr\uparrow}(\sigma_o)/\varepsilon^{tr\uparrow}(T) = \varepsilon^{tr\downarrow}(\sigma_o)/V\varepsilon^{tr\downarrow}(T)$, because the magnitude of the transformation strain is the same for the up and down branches of the loop at fixed *T*.

It can be seen from relations (8) and (9) that, in the case of the simultaneous action of temperature and uniaxial stress only, the stress dependence of the equilibrium transformation temperature, as well as the temperature dependence of the equilibrium transformation stress, introducing the notation $\Delta s=\Delta s^{\downarrow}=-\Delta s^{\uparrow}(<0)$, can be given as

$$T_o(\sigma) = T_o(0) + \sigma V \varepsilon^{tr}(\sigma) / (-\Delta s), \tag{10}$$

and

$$\sigma_o(T) = \sigma_o(0)\varepsilon^{tr}(\sigma_o) / \varepsilon^{tr}(T) - T\Delta s / V\varepsilon^{tr}(T) = [T_o(0) - T]\Delta s / V\varepsilon^{tr}(T), \tag{11}$$

respectivelty. It can be seen that (10) and (11) are the well known Clausius-Clapeyron relations and they are linear only if $\varepsilon^{tr}(\sigma)$ as well as $\varepsilon^{tr}(T)$ are constant. It will be illustrated below that in most of the cases this is not fulfilled.

Now taking the assumptions usual in the treatment of thermoelastic transformations, i.e. assuming that the magnitudes of elastic and dissipative energies and their derivatives are the same in both A/M and M/A transformations; $e(\xi)=e^{\downarrow}(\xi)=-e^{\uparrow}(\xi)$ as well as $d(\xi)=d^{\downarrow}(\xi)=d^{\uparrow}(\xi)$, (8a) and (8b) can be rewritten as

$$\begin{aligned} T^{\downarrow}(\zeta) &= T_o(\sigma) - [e(\zeta) + d(\zeta)] / (-\Delta s) \quad \text{(a)} \\ T^{\uparrow}(\zeta) &= T_o(\sigma) + [d(\zeta) - e(\zeta)] / (-\Delta s). \quad \text{(b)} \end{aligned} \tag{12}$$

Thus

$$T^{\uparrow}(\zeta) - T^{\downarrow}(\zeta) = 2d(\zeta) / (-\Delta s) \tag{13}$$

and

$$T^{\uparrow}(\zeta) + T^{\downarrow}(\zeta) = 2T_o(\sigma) - 2e(\zeta) / (-\Delta s) \tag{14}$$

Furthermore, for the branches of the $\xi(\sigma)$ hysteresis loops

$$\begin{aligned} \sigma^{\downarrow}(\zeta) &= \sigma_o(T) + [e(\zeta) + d(\zeta)] / V\varepsilon^{tr}(T) \quad \text{(a)} \\ &\text{and} \\ \sigma^{\uparrow}(\zeta) &= \sigma_o(T) - [d(\zeta) - e(\zeta)] / V\varepsilon^{tr}(T). \quad \text{(b)} \end{aligned} \tag{15}$$

Thus

$$\sigma^{\downarrow}(\zeta) - \sigma^{\uparrow}(\zeta) = 2d(\zeta) / V\varepsilon^{tr}(T) \tag{16}$$

and

$$\sigma^{\downarrow}(\zeta) + \sigma^{\uparrow}(\zeta) = 2\sigma_o(T) + 2e(\zeta) / V\varepsilon^{tr}(T). \tag{17}$$

It can be seen from eqs. (12)-(17) that, as it was mentioned in the introduction, while the dissipative term can be directly calculated from the hysteretic loops, the elastic and chemical terms appear in sums on the right hand sides of (14) and (17). It is worth noting that the integrals of (13) as well as (16), as it is expected, are nothing else that the area of the thermal and mechanical hystersis loops, respectively.

Nevertheless, relations (10)-(17) allow the determination of the dissipative and elastic energy contributions as the function of ξ at different fixed values of σ as well as T from the thermal and stress induced hysteresis loops, respectively. Thus even the σ and T dependence of E and D can be calculated by integrating the e(ξ) and d(ξ) functions between ξ=0 and ξ=1. It should be noted that the elastic energy contribution can be determined only exclusive the term $T_o(0)$ if its value is not known. The values of Δs can be obtained from DSC measurements (see also below) and the $\varepsilon^{tr}(T)$ and $\varepsilon^{tr}(\sigma)$ values can be read out from the $\varepsilon(\sigma)$ and $\varepsilon(T)$ hysteresis loops, respectively. Thus e.g. the stress or temperature dependence of the elastic energy contribution can be determined, since $T_o(0)$ appears only in the intercept of the e(σ) and e(T) or E(σ) and E(T) functions.

From relations (12) and (15) expressions for the start and finish temperatures as well as stresses can be simply obtained at ξ=0 and at ξ=1:

$$\begin{aligned} M_s(\sigma) &= T_o(\sigma) - \left[d_o + e_o\right]/[-\Delta s] \\ M_f(\sigma) &= T_o(\sigma) - \left[d_1 + e_1\right]/[-\Delta s] \\ A_f(\sigma) &= T_o(\sigma) + \left[d_o - e_o\right]/[-\Delta s] \\ A_s(\sigma) &= T_o(\sigma) + \left[d_1 - e_1\right]/[-\Delta s] \end{aligned} \tag{18}$$

and

$$\begin{aligned} \sigma_{Ms}(T) &= \sigma_o(T) - \left[d_o + e_o\right]/[-V\varepsilon^{tr}] \\ \sigma_{Mf}(T) &= \sigma_o(T) - \left[d_1 + e_1\right]/[-V\varepsilon^{tr}] \\ \sigma_{Af}(T) &= \sigma_o(T) + \left[d_o - e_o\right]/[-V\varepsilon^{tr}] \\ \sigma_{As}(T) &= \sigma_o(T) + \left[d_1 - e_1\right]/[-V\varepsilon^{tr}]. \end{aligned} \tag{19}$$

Here in principle the d_o, d_1, e_o and e_1 can also be σ or T-dependent: in this case e.g. the stress dependence of the start and finish temperatures can be different from the stress dependence of T_o. It can be seen from relations (18) that the simple expression proposed by Tong and Waynman [2] for T_o as $T_o=(M_s+A_f)/2$ can be valid only if e_o is zero. Indeed Salzbrenner and Cohen [1] illustrated that T_o can be calculated only in those cases when the elastic energy contributions to M_s and A_f can be neglected. In their paper the phase transformation was driven by a slowly moving temperature gradient in a single crystalline sample, which resulted in slow motion of only one interface across the specimen (single-interface transformation). This way the elastic energy could easily relax by the formation of the surface relief at the moving (single) phase-boundary. In general experiments for the determination of hysteresis loops, where typically many interfaces move simultaneously and the elastic fields of the different nuclei overlap, this separation is not possible. However, as we have shown in [5], and as it will be illustrated below, in single crystalline samples under relatively slow heating (cooling) rates, from the analysis of the different shapes of the hysteresis curves at low and high stress levels T_o can be determined experimentally as the function of σ.

Finally it is worth summarizing what kind of information can be obtained from the analysis of results obtained by differential scanning calorimeter, DSC. The heats of transformation measurable during both transitions are given by

$$Q^{\downarrow} = \int[\Delta u_c^{\downarrow} + e(\zeta) + d(\zeta)]d\zeta \tag{20}$$

and

$$Q^{\uparrow} = \int[-\Delta u_c^{\downarrow} - e(\zeta) + d(\zeta)]d\zeta. \tag{21}$$

It is worth noting that the heat measured is negative if the system evolves it: thus e.g. the first term in (20) has a correct sign, because it is negative ($\Delta u_c^{\downarrow}<0$). Similarly the dissipative and elastic tems should be positive for cooling (the system absorbs these energies): indeed $e(\xi)$, $d(\xi)>0$, while for heating $e(\xi)^{\downarrow}=-e(\xi)^{\uparrow}=e(\xi)$ and $d(\xi)^{\downarrow}=d(\xi)^{\uparrow}=d(\xi)$.

Now, using the notations $\int\Delta u_c^{\downarrow}=\Delta U_c(<0)$, $\int d(\xi)d\xi= D(>0)$, $\int e(\xi)d\xi= E\,(>0)$

$$Q^{\downarrow} = \Delta U_c + E + D \tag{22}$$

and

$$Q^{\uparrow} = -\Delta U_c - E + D. \tag{23}$$

(In obtaining (22) and (23) it was used that $\Delta u_c^{\downarrow}$ is independent of ξ.) Consequently

$$Q^{\uparrow} - Q^{\downarrow} = -2\Delta U_c + 2E \tag{24}$$

and

$$Q^{\uparrow} + Q^{\downarrow} = 2D. \tag{25}$$

It is important to keep in mind that the last equations are strictly valid only if after a cycle the system has come back to the same thermodynamic state, i.e. *it does not evolve from cycle to cycle*. Furthermore, it can be shown [12] that these are only valid if the heat capacities of the two phases are equal to each other: $c_A \cong c_M$, which was the case in our samples (see also below).The DSC curves also offer the determination of Δs. Indeed from the Q versus T curves, taking the integrals of the $1/T$ curves by $Q^{\downarrow}$ or $Q^{\uparrow}$ between M_s and M_f, as well as between A_s and A_f, respectively, one gets the $\Delta s^{\downarrow}$ as well as $\Delta s^{\uparrow}$ values. If, again, the $c_A \cong c_M$ condition fulfils, then $\Delta s^{\downarrow} \cong \Delta s^{\uparrow}$ [12].

Finally, it is possible, by using the DSC curve [I6], to obtain the volume fraction of the martensite, ξ, as a function of temperature (both for cooling and heating) as the ratio of the partial and full area of the corresponding curve ($A_{Ms\text{-}T}$ and $A_{Ms\text{-}Mf}$, respectively: see also Figure 2):

$$\varsigma\left(T^{\downarrow}\right)=\frac{A_{Ms-T}}{A_{Ms-Mf}}=\int_{M_s}^{T}\frac{dQ^{\downarrow}}{T}\Big/\int_{M_s}^{M_f}\frac{dQ^{\downarrow}}{T}. \tag{26}$$

Similar relation holds for the $\xi\ (T^{\uparrow})$ curve (obviously in this case the above integrals run between A_s and T as well as A_s and A_f, respectively). The denominator is just the entropy of this transformation.

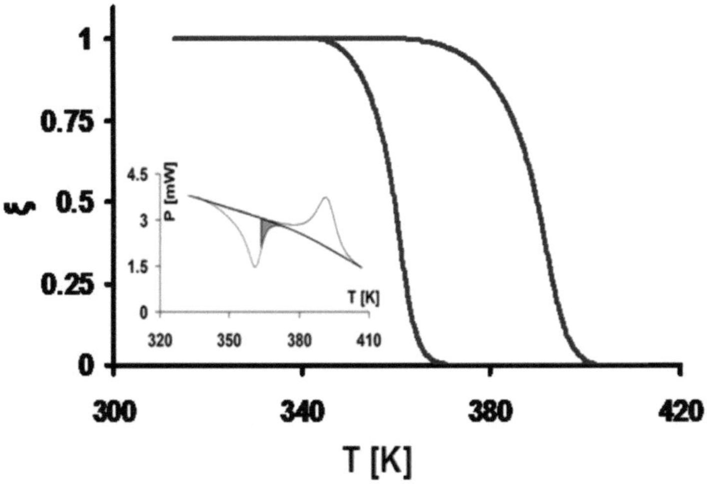

Figure 2. DSC curve measured at zero stress (a) and the $\xi\,(T)$ hysteresis curve (b): the dashed area (on the cooling down curve in a)) can be transformed to the nominator of equation (26); see also the text and [17].

3. Analysis of experimental data

3.1. Stress and temperature dependence of the transformation strain

As it was mentioned in the previous section it is generally expected that the transformation strain depends on the martensite variant structure developed. Since for thermal hysteresis loops this structure can vary from the randomly oriented structure to a well oriented single variant structure with increasing uniaxial stress, ε^{tr} should increase with σ. Figure 3b shows this function for single crystalline CuAl(11.5wt%)Ni(5.0wt%) alloy (the applied stress was parallel to the [110] direction), as determined from the saturation values of the ε~T loops shown in Figure 3a [18]. In this alloy (i.e. at this composition) the β (austenite) to β′(18R, martensite) transformation takes place. Figure 4a shows the temperature dependence of ε^{tr}, in the same alloy, as determined form the σ~ε loops shown in Figure 4b [18]. It can be seen that ε^{tr} increases with increasing temperature and saturates at the same maximal value which is obtained from the ε^{tr} versus σ plot and is approximately equal to the maximal possible transformation strain, ε^{tr}_{max}, corresponding to the estimated value for the case when a single crystal fully transforms to the most preferably oriented martensite [19].

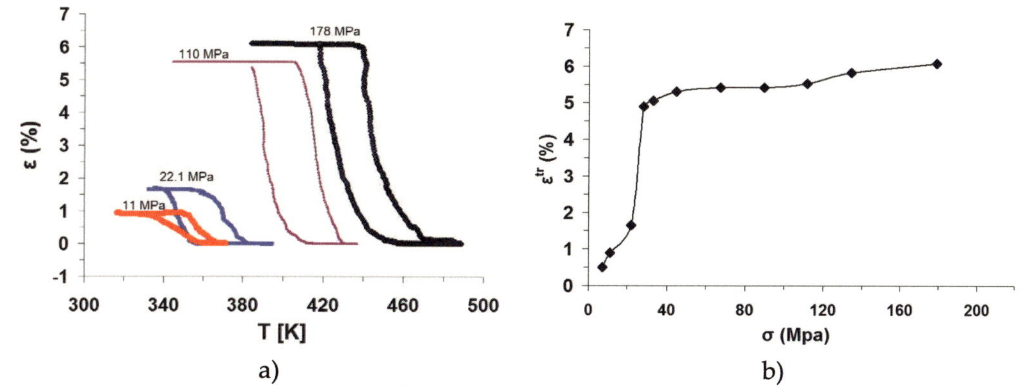

Figure 3. a) Thermal hysteresis loops (ε versus T curves) at four different uniaxial stress levels, b) Transformation strain as function of stress (ε^{tr} is the maximal of value of ε in a) for β/β' transformation in single crystalline CuAl(11.5wt%)Ni(5.0wt%) alloy [18].

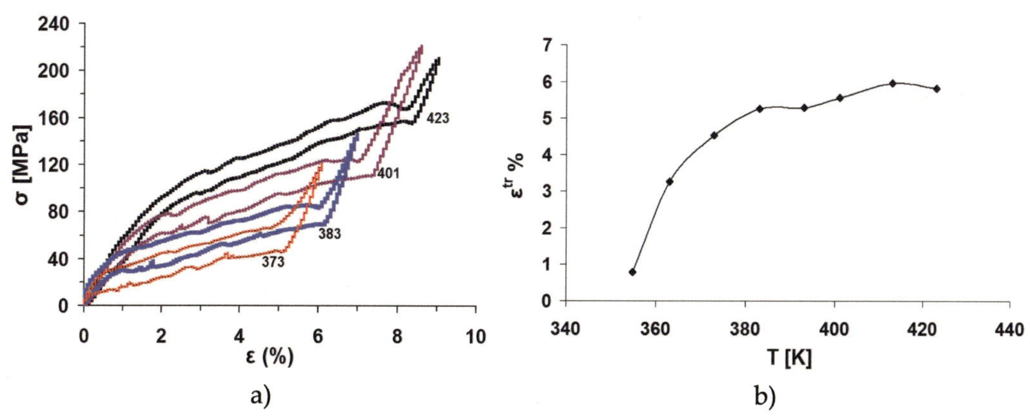

Figure 4. a) σ versus ε curves at four different temperatures, b) transformation strain as the function of the temperature (read out from curves like shown in a) in single crystalline CuAl(11.5wt%)Ni(5.0wt%) alloy for β/β' transformation [18].

Figure 5 shows the stress dependence of the transformation strain for the β to orthorhombic γ(2H) phase transformation obtained in CuAl(17.9w%)Ni(2.6 w%) single crystalline alloy in [5]. It can be seen that it has S shape dependence with a saturation value of 0.075. It is interesting that in this case ε^{tr} has a finite (remanent) value even at σ=0.

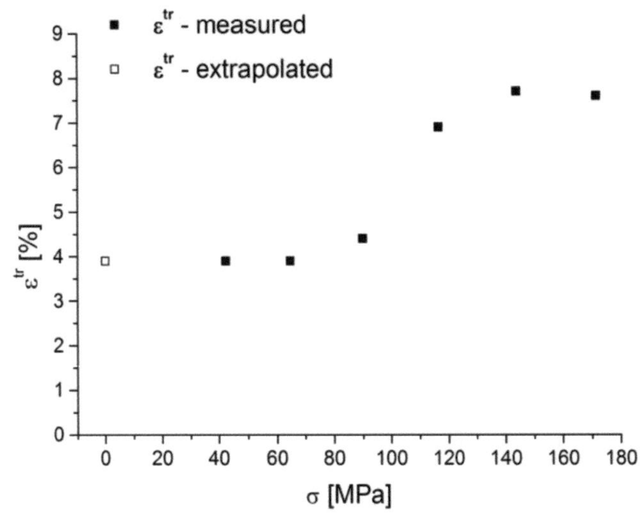

Figure 5. Stress dependence of ε^{tr} in CuAl(17.9w%)Ni(2.6 w%) single crystalline alloy for β/γ transformation [5].

As it was analyzed in detail in [19], from the above curves the η dependence of ε^{tr} can be constructed using the relation introduced in [8]:

$$\varepsilon^{tr} = \varepsilon_T + (\varepsilon_s - \varepsilon_T)\eta, \tag{27}$$

where ε_T and ε_σ are the transformation strains when fully thermally induced multi variant structure forms (η=0), as well as when the martensite consists of a fully ordered array of stress preferred variants (single variant state, η=1), respectively. Thus ε^{tr} can be very small or even close to zero for the formation of the thermally induced (randomly oriented) martensite variants (usually there is a small resultant (remanent) strain in single crystalline samples). On the other hand during the formation of stress induced martensite a single variant structure can form (η=1) i.e. $\varepsilon^{tr}=\varepsilon^{tr}_{max}=\varepsilon_\sigma$. On the basis of the experimental curves shown in Figure 3b, 4b and 5 as was well as of relation (27) it can be concluded that a fully ordered single variant martensite structure develops above 140 MPa for the β/γ phase transformation, while for the β/β' transformation η is about 80% already for 28 MPa and then gradually increases up to 100% in the 40 - 178 MPa interval. As regards the temperature dependence of η, it can be seen from Figure 4a that (according to eq. (27) $\varepsilon_T \cong 0$ and ε_σ=0.061) η monotonously increases from about 10% up to 100% between 350 and 430 K.

Thus it can be concluded that the transformation strain depends both on the uniaxial stress and on the temperature and this dependence is related to the change of the martensite variant distribution with increasing field parameters. Then it is plausible to expect that the Clausius-Clapeyron type relations (see eqs. (10) and (11)) should also be non linear. Furthermore, the elastic and dissipative energy contributions should also be influenced by the martensite variant distribution. These points will be discussed in detail in the following sections.

3.2. Stress dependence of the equilibrium transformation temperature

In reference [5] we have investigated the thermal hysteresis loops in CuAl(17.9w%)Ni(2.6 w%) single crystalline alloy at different uniaxial stresses (applied along the $[110]_A$ axis). Very interesting shapes were obtained (see Figure 6): the ε~T loops had vertical parts, indicating that at these parts there were no elastic energy contributions (see also Figure 1c), allowing the determination of T_o from the start and finish temperatures (see also eqs. (18)) either using the Tong-Waymann formula, $T_o=(M_s+A_f)/2$, (see the curve at 171.5 MPa in Figure 6) or $T_o=(M_s+A_s)/2$ (see e.g. the curve at 42.4 MPa in Figure 6). Thus it was possible (using also relation (10) and the value of the entropy, $\Delta s=-1.169\cdot10^5 JKm^{-3}$, determined also in [5] and the stress dependence of ε^{tr} shown in Figure 5) to determine the stress dependence of T_o as it is shown in Figure 7. It can be seen that this is indeed not a linear function.

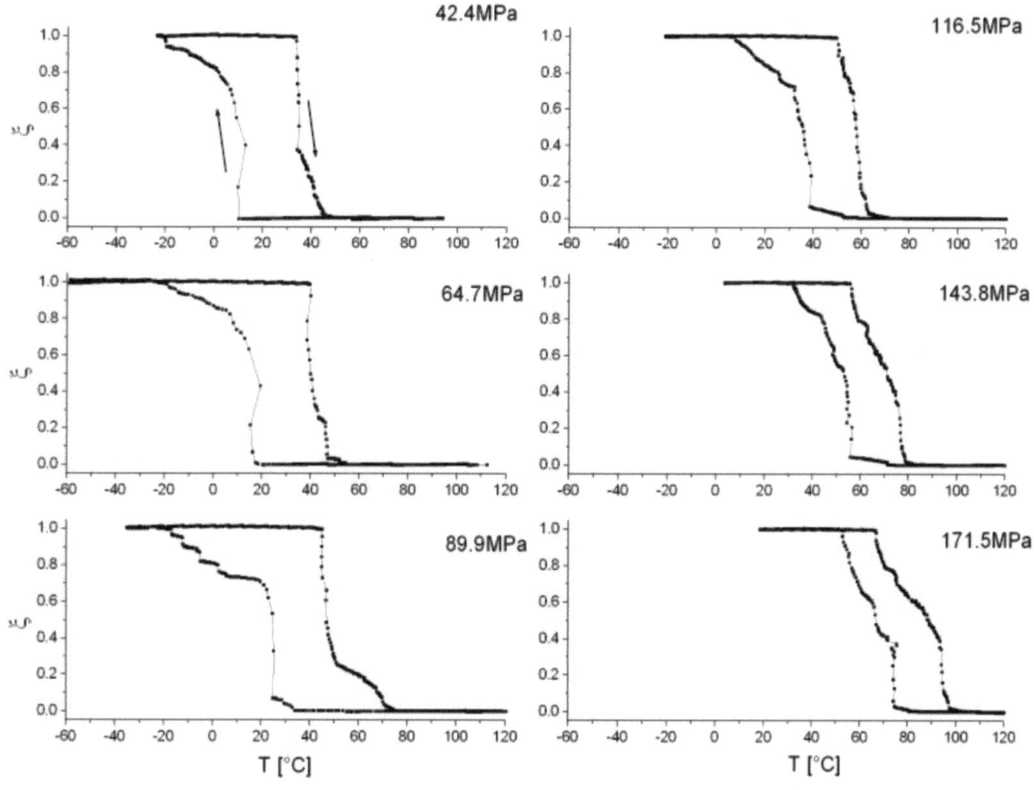

Figure 6. Thermal hystersis loops at different stress levels in CuAl(17.9w%)Ni(2.6 w%) single crystalline alloy [5].

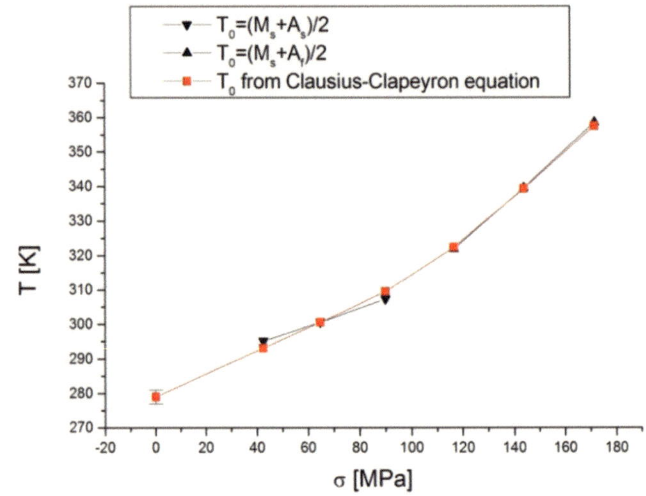

Figure 7. Stress dependence of T_o in in CuAl(17.9w%)Ni(2.6w%) single crystalline alloy [5].

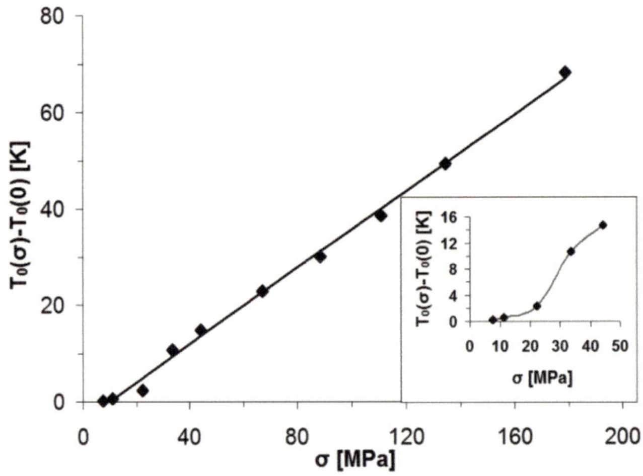

Figure 8. Stress dependence of T_o in single crystalline CuAl(11.5wt%)Ni(5.0wt%) alloy [18].

Figure 8 shows the stress dependence of T_o for the β/β transformation. In this case the determination of absolute values of T_o was not possible, but the $T_o(\sigma) - T_o(0)$ difference could be calculated using the measured Δs value and the $\varepsilon^{tr}(\sigma)$ curve (Figure 3b). It can be seen that this function can be approximated by a straight line in the entire stress interval. But, as it is illustrated in the insert of this figure, if we plot this function only at low stresses then an S-shape dependence appears. Thus it can be concluded, in contrast to the very frequently used approximation in the literature [9,20,21] about linear Calusius-Clapeyron relations, that the σ dependent ε^{tr} usually leads to nonlinear dependence [18,19]. Of course in special cases, i.e. when the dependence of ε^{tr} in the investigated range is week, or the stress interval wide enough to have many points belonging to the saturation value of ε^{tr} a linear fit with an effective slope can be made, like in Figure 8. The slope of this straight line is 0.90 K/MPa, which corresponds to an effective constant ε^{tr} value in equation (10) equal to 0.065 (Δs=-7.2x10^{4}J/Km^{-3} [18]), which is a bit larger that ε^{tr}_{sat}=0.061 [18,19].

Closing this section Figure 9 shows the stress dependence of the transformation strain in polycrystalline Cu-20at%Al-2.2at%Ni-0.5%B alloy [6,22] for β/β′ transformation. It can be seen that here ε_T is zero. Indeed, quite frequently in polycrystalline samples (see also [14,15]) ε_T is zero or close to zero and it can also happen that the saturation can not be reached in the σ interval investigated (as it is the case here as well).

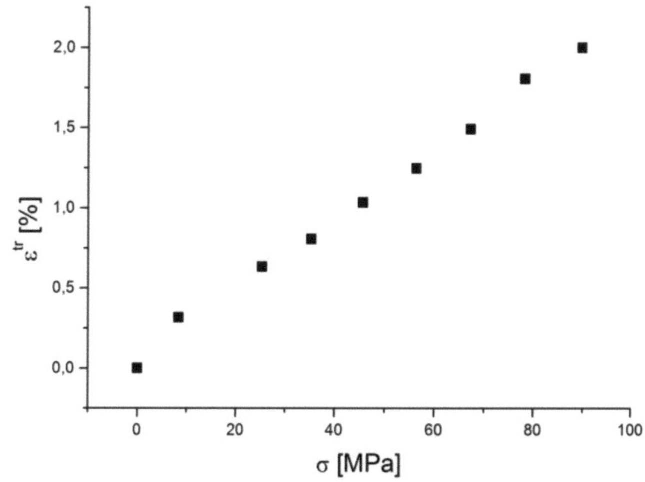

Figure 9. $\varepsilon^{tr}(\sigma)$ *function for β/β′* transformation *in polycrystalline samples [6, 22].*

3.3. Dependence of the derivatives of the elastic and dissipative energy contributions on the martensite volume fraction

As it was pointed out in Section 2 equations (13), (14) and (16), (17) offer the possibility to calculate the dependence of *d* and the $2T_o(\sigma)$ $-2e(\xi)/(-\Delta s)$ terms (or the *e* term directly if T_o is known) on the transformed martensite volume fraction. In the case of CuAl(17.9w%)Ni(2.6w%) single crystalline alloy we could determine both the equilibrium transformation temperature and the entropy thus Figure 10 shows the $d(\xi)$ as well as the $e(\xi)$ function, respectively for 171.5 MPa (high stress limit). It can be seen that indeed the elastic energy contributions is zero up to about ξ_c=0.37 and then significantly increases with increasing ξ (see also Figure 6) indicating that there is an elastic energy accumulation in this stage. Furthermore, since we have different shapes of the hysteresis loops at low and high stress limits (see also Figures 6), Figure 11 shows the $e(\xi)$ function at 42.4 MPa for the cooling down process. It is worth mentioning that a detailed analysis (see [5]) shows that the unusual shape of the loop at this stress level indicated (see Figure 12 which shows the inverse of the $T(\xi)$ loop obtained at 42.4 MPa: the sums and differences of the cooling and heating branches give the ξ dependence of the elastic and dissipative terms, respectively) that the elastic energy accumulation was practically zero up to about ξ=0.63 during cooling and again zero for heating but, surprisingly now from ξ=1 down to ξ=0.37.

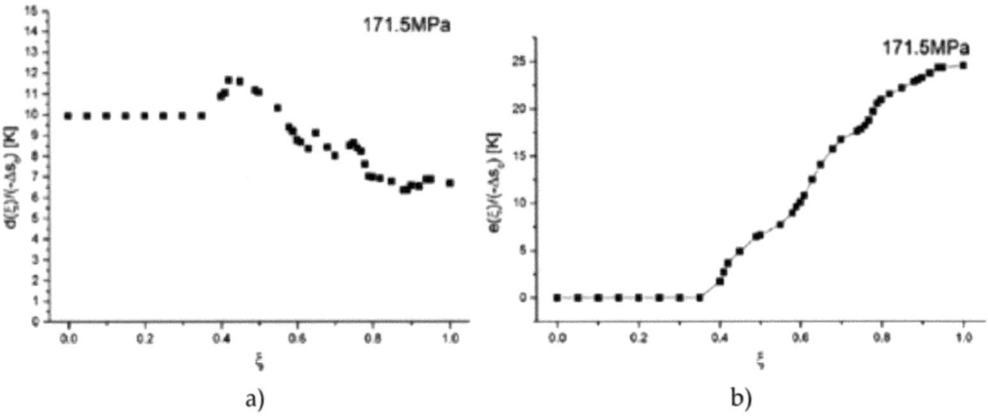

Figure 10. Derivative of the dissipative (left) and elastic energy (right) contributions versus transformed fraction in CuAl(17.9w%)Ni(2.6w%) single crystalline alloy for β/γ transformation at 171.5 MPa (high stress limit) [5].

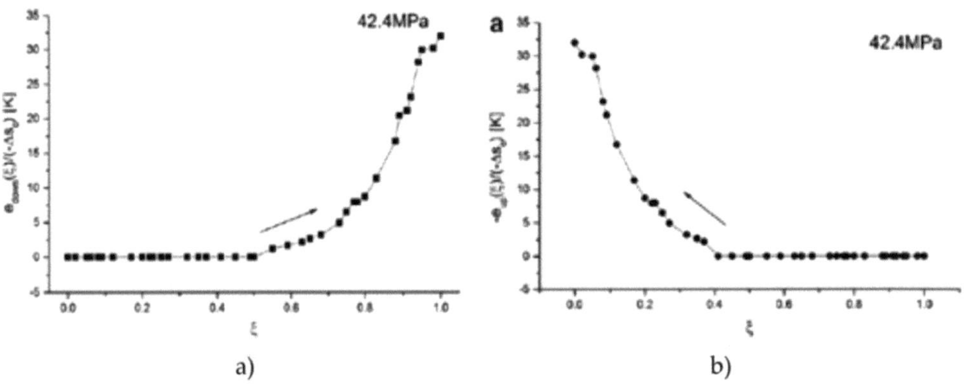

Figure 11. Derivative of the elastic energy versus the transformed fraction in CuAl(17.9w%)Ni(2.6w%) single crystalline alloy for β/γ transformation at 42.4 MPa (low stress limit) for cooling down (left) and heating up (right; in obtaining this curve a mirror transformation was made i.e. $-e^{\uparrow}(\xi=0)=e^{\downarrow}((\xi=1)$ and $-e^{\uparrow}(\xi=0.37)=e^{\downarrow}((\xi=0.63))$ [5].

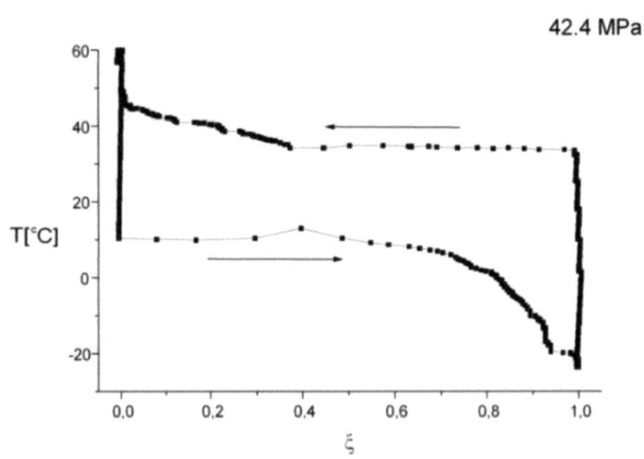

Figure 12. Inverse of the thermal loop shown on Fig. 6.

The above behaviour can be understood as follows [5]: under high stress levels the stress will prefer the nucleation of special variant(s), which can freely grow without the accumulation of elastic energy at the beginning and during cooling the relaxation of the stress starts from $\xi=1$ and after a certain value the elastic contribution will be zero. This is what was usually observed in martensitic transformations and can be described as "the first plate of martensite to form during cooling is usually the last plate of martensite to revert on heating" [1]. Thus in this case obviously after $\xi>\xi_c^{\downarrow}$ the elastic fields of the growing martensite variants will overlap (or in addition to the single growing variant, new nuclei can also form) and accumulation of the elastic energy takes place. On heating the reverse phenomenon (i.e. first the last martensite plates start to revert and the relaxation of the stored elastic energy between $\xi=1$ and $\xi=\xi_c^{\downarrow}$ takes place) can be observed. On the other hand curves at low stress levels showed different features. Indeed the multiple interface transformation takes place in the form as described above only in bulk samples and as stated in [1] "for other shapes of the same crystal (say, thin discs) the reverse transformation may nucleate competitively at separate places". Indeed in [5] the samples had a form of rod with a relatively small cross-section. In this case there are no preferred martensite variants (if the stress level is too low and is in the order of magnitude of the internal random stress field) and the first martensite nuclei can appear at easy nucleation places (e.g. tips, edges). Nevertheless, at the beginning (around M_s) of cooling down, there is no change in the elastic energy (i.e. e is approximately zero) up to a certain value of $\xi_c^{\downarrow}$ (either because the transformation takes place in a single interface mode, or because the elastic fields of the formed nuclei does not overlap yet). Obviously, for $\xi>\xi_c^{\downarrow}$ the elastic fields of the martensites formed start to overlap and accumulation of the elastic energy takes place. Thus this forward part of the transformation is very similar to that observed at high tensile stresses. In the reverse process the heating up branch of the hysteresis curve indicates that the first austenite particles may nucleate competitively at easy nucleation places (where the first martensite nuclei were formed during cooling) and thus at A_s the change in the elastic energy can be negligible. Indeed, as optical microscopic observations confirmed [5], the formation of surface relief at low stress level (at about $\sigma=0$) in the backward transformation usually started at places where the formation of the first martensite plates occurred (and not at places where their formation finished). Thus Figure 11 (on the right) shows the $e(\xi)$ for the heating up branch, but by using a mirror transformation (for the details see [5]).

Figure 13 shows the $d(\xi)=d^{\downarrow}(\xi)=d^{\uparrow}(\xi)$ as well as the $e(\xi)=e^{\downarrow}(\xi)=-e^{\uparrow}(\xi)$ functions in single crystalline samples for β/β' transformations [18], respectively. Since in this case we were not able to determine T_o the elastic energy derivative contains also the constant term $2T_o(\sigma)\Delta s$ (see eq.(14)).

In Figure 14 the $d(\xi)=d^{\downarrow}(\xi)=d^{\uparrow}(\xi)$ as well as the $e(\xi)=e^{\downarrow}(\xi)=-e^{\uparrow}(\xi)$ functions are shown for polycrystalline Cu-20at%Al-2.2at%Ni-0.5%B alloy (β/β' transformation) [22]. Here again the elastic energy derivative contains the constant $2T_o(\sigma)\Delta s$ term.

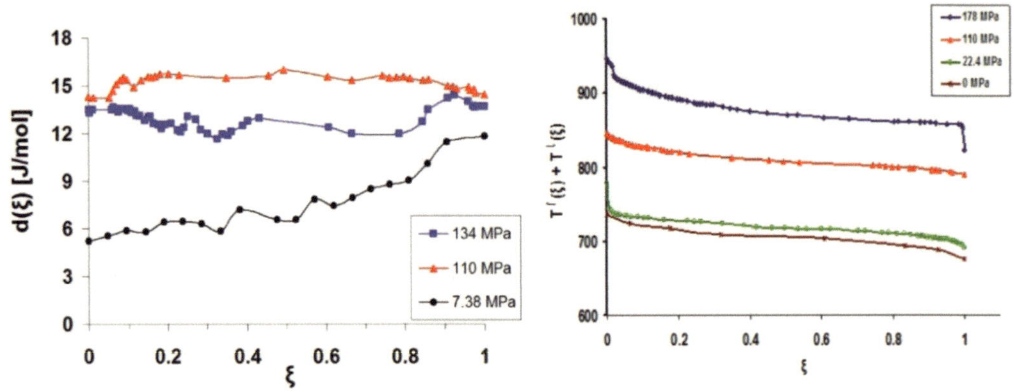

a) b)

Figure 13. Dissipative (left) and elastic (right) energy terms versus the transformed martensite fraction *for* β/β' transformation *in single crystalline samples [18], respectively. On the right only the difference of equations (6) are shown because* $T_o(\sigma)$ *is not known (see also the text).*

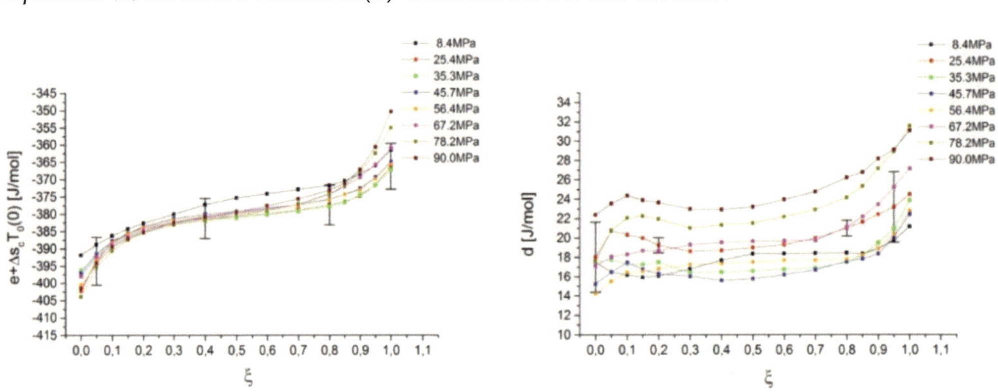

a) b)

Figure 14. Elastic (left) and dissipative (right) energy terms versus ξ at different stress levels in polycrystalline *Cu-20at%Al-2.2at%Ni-0.5%B alloy* for β/β' transformation [22].

3.4. Stress and temperature dependence of the elastic and dissipative terms

We have seen that the relations presented in Section 2 allow calculating the stress as well as temperature dependence of the derivatives of the elastic or dissipative energies, at a fixed ξ value, or their integrals, i.e. the E and D quantities, from the ξ~T, as well as from the ξ~σ loops, respectively. Let us see these functions for the there alloys investigated.

In the *single crystalline CuAl(17.9w%)Ni(2.6w%) samples* (β/γ transformation) the dissipative energy contributions were calculated from the parallel parts of the loops (see Figure 6), using that d is independent of ξ here. These values can be seen in Figure 15 as the function of the applied stress [5, 22]. It shows a slight maximum at around 90 MPa, i.e. there are increasing and decreasing tendencies in the low and the high stress range, respectively.

Figure 16 shows the full dissipated energy and stored elastic energy in martensitic state as the function of applied stress. It can be seen that the dissipative energy slightly decreases while the elastic one increases with increasing stress. This is similar to the behaviour observed in NiTi single crystals in [23].

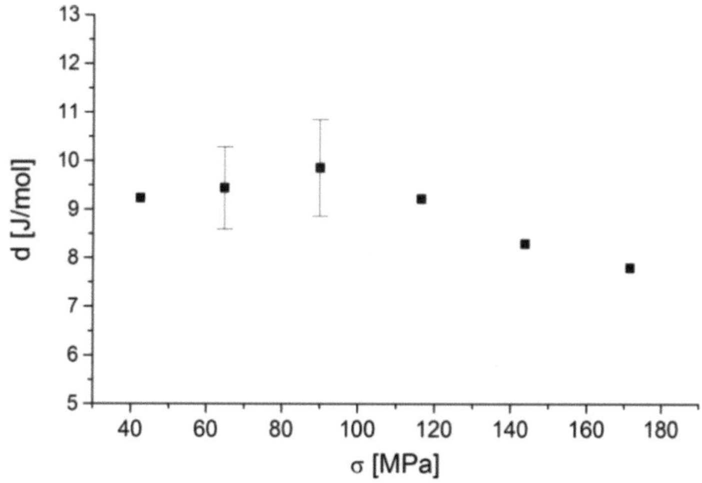

Figure 15. Stress dependence of the derivative of the dissipative energy calculated form the ξ intervals of the thermals loops where the two branches were parallel to each other [5,22] in single crystalline CuAl(17.9w%)Ni(2.6w%) samples (β/γ transformation.

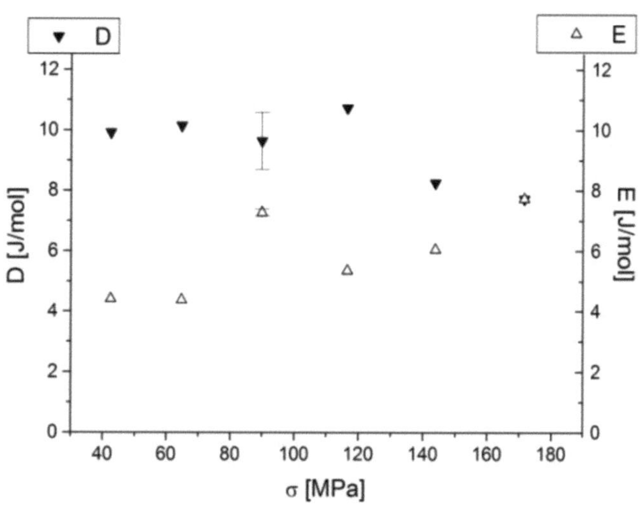

Figure 16. Stress dependence of the integral values of the dissipative and elastic energies [5,22] in single crystalline CuAl(17.9w%)Ni(2.6w%) samples (β/γ transformation).

In *single crystalline CuAl(11.5wt%)Ni(5.0wt%) alloys (β/β' transformation)* the stress dependence of e and d quantities at fixed values of ξ (at ξ=1 and ξ=0, denoted by indexes 1 and 0, respectively) is shown in Figure 17, while Figure 18 illustrates the temperature dependence of them. Furthermore in Figure 19 and 20 the total dissipative and elastic energies are shown as the function of σ as well as T. It can be seen from Figure 17 that, although the scatter of points is rather high, the d_i $(i=1, 0)$ terms can have a maximum at around 60 MPa, while their average value at the low and high stress values is 7 J/mol [18]. On the other hand the elastic energy term has definite stress dependence with the slopes -0.25 and -014 J/molMPa for e_0 and e_1, respectively. Furthermore, both the elastic and dissipative terms have linear temperature dependence (Figure 18) with the following slopes: $\partial e_0/\partial T=-0.50J/molK$, $\partial e_1/\partial T=-0.18J/molK$, and $\partial d_0/\partial T\cong\partial d_1/\partial T=-0.028J/molK$ [18, 24]. Thus it is not surprising that in Figure 19 the dissipative energy D has a maximum at about 60 MPa and the elastic energy, E, has linear stress dependence (decreases with increasing stress), while in Figure 20 the D versus T function is almost constant and E has a negative slope too.

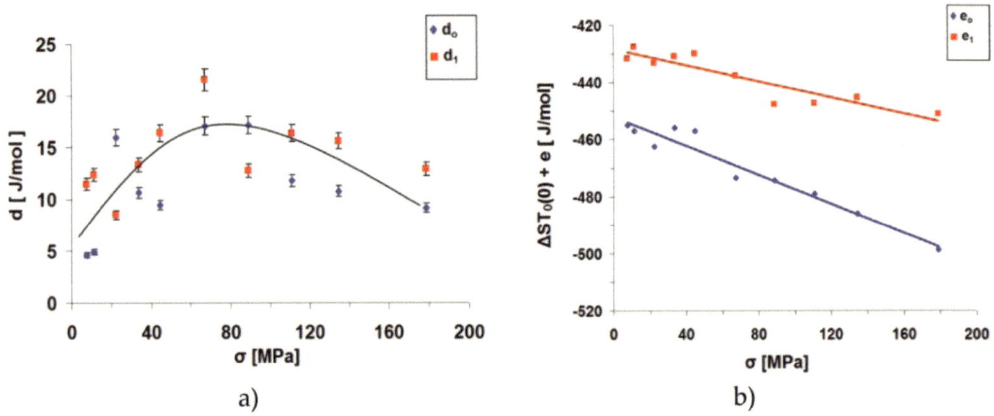

Figure 17. Stress dependence of the of the derivatives of the dissipative (left) and elastic (right) energies at ξ=1 and ξ=0 in single crystalline CuAl(11.5wt%)Ni(5.0wt%) alloys (β/β' transformation) [18].

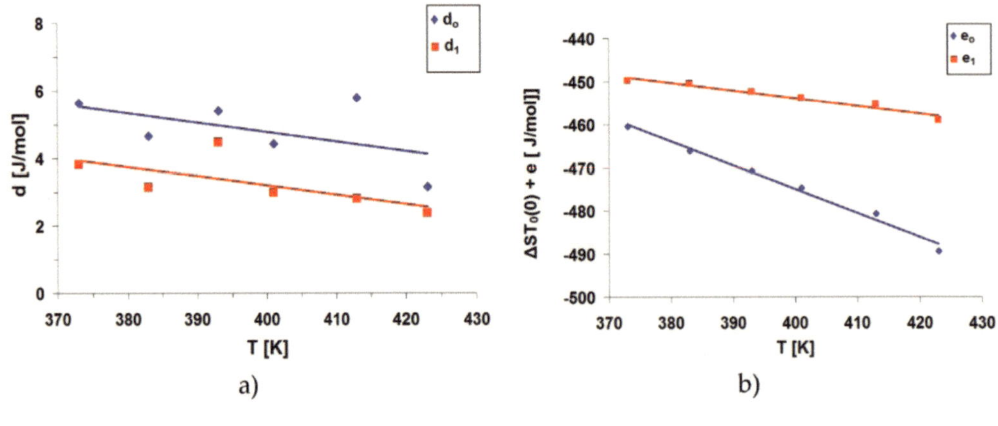

Figure 18. Temperature dependence of the of the derivatives of the dissipative (left) and elastic (right) energies at ξ=1 and ξ=0 in single crystalline CuAl(11.5wt%)Ni(5.0wt%) alloys (β/β' transformation) [18, 24].

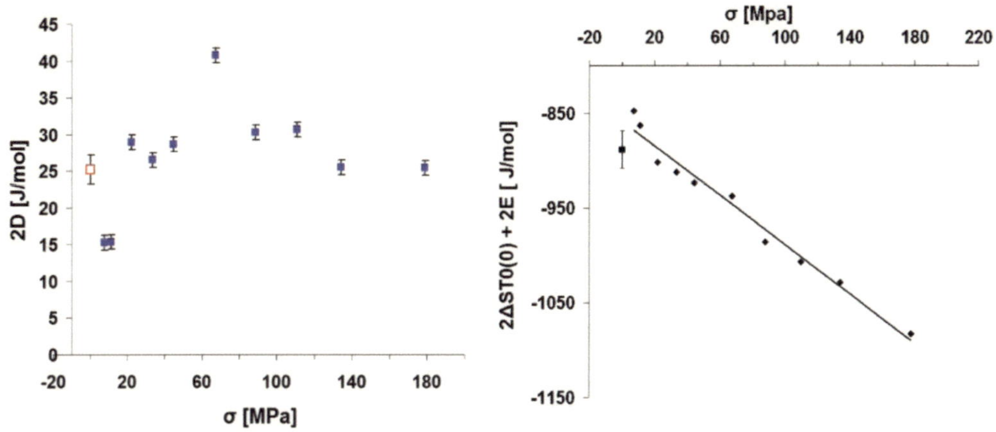

a) b)

Figure 19. Total dissipative (left) and elastic (right) energies as the function of stress in single crystalline CuAl(11.5wt%)Ni(5.0wt%) alloys (β/β' transformation) [18].

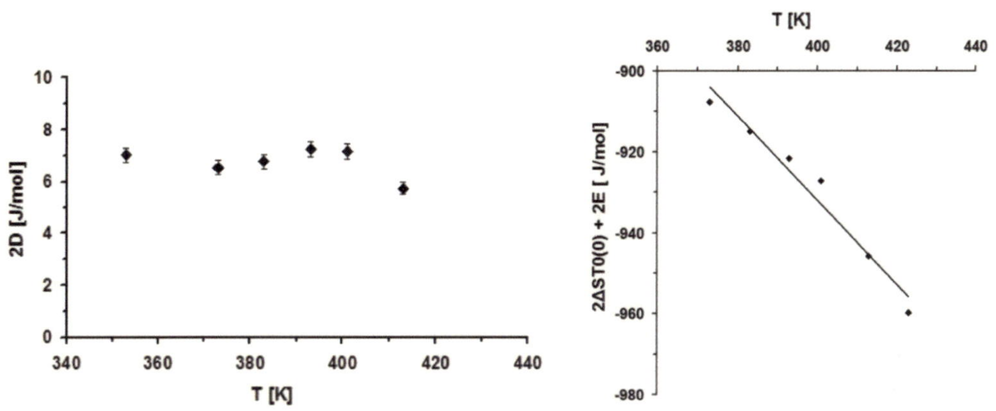

a) b)

Figure 20. Total dissipative (left) and elastic (right) energies as the function of temperature in single crystalline CuAl(11.5wt%)Ni(5.0wt%) alloys (β/β' transformation) [18].

The values obtained for the d_o and d_1 (and D) quantities are almost the same values in both sets, but their value is lower for the $\sigma \sim \varepsilon$ loops by a factor of 3. Nevertheless, the average value on the d_i versus σ plots at low and high stresses (~7J/mol) is close to 4 J/mol obtained from the $d_i(T)$ functions. Furthermore, since at higher temperatures higher stress is necessary to start the transformation, it is also plausible that the negative slope of the second part on Figure 17 should correspond to a negative slope on the $d_i(T)$ functions. Indeed there is a slight decreasing tendency with increasing T on Figure 18. Unfortunately, the accuracy of our present results does not allow a deeper and proper analysis of the field dependence of the dissipative terms. In addition, the details of the transformation (and thus the magnitude of d_i) can be different for stress and temperature induced transformations as well as can also depend on the prehistory of the samples (not investigated here).

In *polycrystalline* Cu-20at%Al-2.2at%Ni-0.5%B samples *(β/β′ transformation)* [3,22] Figures 21 and 22 show the stress dependence of the d_i, e_i as well as D and E quantities, respectively.

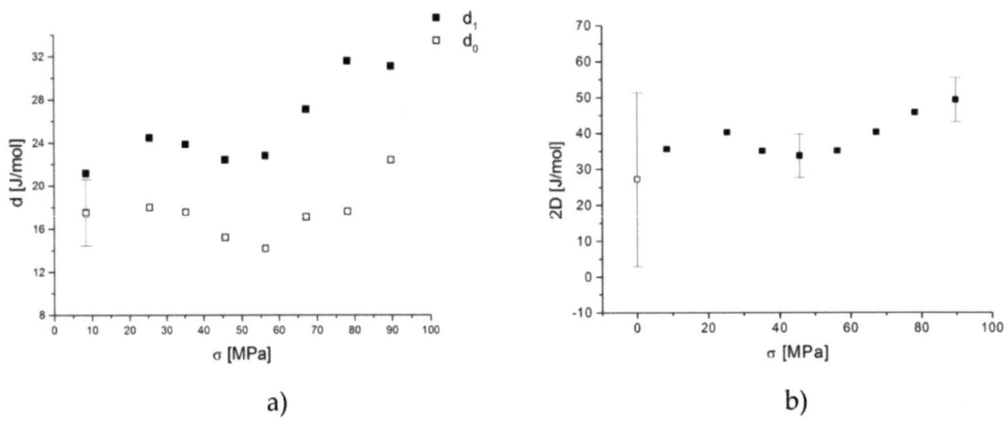

Figure 21. Stress dependence of the of the derivatives of the dissipative (left) and elastic (right) energies at ξ=1 and ξ=0 in polycrystalline *Cu-20at%Al-2.2at%Ni-0.5%B samples* (*β/β′* transformation) [3, 22].

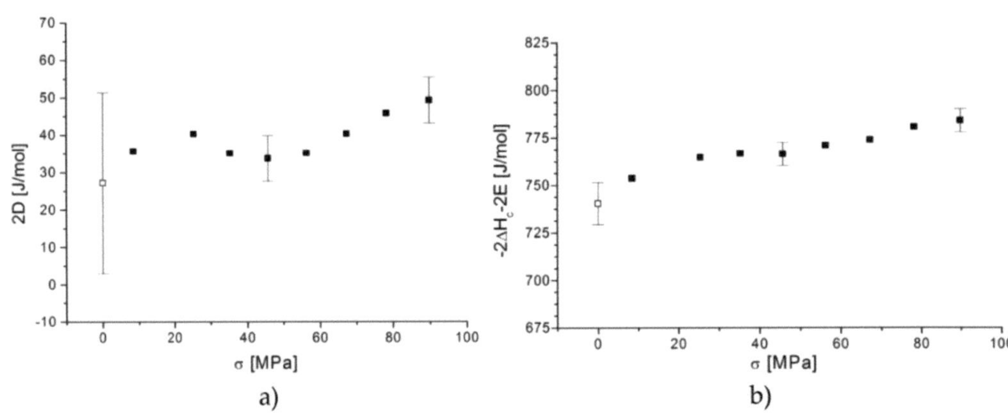

Figure 22. Stress dependence of the dissipative (left) and elastic (right) energies at ξ=1 and ξ=0 in polycrystalline *Cu-20at%Al-2.2at%Ni-0.5%B samples* (*β/β′* transformation) [3, 22].

Closing this subsection it is worth mentioning two more aspects. One is the self-consistency of our analysis. The dots at σ=0 in Figures 19 and 22 show the values calculated from the DSC curves, according to the relations (24) and (25). Thus e.g. $Q^{\downarrow}+Q^{\uparrow}=2D=25J/mol$ ($Q^{\downarrow}$=-331.6 J/mol, $Q^{\uparrow}$= 357.6 J/mol [18]) in Figure 19. It can be seen that these dots fit self-consistently within the experimental errors to the other dots calculated from the independent (hysteresis loops) measurement. The another point is related to the connection between the stress and temperature dependence of ε^{tr}(i.e. the change of the martensite variant structure) and the stress and temperature dependence of the characteristic parameters of the hysteresis loops in single crystalline samples. Although this point will be analyzed in detail in the next subsection too, it is worth summarizing some qualitative correlations: i) as it can be seen from Figure 5 as well as Figures 15 and 16 the E and D quantities change in the same stress interval where ε^{tr} for the β/γ transformation, ii) a very similar relation can be observed between ε^{tr} (Figure 3b) and d as well as D for β/β' transformation (Figures 17 and 19).

3.5. Stress and temperature dependence of the start and finish temperatures and stresses, respectively

3.5.1. Stress dependence of the start and finish temperatures

It is worth investigating whether the commonly used assumption in the literature (see e.g. [9, 25, 26]) that the slopes of the start and finish temperatures and the slope of the $T_o(\sigma)$ are approximately the same or not. From the relations, presented in Section 2, it is clear that i) strictly even the linear σ dependence of T_o is not fulfilled in general (see e.g. Figure 3b which illustrates that ε^{tr} is not constant), ii) the σ dependence of the elastic and dissipative terms (e_i, d_i, $i= 0,1$) as compared to the $T_o(\sigma)$ function, can also give a contribution to the stress dependence of the start and finish temperatures (see relations (18)). Such an analysis was carried out for the results obtained in single crystalline CuAl(11.5wt%)Ni(5.0wt%) alloys (β/β' transformation) in [18] and will be summarized here. As we have already seen in Figure 8 the $T_o(\sigma)$-$T_o(0)$ function can be approximated by a straight line, neglecting the small deviations in the interval between 0 and 50 MPa. In fact this slight S-shape part up to 50 MPa is the consequence of the stress dependence of ε^{tr}(see the insert in Figure 8). The straight, line fitted in the whole stress range, gives the slope 0.39 ±0.05 K/MPa. At the same time the slopes of M_s and A_f as well as M_f, and A_s (as shown in Figure 23, on the left) are almost the same: 0.59 as well as 0.50 K/MPa, respectively. Thus these differ from the one obtained for the slope of $T_o(\sigma)$. It should be decided whether this difference comes from the stress dependence of d_i or e_i parameters or from both. As it can be seen in Figure 17, although the d_i function indicates a maximum at around 60 MPa, from the point of view of the slope of this function in the whole stress interval, one can assume that within the scatter of the measured points they are independent of the stress. On the other hand the e_0 and e_1 parameters have a linear stress dependence with the slopes (see also above) -0.25 and -014 J/molMPa for e_0 and e_1, respectively. Dividing these by the value of Δs (=–1.26 J/Kmol [18]) the elastic energy contribution to the slope of the start and finish temperatures (see relations (18)) will be - 0.20 and - 0.11K/MPa, respectively. Thus the differences in the slopes of the start and finish temperatures and the equilibrium transformation temperature are caused by the stress dependence of the derivative of the elastic energy contribution.

Finally it is worth mentioning that since both the stress dependence of $T_o(\sigma)$ and the elastic terms can be relatively well fited by straight lines, it is not surprising that in the literature frequently a linear relation is found for the stress dependence of the start and finish temperatures.

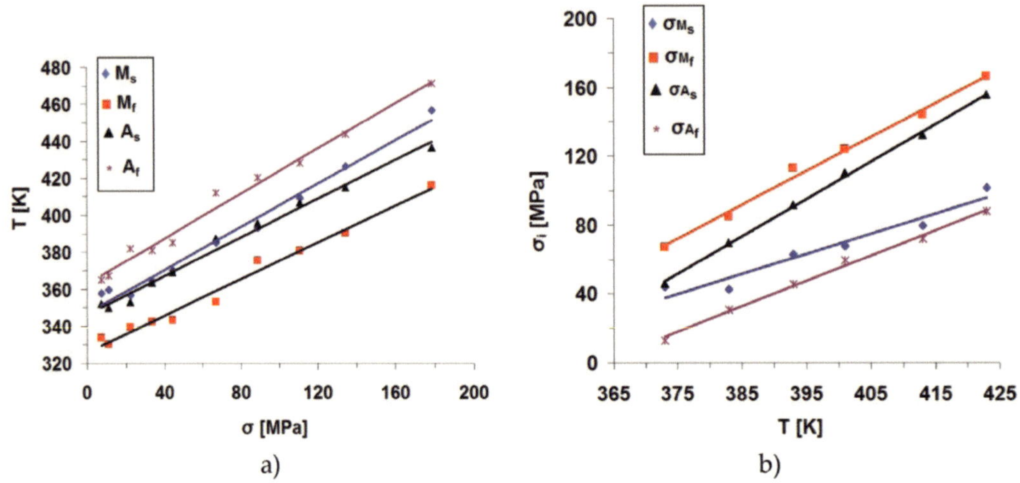

Figure 23. Stress dependence of the start and finish temperatures (left) and temperature dependence of the start and finish stresses (right) in single crystalline CuAl(11.5wt%)Ni(5.0wt%) alloys (β/β' transformation) [18].

3.5.2. *Temperature dependence of the start and finish stresses [24]*

In many papers about the relations between the start/finish stresses and the test temperature, *T*, in martensitic transformations of shape memory alloys it is assumed that e.g. the temperature dependence is the same as that of the $\sigma_o(T)$ function (σ_o is the equilibrium transformation stain). As we have seen the linearity of this (or the $T_o(\sigma)$ relation) Clausius-Clapeyron-type relation would be fulfilled only if the transformation strain, ε^{tr}, would be constant. Furthermore, it was illustrated in the previous section that relations between the start and finish temperatures versus stresses can contain stress dependent elastic and dissipative energy contributions. Thus even if these relations are approximately linear their slopes can be different from each other and from the slope of the $T_o(\sigma)$ function. The situation is very similar when one considers the $\sigma_o(T)$ as well as temperature dependence of the start and finish stresses.

In practice σ_{Ms} and σ_{As} are the most important parameters in thermomechanical treatments. Let us consider isothermal uniaxial loading tests carried out at temperatures $T>A_f$. In this case σ_{Ms} means the critical stress for the formation of stress induced martensite variants. In order to get expression for $\sigma_{Ms}(T)$ let us take the first relations of (18) (at $\sigma=0$) and (19) and make the use of (11) [24]:

$$\sigma_{Ms}(T) = -(\Delta s / V\varepsilon^{tr}(\sigma_o(T)))\left[T - M_s(0)\right] + \left[1/V\varepsilon^{tr}(\sigma_{Ms})\right]\left[d_o(\sigma_{Ms}) + e_o(\sigma_{Ms})\right] - \\ -\left[d_o(0) + e_o(0)\right]\left[1/V\varepsilon^{tr}(\sigma_o(T)\right]. \tag{28}$$

Note that in the relations used in obtaining (28) the transformation strain and the transformed fraction derivatives of the dissipative and elastic terms were considered stress dependent. It can be seen that relation (28) will have the form usually found in the literature (see e.g. [10,27]) only if the sum of the last two second terms is zero and, even in this case, it will have a linear temperature dependence only if $\varepsilon^{tr}(\sigma_o(T))$ is constant. Similar relations can be obtained for the other start and finish stresses. In the case of σ_{Mf} the sum of d_1 and e_1 appears and in the second term they should be taken at σ_{Mf}, while for σ_{Af} and σ_{As} the e_o-d_o as well as e_1-d_1 differences will be present. For example;

$$\begin{aligned}\sigma_{As}(T) &= -(\Delta s / V\varepsilon^{tr}(\sigma_o(T)))\left[T - A_s(0)\right] + \left[1/V\varepsilon^{tr}(\sigma_{As})][e_1(\sigma_{As}) - d_1(s_{As})\right] - \\ &-\left[e_1(0) - d_1(0)][1/V\varepsilon^{tr}(\sigma_o(T)\right]].\end{aligned} \tag{29}$$

One can recognize from (28) or (29) that interestingly if the contributions from the elastic and dissipative contributions are neglected the slopes of all start and finish stresses versus temperature have the same value (or have the same curvature).

Now the analysis of the experimental data obtained in single crystalline CuAl(11.5wt%)Ni(5.0wt%) alloys (β/β' transformation) resulted in the following results [24]. First it is interesting to recognize a correlation between the stress and temperature dependence of ε^{tr}: it can be seen from Figure 4a that e.g. at 373 K the martensite start stress is about 30 MPa and on the curve shown in Figure 3b this leads to about 4% ε^{tr} value, which is approximately the same as was observed at this temperature ((see Figure 4b). *Thus the transformation strain has indirect temperature dependence and it is the result of its σ-dependence.* It is easy to understand the above indirect temperature dependence: since in expression (2) the elastic and thermal terms play equivalent roles with opposite sings in the thermoelastic balance [8,9] at higher temperatures higher stress is necessary to start the transformation and the martensite structure formed will be more oriented at this higher temperature: η and thus ε^{tr} will be larger.

Next, let us see whether the slopes of the start and finish stresses versus temperature are the same or not. It can be seen in Figure 23 (on the right) that the functions can be approximated by straight lines and Table 1 contains their slopes. However, while the slopes of σ_{Ms}(T) and σ_{Af}(T) as well as σ_{Mf}(T) and σ_{As}(T) are the same the slopes of these two groups differ from each other more than the estimated error (about 0.05 MPa/K [18]).

In (28) and (29) both d_o and d_1 terms has a very moderate temperature dependence with the same slopes of (Figure 18) -0.028J/molK (leading to a small contribution to the slope of the temperature dependence of the start/finish temperatures as -0.064MPa/K) while *$e_o(\sigma_{Ms}(T))$ depends on temperature* (see Figure 23: $\partial e_o/\partial T$=-0.50 J/molK, $\partial e_1/\partial T$=-0.18 J/molK [18, 24]). Furthermore the $\varepsilon^{tr}(\sigma_o(T))$ and $\varepsilon^{tr}(\sigma_{Ms}(T))$ functions should be considered in the temperature interval 373-425K (Figures 23 and 4b) i.e., as an average value, one can take $\varepsilon^{tr}(\sigma_o(T))\cong\varepsilon^{tr}(\sigma_{Ms}(T))\cong 0.055$. Thus the terms containing $1/V\varepsilon^{tr}$ will be approximately constant $1/V\varepsilon^{tr} \cong 2.3\text{x}10^6$ mol/m^3 (a bit larger than the value belonging to ε^{tr}_{max}: 2.1x10^6 mol/m^3, $V=7.9\text{x}10^{-6} m^3/mol$ [18]).

Thus, one can estimate the contributions of the 1st, 2nd and 3rd terms in (28) and (29) to the slope of σ_{Ms} and σ_{As} *vs.* T functions (Table 1). The slope of the third term is 0 ($\varepsilon^{tr}(\sigma_o(T))\cong\varepsilon^{tr}(\sigma_{Ms}(T))\cong const.$) and from *the second term the elastic term gives determining contribution to the slope*. This also explains why the slopes of σ_{Ms} and σ_{Af} as well as σ_{Mf} and σ_{As} are similar, because they contain the different temperature derivatives of e_o and e_1, respectively.

Experimental vales [18]	σ_{Ms} vs. T	σ_{Mf} vs. T	σ_{Af} vs. T	σ_{As} vs. T
Slope in MPa/K	1.6	2.0	1.5	2.2

Estimated (parts)	Eq. (11)	1st term in (28) and (29)	2nd term in (28), e_o	2nd term in (29) e_1	3rd term in (28) or (29) d_0=d_i
Slope in MPa/K	2.59	2.83	-1.15	-0.41	-0.06

Estimated (whole)	σ_{Ms} vs. T	σ_{Mf} vs. T	σ_{Af} vs. T	σ_{As} vs. T
Slope in MPa/K	1,6	2.3	1.7	2.5

Table 1. Experimental and estimated values of the slopes of the start and finish stresses versus T [24].

It can be seen from Table 1 that taking all the contributions into account the agreement between the estimated and experimental values is very good.

Finally a comment, similar to that given at the end of Section 2.5.1., can be made here too: since both the $\sigma_o(T)$ and the temperature dependence of the elastic terms (giving the determining contribution to the T dependence) can be well approximated by straight lines, the linear relations between the start and finish stresses and the test temperature can be frequently linear.

3.6. Effect of cycling

After the illustration of the usefulness of the above model in the calculation of the elastic and dissipative energy contributions from hysteresis loops of thermal and mechanical cycling in this section the results on the effect of number of the above cycles on the energy contributions will be summarized.

In [17] the effect of thermal and mechanical cycling on β/β' phase transformation in CuAl(11.5W%)Ni(5.0W%) single crystalline shape memory alloy was studied. The σ~ε and ξ-T hysteresis loops were investigated after different numbers of thermal and mechanical cycles. The σ~ε loops were determined at fixed temperature (373 K) and the ξ-T loop under zero stress was calculated from the DSC curves measured.

Figure 24 (left) shows the ξ -T loops, calculated from the DSC curves, after different numbers of cycles, N, and the N dependence of the start and finish temperatures (right). Figure 25 illustrates the N dependence of the start and finish stresses, while in Figures 26 and 27 the N dependence of the calculated dissipative and elastic energies are shown as calculated form the thermal and mechanical cycling.

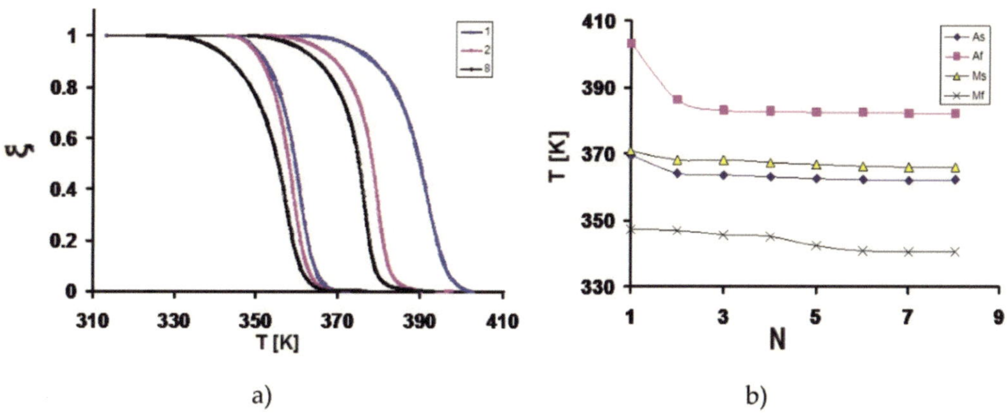

Figure 24. ξ -T loops (left), calculated from the DSC curves, after different numbers of cycles, N, in CuAl(11.5W%)Ni(5.0W%) single crystalline alloy and the N dependence of the start and finish temperatures (right) [17].

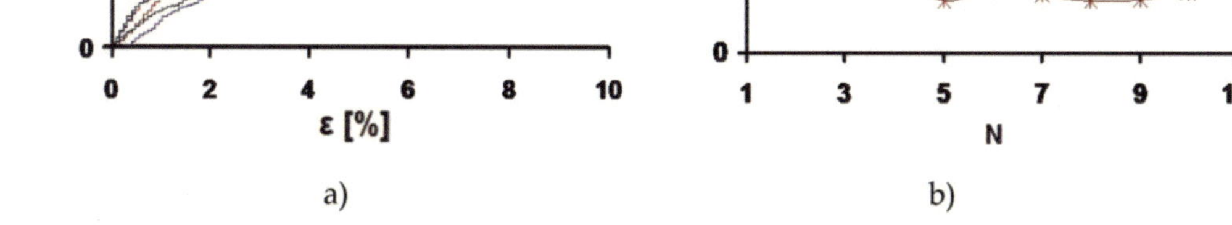

Figure 25. $\sigma \sim \varepsilon$ loops (left) after different numbers of cycles, N, and the N dependence of the start and finish stresses in CuAl(11.5W%)Ni(5.0W%) single crystalline alloy (right) [17].

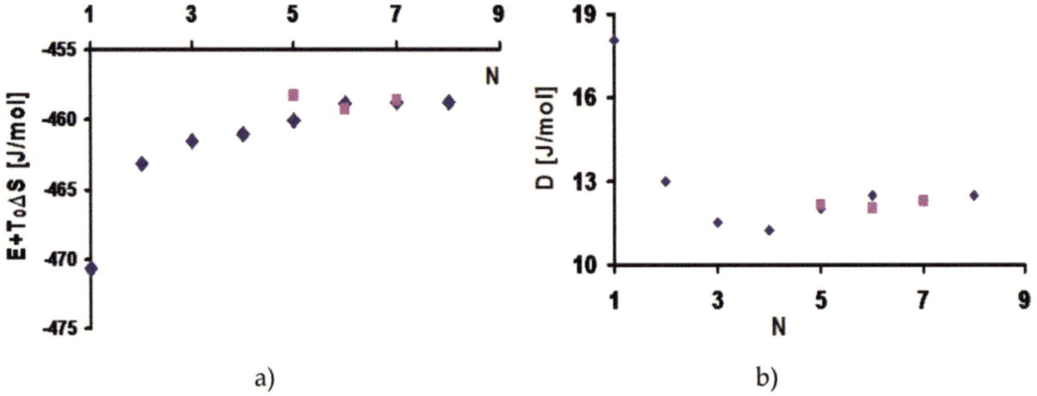

Figure 26. Cycle number dependence of the total elastic energy (left) and the total dissipative energy (right) for thermal cycles (♦ obtained from the ξ-T loops, ■ obtained from the heats of transformation) in CuAl(11.5W%)Ni(5.0W%) single crystalline alloy (right) [17].

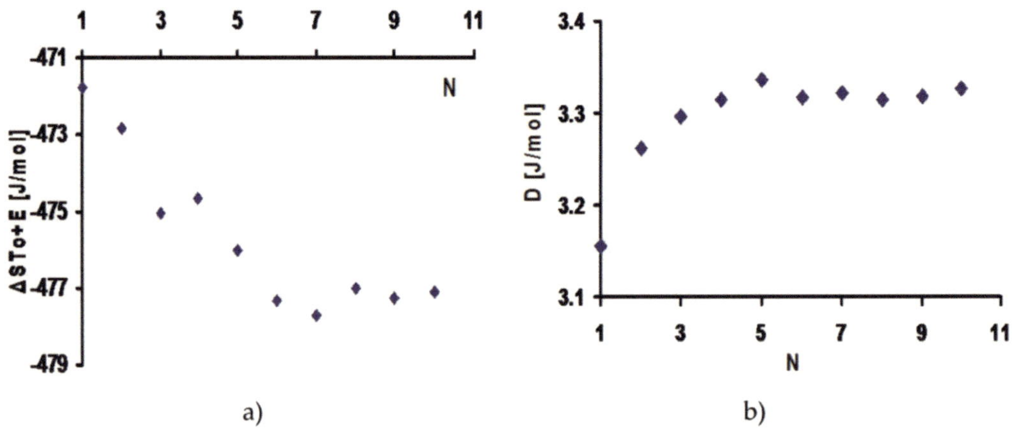

Figure 27. Cycle number dependence of the total elastic energy (left) and the total dissipative energy (right) for mechanical cycles in CuAl(11.5W%)Ni(5.0W%) single crystalline alloy (right) [17].

From the results presented in Figures 24-27 the following conclusions can be drawn [17]:

i. Both the thermal and mechanical cycling causes some changes in the hysteresis loops: after a fast shift in the first few cycles the stress-strain and strain-temperature response stabilize.

ii. In thermal cycling the elastic energy, E, as well as the dissipative energy, D (per one cycle), increases as well as decreases, respectively with increasing number of cycles, while in mechanical cycling there is an opposite tendency. These changes are inevitably related to the change in the martensite variant structure during cycling.

iii. In thermal cycling, where self-accommodated martensite variant structure develops, with increasing numbers of N, due to some "learning process in nucleation of similar variants" at different places, the marensite variant structure stabilizes and interestingly in this process E increases (by about 2.5%) and D decreases (by about 50 %).

iv. In mechanical cycling it is expected that the learning process can lead to an increased number of nucleation of preferentially oriented (according to the direction of the applied uniaxial stress) martensite variants. This decreased E and increased D by about 1 % and 6% respectively.

v. In general there are two energy dissipative processes [23]: the first is related to the frictional interfacial motion, while the second is due to the dissipation of the stored elastic energy when the coherency strains at the martensite/austenite interface relax. Assuming the first contribution independent of N, the increase/decrease of E can be accompanied by a decrease/increase in D, but for a deeper understanding detailed microscopic investigation of the variant structure and the interfaces, similarly as e.g. was done in [23], is necessary.

4. Conclusions

The analysis of extended experimental data obtained in poly- and single crystalline Cu based alloys provided the following main conclusions:

1. It has been illustrated that the transformation strain, ε^{tr}, depends on both the uniaxial stress and temperature in measurements carried out in single crystalline samples at different constant stress and temperature values, respectively. In both functions the saturation values were the same corresponding to the maximal possible transformation strain, ε^{tr}_{max}, estimated for the case when a single crystal fully transforms to the most preferably oriented martensite. This behaviour was interpreted by the change of the martensite variant structure as the function of the parameter, η, the volume fraction of the stress induced (single) variant martensite structure. In the $\varepsilon^{tr} = \varepsilon_T + (\varepsilon_\sigma - \varepsilon_T)\eta$ relation ε_T and ε_σ are the transformation strains when fully thermally induced multi variant structure forms (η=0), as well as when the martensite consists of a fully ordered array of stress preferred variants (single variant state, η=1), respectively. It has been illustrated that ε_T can be either zero or can have a finite value (remanent strain) depending on the details of the variant structure (and thus on the prehistory of the sample).
2. The stress and temperature dependence of ε^{tr}(or η) is reflected in deviations from the Clausius-Clapeyron-type relations. Indeed it was demonstrated that the equilibrium transformation temperature, T_o, was not a linear function of the stress in single crystalline alloys.
3. Using relations for the ξ~T and σ~ε (ε~ξ) loops (ξ is the transformed martensite volume fraction) the ξ dependence of the ξ derivatives of the elastic and dissipative energies, ($e(\xi)$ and $d(\xi)$) could be determined. The integrals of these functions gave the elastic, E, and dissipative, D, energies per on cycle. Thus it was also possible to determine their dependence on the stress and temperature. Note that the σ or T dependence of the elastic energy can be calculated only exclusive of a constant term containing the product of the entropy and $T(\sigma=0)$ (see eqs. (10), (13) and (16)). In the CuAl(17.9w%)Ni(2.6w%) single crystalline alloy, by the analysis of the peculiar shapes of the ξ~T loops even the determination of the equilibrium transformation temperature and its σ dependence was possible. It was also demonstrated that our procedure is self-

consistent since e.g. at zero stress the D and E quantities were also calculated from independent measurements (DSC curves) and the results were in very good agreement with those values obtained form the integrals of the $e(\xi)$ and $d(\xi)$ functions.

4. It was shown that the stress and temperature dependence of ε^{tr}(or η) is also reflected in the shape of the $D(\sigma)$, $D(T)$, $E(\sigma)$ and $E(T)$ functions, since these terms should be plausibly dependent on the martensite variant structure developed.
5. It was illustrated that both the stress dependence of the start and finish temperatures as well as the temperature dependence of the start and finish stresses in general can be approximated by straight lines. This is due to the facts that the $T_o(\sigma)$, $\sigma_o(T)$ functions, in a wider interval of their variables, can be linear as well as the elastic energy contributions (giving dominating contributions to the σ or T dependence) can also be fitted by a linear functions. On the other hand, the slopes of the start and finish parameters as well as the slopes of the $T_o(\sigma)$ or $\sigma_o(T)$ can be definitely different from each other.

It was shown that the number of thermal and mechanical cycling, N, has effected the values of E and D: in thermal cycling E increased, while D decreased with N. During mechanical cycling an opposite effect was observed.

Author details

Dezso L. Beke, Lajos Daróczi and Tarek Y. Elrasasi
Department of Solid State Physics, University of Debrecen, Debrecen, Hungary

Acknowledgement

This work has been supported by the Hungarian Scientific Research Found (OTKA) project No. K 84065 as well as by the TÁMOP-4.2.2/B-10/1-2010-0024 project which is co-financed by the European Union and the European Social Fund.

5. References

[1] Salzbrenner R J, Cohen M (1979) On the thermodynamics of thermoelastic martensitic transformations. Acta Metall. 27: 739-748.
[2] Tong H C, Wayman C M (1974) Characteristic temperatures and other properties of thermoelastic martensites. Acta Metall. 22: 887-896.
[3] Palánki Z, Daróczi L, Beke D L (2005) Method for the determination of non-chemical free energy contributions as a function of the transformed fraction at different stress levels in shape memory alloys. Mater. Trans. A46: 978-982.
[4] Daróczi L, Beke D L, Lexcellent C, Mertinger V (2004) Effect of hydrostatic pressure on the martensitic transformation in CuZnAl(Mn) shape memory alloys. Scripta Mat. 43:691- 697.
[5] Palánki Z, Daróczi L, Lexcellent L, Beke D L (2007) Determination of the equilibrium transformation temperature (T_o) and analysis of the non-chemical energy terms in a CuAlNi single crystalline shape memory alloy. Acta Mat. 55: 1823-1830.

[6] Daróczi L, Palánki Z, Szabó S, Beke D L (2004) Stress dependence of non-chemical free energy contributions in Cu–Al–Ni shape memory alloy. Material Sci. and Eng. A378: 274-277.

[7] Beke D L, Daróczi L, Lexcellent C, Mertinger V (2004) Determination of stress dependence of elastic and dissipative energy terms of martensitic phase transformations in a NiTi shape memory alloy. Journal de Physique IV (France) 115: 279-285.

[8] Beke D L, Daróczi L, Palánki Z (2008) On relations between the transformation temperatures, stresses, pressures and magnetic fields in shape memory alloys. in S. Miyazaki editor. *Proc. of Int. Conf. on Shape Memory and Superleastic Technologies,* 2007, Tsukuba, Japan, (ASM International, Materials Park, Ohio) pp. 607-614.

[9] Planes A, Castan T, Ortin J, Delaey L (1989) State equation for shape memory alloys: Application to Cu Zn Al. J. Appl. Phys. 66(6): 2342-1348

[10] Delaey L (1991) Diffusionless transformations. in Cahn R W, Haasen P, Kramer E J, editors. Materials Science and Technology – A Comprehensive Treatment Vol. 5., Chp.5 in Haasen P, editor. Phase Transformations in Materials: Weinheim, VCH, p. 339

[11] Tanaka K, Sato Y (1986) Analysis of Superplastic Deformation During Isothermal Martensitic Transformation. Res. Mech. 17: 241-252.

[12] Ortin J, Planes A (1989) Thermodynamic analysis of thermal measurements in thermoelastic martensitic transformation. Acta metall. 36: 1873-1889.

[13] J. van Humbeck and R. Stalmans (1998) Characteristics of shape memory alloys. Chp. 7 in Osuka K and Wayman C editors. Shape memory materials, Cambridge, Cambridge University Press, p. 151.

[14] Leclercq S, Lexcellent C (1996) A general macroscopic description of the thermomechanical behavior of shape memory alloys. J.Mech.Phys.Solids, 44(6): 953-980.

[15] Lexcellent C, Boubakar M L, Bouvet Ch, Calloch S (2006) About modelling the shape memory alloy behaviour based on the phase transformation surface identification under proportional loading and anisothermal conditions. Int. J. of Sol. and Struct. 43: 613-626.

[16] Planes A, Macquron J L, Ortin J (1988) Energy contributions in the martensitic transformation of shape-memory alloys. Phil. Mag. Let. 57(6): 291-298.

[17] Elrasasi T Y, Dobróka M M, Daróczi L, Beke D L. (2012) Effect of thermal and mechanical cycling on the elastic and dissipative energy in CuAl(11.5W%)Ni(5.0W%) shape memory alloy. J of Alloys and Comp. (in print)

[18] Elrasasi T Y, Daróczi L, Beke D L (2010) Investigation of thermal and stress induced hystersis curves in CuAl(11.5 wt%)Ni(5.0 wt%) single crystalline shape memory alloy. Intermetalics 18: 1137, 2010

[19] Beke D L, El Rasasi T Y, Daróczi L (2009) On the temperature and stress dependence of transformation strain in single crystalline Cu-Al-Ni shape memory alloys. ESOMAT Proceedings. Available: http://dx.doi.org./10.1051/esomat/200902002.

[20] Chernenko V A, L'vov V A (1996) Thermodynamics of martensitic transformations affected by hydrostatic pressure. Phil. Mag. A73: 999-1008.

[21] Johari, G P, McAnananma J G, Sartor G (1996) Effect of hydrostatic pressure on the thermoelastic transformation of Ni-Ti alloy and the entropy of transformation. Phil. Mag. B74: 243-257.

[22] Palánki Z, Theses, (2008) Departement Mecanique, Universite de Franche-Comte, France.

[23] Hamilton R F, Sehitoglou H, Chumljakov Y, Maier H J (2004) Stress dependence of the hysteresis in single crystal NiTi alloys. Acta Mater. 52: 3383-3402.

[24] Elrasasi T Y, Daróczi L, Beke D L (2010) On the relation between the matensite start stress and the temperature in single crystalline Cu-11.5wt%Al-5.0wt%Ni shape memory alloy. Mat. Sci. Forum, 659: 399-404.

[25] Rodriguez-Aseguinolaza J, Ruiz-Larrea I, No M L, Lopez-Echarri A, San Juan J (2008) A new quantitative approach to the thermoelastic martensitic transformation: The density of elastic states. Acta Mater. 56: 6283-6290.

[26] Otsuka K, Sakamoto H, Shimizu K (1979) Successive stress-induced martensitic transformations and associated transformation pseudoelasticity in Cu-Al-Ni alloys. Acta Metall. 27: 585-601.

[27] Kato H, Miura S (1995) Thermodynamical analysis of the stress-induced martensitic transformation in Cu-15.0 at.% Sn alloy single crystals. Acta metall. Mater. 43(1): 351-360.

Chapter 8

Micromechanical Behavior of CuAlBe Shape Memory Alloy Undergoing 3-Point Bending Analyzed by Digital Image Correlation

R.J. Martínez-Fuentes, F.M. Sánchez-Arévalo, F.N. García-Castillo, G.A. Lara-Rodríguez, J. Cortés-Pérez, A. Reyes-Solís

Additional information is available at the end of the chapter

http://dx.doi.org/10.5772/48772

1. Introduction

The study and development of new materials have been used to forge new technology. Shape memory alloys (SMA's) have been cataloged as new materials which applications are in fields like the construction industry where these materials have been used as mechanical elements of damping; in medicine, SMA's have been used for biomechanical prosthesis in order to correct the position of bones and also to make surgical instruments. In these applications, because of their mechanical behavior, SMA's can substitute more efficiently, conventional materials. To get a better understanding of SMA's, new investigations are needed.

In recent years, the scientific and technologic community has been interested in the study of "exotic materials" like shape memory alloys (SMA's). Shape memory alloys are materials that can recover their original shape, after being elastically or pseudo-plastically deformed, by increasing their temperature; all these associated to a martensitic transformation [1,2]. The martensitic transformation is defined as a first-order displacive process, where a body center cubic parent phase (austenitic phase) transforms by a shearing mechanism into a monoclinic or orthorhombic martensitic phase [1]. The shape memory effects are useful in replacing conventional materials and developing new applications in science and industry. The SMA's present some associated effects like single, double shape memory effects and superelastic effect. All these effects are well known and they have been reported in literature for several authors [3-7].

The superelastic effect is one of the reasons that have encourage a continuous effort to understand, predict and explode the shape memory behavior of these materials. Since the

CuAlBe system appeared in 1982; it is considered one of the first studies that were done in Cu-Al-Be shape memory alloys by Higuchi et al. These works studied the influence of the thermal stability of Cu-Al-Be alloy which had a nearly eutectoid composition; Higuchi et al. observed small changes in the temperatures of transformation and also realized that the austenite phase was not decomposed up to 300 ºC. With the thermal cycling also were observed the displacement variations; this was confirmed with a coil with a constant load; increasing the coil temperature, small displacement were register; thus the two way shape memory effect was reported in Cu-Al-Be [8,9].

After a while other interesting works concerning with this alloy were developed. Belkahla et al. reported the elaboration and characterization of a low temperature of the Cu-Al-Be shape memory alloy; obviously this work was based in Higuchi's research. The main contributions of this work were the quasi-binary Cu-Al phase diagram with the addition of 0.47 wt.% of beryllium and the M_s analytical expression, as a function of elements composition, to determine the critical temperature that indicates the start of the martensitic transformation. This study confirmed that the addition of small concentration of beryllium Cu-Al system decreased the eutectoidal temperature around 50 ºC. In addition to the temperature decrement, a new ternary domain was observed; here the phase α, β and γ_2 are present [10].

Jurado et al. were concerned about the order-disorder phase transition in Cu-Al-Be system alloys; in this case the studied ally was close to the eutectic composition. The main contribution of Jurado et al. was to reveal the effect of beryllium atom on the ordering behavior of the Cu-Al based alloys. Nowadays the X-Ray diffraction measurements reported by Jurado et al. are very useful to identify the involved phases in this system [11].

If order-disorder behavior takes place in CuAlBe system, it is obvious that this material can exhibit different mechanical response. The difference in mechanical behavior is due to the high anisotropy of this material. As a matter of fact, the Cu-Al-Be system presents three kinds of anisotropy. The first type is due to the austenitic phase and it is known as an inherent anisotropy. The second type is known as transformational anisotropy; this depends on the applied stress level or even the test temperature, preferential crystal orientation due to manufacturing process of the material. The last kind anisotropy is related with the mixture of austenite and martensite phases; the proportions of both phases, in the alloy, will change the mechanical response of this material. In order to clarify the mechanical response of this material some studies on the thermomechanical behavior has been done in monocrystals of Cu-Al-Be. These studies were able to determine the metastable phase stress vs. temperature diagram (σ-T). With this diagram is possible to get the critical stress or transformation stress value if the martensite phase is induced by stress [12].

Siredey and Eberhardt presented other interesting result, on the fatigue behavior of Cu-Al-Be monocrystals. A model to explain the fatigue mechanism was proposed. This model was based on the assumption that there are different zones where the martensite phase gets reordering or other diffusional phenomena, which vary the expected behavior inside the material; so the M_s temperature can change locally and it will modify the global behavior of the material [13]. As it was previously mentioned the X-ray and differential screw calorimetric studies represent a suitable way to characterize shape memory materials. Balo et al. used these techniques to show the influence of heat treatments and beryllium content for this alloy. In this study can be observed the indexed X-ray pattern for martensite phase in addition the lattice parameters for 18R martensite were reported too [14,15].

All previous research has motivated the improvement of the Cu-Al-Be system in order to spread its application in different industry branches. One aspect to improve in this system is the mechanical properties that depend on the alloy's microstructure. In other words the involved phases and grain size have to be customized. In order to modify the mechanical response of Cu-A-Be system several grain refiners have been used. Ultrasonic and mechanical tensile testing has proved the influence of those refiners by different authors [16,17].

More specific studies, to understand the mechanical behavior of SMA, were done using non-conventional techniques to mimic the martensitic transformation in the superelastic regime for both monocrystal and polycrystal CuAlBe undergoing tension [18-20]. As a result, scientists have proposed transformation criterions, constitutive equations and also new tools to be familiar with the involve mechanisms during the martensitic transformation [21-26]. As it can be notice, considerable efforts have been done to understand the complex behavior of SMA's; mainly, because of their properties will eventually lead to replace conventional materials with these alloys.

Although there are several works trying to explain the mechanical behavior of SMA's nobody has studied the stress-induced martensitic transformation and its granular interaction in 2D-confined polycrystalline sample of CuAlBe undergoing 3-point bending by Digital Image Correlation. That is why the objective of this work is based on getting a practical methodology to understand the micro and macromechanical behavior of poly and monocrystalline Cu-Al 11.2 wt.%-Be 0.6 wt% , Cu-Al 11.2 wt.%-Be 0.5 wt% (respectively) undergoing a stress-induced martensitic transformation by a 3-point bending using digital image correlation.

Taking in to account the good thermal stability, excellent shape memory properties, temperature transformation wide ranges, damping capacity and low cost of production, the Cu-Al-Be system has become in a excellent alternative to take advantage of the shape memory effects; that is why several works will be conducted to get a full understanding on the Cu-Al-Be properties.

2. Experimental details

The experimental section describes the elaboration and characterization of the material. The first part of this section is focused in the elaboration and structural characterization of the material. The second part is dedicated to the mechanical arrangement that makes possible the simultaneous state of stress (tension and compression) in the sample.

2.1. Material

As it was previously mentioned, the composition was Cu-Al 11.2 wt%-Be 0.6 wt%, and Cu-Al 11.2 wt.%-Be 0.5 wt% which is close to the eutectoidal composition [10]. An induction furnace (Leybold-Heraeus) was employed to elaborate the alloy by a melting process; this furnace has a controlled atmosphere and in our case argon gas was used. From the castings, suitable slices were cut and then hot rolled to obtain thin sheets. The length, width and thickness of the sheets were 280, 57 and 0.7-0.9 millimeters, respectively. These dimensions were reached after a 191 % hot-rolled (800 °C) deformation process. The hot-rolled process was carried out in an oven (Sola Basic-Lindberg model 847) and a roll machine (Fenn Amca International). Subsequently, the sheets were subjected to a heat treatment, called betatization, to reveal the shape memory effects; the sheets were heated at 750 °C during 15 minutes and then water-quenched to 95 °C during 20 minutes [24]. Then the samples were studied by X-ray diffraction (Bruker AXS modelo D8 Advance) in order to detect the phases involved in the alloy; finally Critical transformation temperatures were obtained by DSC 2910 Modulated de TA Instrument. After this, the sheets were cut in rectangular samples according to the beam theory.

2.2. Three-Point bending test

Bending tests were carried out on a servohydraulic loading device (MTS 858 MiniBionix axial). To acquire images an optical microscope was coupled to a CCD camera. The modular microscope works as an infinity-corrected compound microscope with magnifications of 2X. To control the MiniBionix MTS a 407 MTS controller was used while data and images acquisition were acquired by a National Instruments PXI-1002 chassis and PXI-boards (6281, 8331 and 1402) connected to a PC. White light illumination (150 W quartz halogen light source) was used to observe the microestructural behavior of the SMA under bending (see Fig. 1).

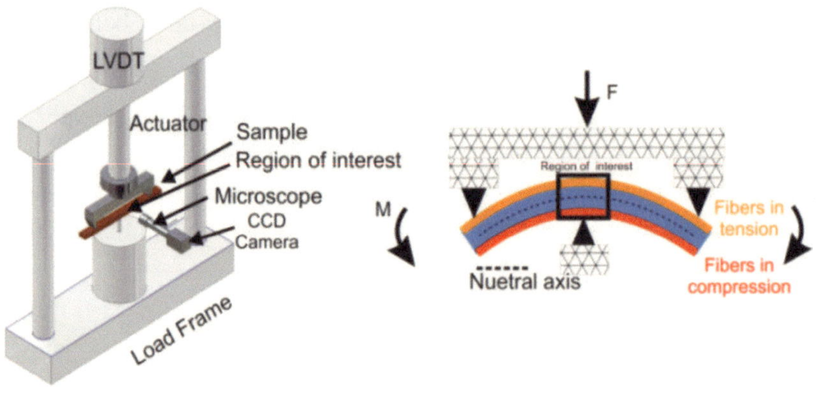

Figure 1. Experimental setup

In order to have a better understanding of CuAlBe mechanical behavior under bending a single-crystal of CuAlBe was also studied under the same conditions. Hence it was possible to compare the mechanical behavior of monocrystalline and polycrystalline samples under bending. From the image sequence were determine the displacement vector fields at the region of interest, using digital image correlation, associated to bending test for both cases.

With the series of images acquired during the bending test, displacement vector fields were calculated from pairs of images. The Willert and Gharib algorithm [27] was used to calculate the displacement field $u_k(x_k, y_k)$ and $v_k(x_k, y_k)$, where u and v represent the displacements of an analysis object or region of interest in the x and y directions respectively [28, 29]. The x and y represent the position coordinates of the analysis object in every image; subindex k indicates the corresponding object, which is defined as a 64x64 pixels subregion of interest. Hence was possible to observe the mechanical behavior in tension and compression by grain simultaneously.

3. Results

3.1. X-ray diffraction

The x-ray spectra are shown in Figure 2. Here it can be observed the all the present phases in the monocrystalline and polycrystalline samples corresponds to those reported phases for this material (Austenite β and Martensiteβ', γ_2) [11]. It is obvious that the monocrystal presents only a peak that corresponds to (440) direction; while the hot-rolled polycrystalline sample presents a preferential orientation in (111) and (220) directions; this confirms the existence of global crystallographic texture in the polycrystalline sample.

This result also confirmed the possibility to get the direct transformation by stress (Austenitic-Martensitic); in other words the superelastic effect in this alloy will be observed when the stress-induced martensitic transformation appears during the mechanical test. The β phase or austenite phase is a supercell DO_3 wich lattice has a higher order of symmetry than β' martensite phase; this martensite is also known as 18R martensite. The lattice correspondence between martensite and austenite was reported by Zhu et al [30]. In this transformation more than one martensite variant can be induced from one austenite. In addition it has to be pointed out than martensite variants have identical crystal lattice but they can appear in different orientation. The relationship of microstructures is essential to get a better understanding of the mechanical response of this smart material. Once the x-ray analysis confirmed the existence of monocrystalline and polycrystalline samples, the differential screw calorimetric studies were done.

3.2. Differential screw calorimetric analysis

The transformation temperatures M_s, M_f, A_s, A_f, and the difference M_s-M_f, A_f-A_s are considered as critical factors in characterizing shape memory behavior. There is a strong dependence between transformation temperatures and the alloy's composition and its processing, this is based on microstructural defects, degree of order in the parent phase, and grain size of the parent phase. The mentioned factors can modify the transformation temperatures by several degrees. When the martensitic transformation takes place,

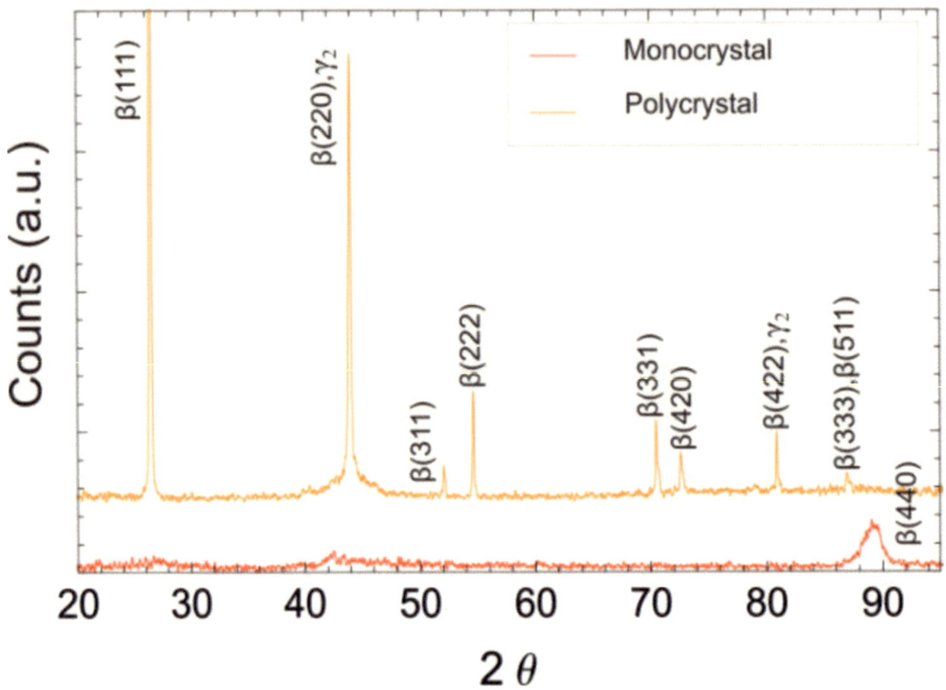

Figure 2. X-ray spectra for mono and polycrystalline samples of CuAlBe shape memory alloy

numerous physical properties are modified. During the transformation, a latent heat associated with the transformation is absorbed or released based on the transformation direction. The forward, austenite- to-martensite (A-M) transformation is accompanied by the release of heat corresponding to a change in the transformation enthalpy (exothermic phase transformation). The reverse, martensite-to-austenite (M-A) transformation is an endothermic phase transformation accompanied by absorption of thermal energy. For a given temperature, the amount of heat is proportional to the volume fraction of the transformed material. The two phases also have different resistance due to their different crystallographic structures, so the phase transformation is associated with a change in the electrical resistivity [31].

Figure 3 shows the DSC curves for both kind of samples mono and polycrystal. In this figure can be realized those discontinuities in the heat flow *vs.* temperature curve which corresponds to exothermic reaction during the direct transformation (Austenite to Martensite phase change); additionally the inverse transformation (Martesnite to Austenite phase change) is located in the peak that reveals an endothermic reaction. These peaks show the four critical temperatures of transformation in these smart materials. The temperatures were labeled in figure 3 as follows: M_s corresponds to the beginning of the martensitic transformation; M_f indicates the end of martensitic transformation. In the same way A_s and A_f indicate the start and the end of the inverse transformation. All these temperatures were determined at 10% and 90% of the peak's areas that defines each transformation; they were found using the Universal Analysis 2000 software of TA Instrument. The critical temperatures were summarized in table 1.

Temperatures (ºC)	Monocrystal	Hot-rolled Polycrystal
M_s	0	-69
M_f	-18	-82
A_s	-8	-81
A_f	5	-66

Table 1. Critical temperatures of transformation of CuAlBe system.

After X-ray analysis and DSC studies, the samples were prepared to observe the microstructure of the mono and polycrystalline samples under bending.

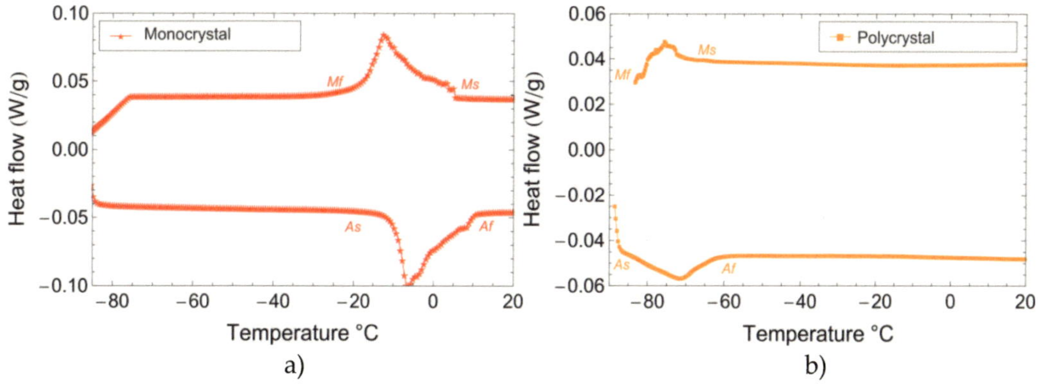

Figure 3. Calorimetric analysis to determine the four critical temperatures that defines the shape memory effect in the CuAlBe alloy. a) Monocrystal Sample and b) Hot-rolled Polycrystal Sample

Monocrystalline and polycrystalline samples of CuAlBe were tested in 3-point bending. According to the experimental setup showed in Figure 1, the samples were focused with the optical microscope coupled to the CCD camera at the center of the sample (*l*/2); where *l* represents the support span. In order to observe the stress-induced martensitic transformation through the regions of interest, the samples were previously polished and chemically etched to reveal the microstructure of each sample (Figure 4). All tested specimens were metallographically prepared and chemically etched with a solution of ferric chloride (2g $FeCl_3$ + 95 ml Ethanol + 2ml HCl) before the mechanical test.

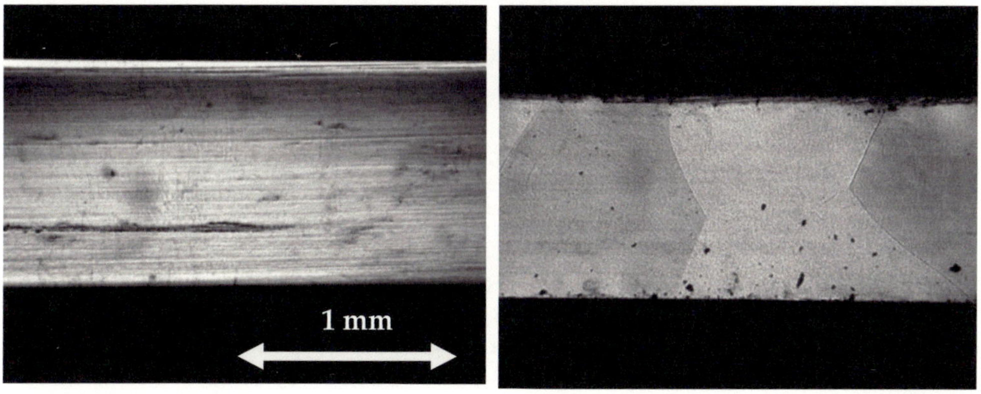

Figure 4. Microstructural details of CuAlBe samples in austenitic phase: a) Monocrystal and b) Polycrystal

The Figure 4(a) shows the monocrystal; here it is evident that there are not any grain boundaries while Figure 4(b) shows four grains in a serial arrangement. These two images were taken at the same loading conditions. It is clear that the polycrystal is a little bit thinner than the monocrystal. The thicknesses were 0.9 and 0.7 mm for monocrystal and polycrystal respectively.

3.3. Mechanical behavior under 3-point bending

From the 3-point bending test, the force *vs.* neutral axis displacement curves for monocrystal and polycrystal were obtained (Figure. 5). In the case of monocrystal is observed the typical reversible hysteretic loop with a second slope close to cero. The polycrystal showed the same reversible hysteretic loop but the second slope was higher than monocrystal's second slope; as it was expected. This is due to the transformational deformation that is directly associated to stress-induced martensitic transformation, which depends on the applied force direction and the crystals orientations. Now taking in to account the *Ms* temperature for monocrystal and polycrystal and the equation of Claussius Clappeyron $\sigma_c = 1.97MPa(T - M_s)$ which relates the *Ms* with the critical transformation stress σ_c, this transformation stress can be easily calculated considering that *T* corresponds to the test temperature 20ºC; furthermore the critical stresses were around 175 and 40 MPa for polycrystal and monocrystal respectively. This last result is good agreement with the stress transformation values for label B and E in the stress vs. neutral axis displacement curve for both samples.

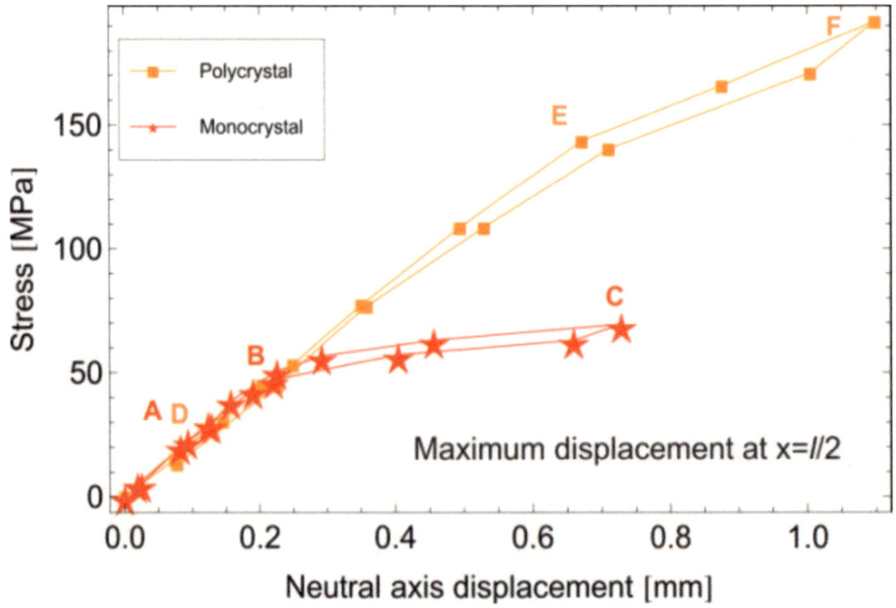

Figure 5. Force vs. Neutral axis displacement curves obtained from 3-point bending arrangement

The Figure 5 shows six labels that indicate the associated images in both cases monocrystal and polycrystal samples; the monocrystal images correspond from letter A to C and polycrystal from D to F. This six labels match with the images presented in Figure 6. These are labeled with the same letters for each case. As it can be observed, the images from A to C showed the austenitic phase (Figure 6A), the beginning of martensitic transformation (Figure 6B) and almost a total martensitic state for the monocrystal under bending (Figure 6C). The polycrystalline sample showed the austenitic phase in all grains (Figure 6D), then the martensitic plates appear in tension and some martensitic plates in compression almost simultaneously (Figure 6E); finally the image (Figure 6F) shows the central region plenty of martensitic plates. It has to be pointed out, that the martensitic plates in the monocrystal sample grow first up in compression and subsequently in tension; in the polycrystalline case happened the opposite.

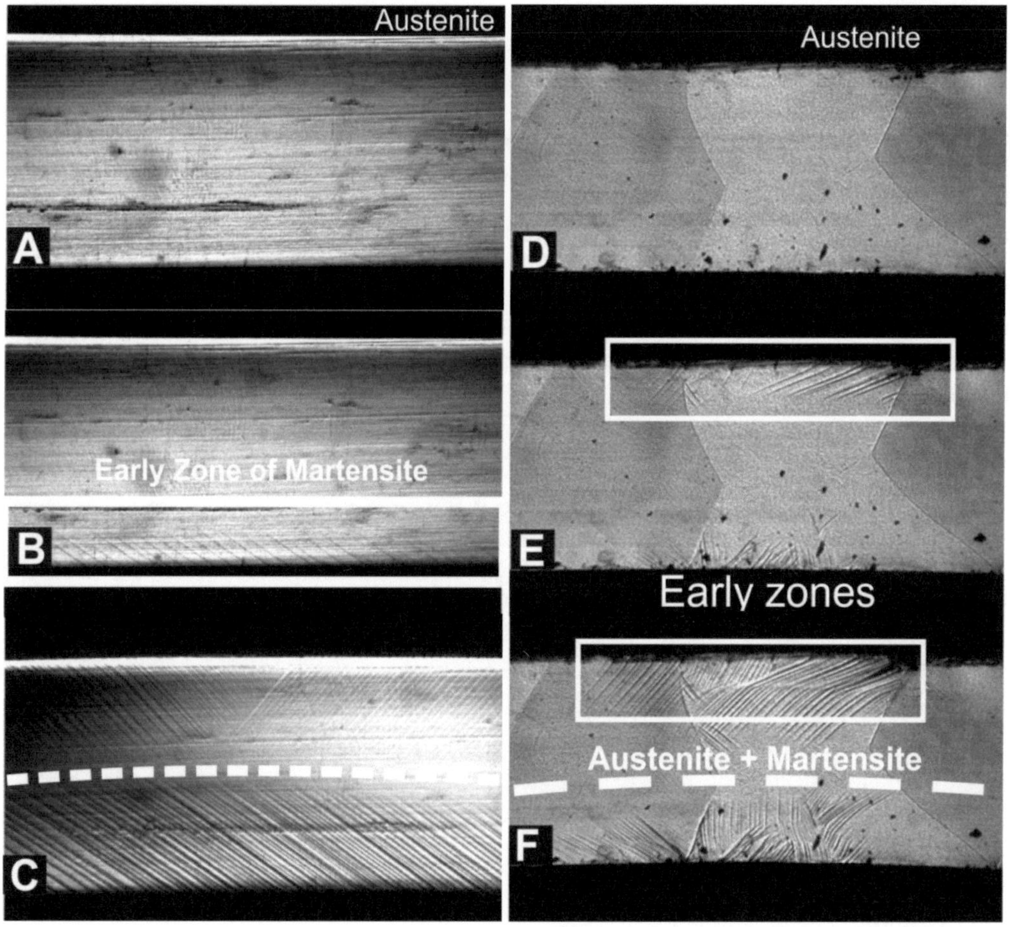

Figure 6. Mono and polycrystalline samples of CuAlBe under 3 point bending: a. Monocrystalline austenitic phase; b. Beginning of the martensitic transformation; c. Almost a total martensitic state for the monocrystal under bending; d. Austenitic phase in the polycrystal; e. Growth of the martensite plates in tension and compression; f. Several variants of martensite appear in the same grain

Another interesting observation was about the martensitic variants that grew up in both cases. The monocrystalline sample showed two variants according to the established angles respect to the horizontal line. In tension appeared a variant with two equivalent directions;

at left hand this variant was around 45º and at right hand 48º. In compression the stress-induced martensitic plates showed just a single direction that it was around 35º (Figure 7a). In this case should have appeared two variants close to 45º; nevertheless, a single variant grew up; it may due to the interaction with the fulcrum.

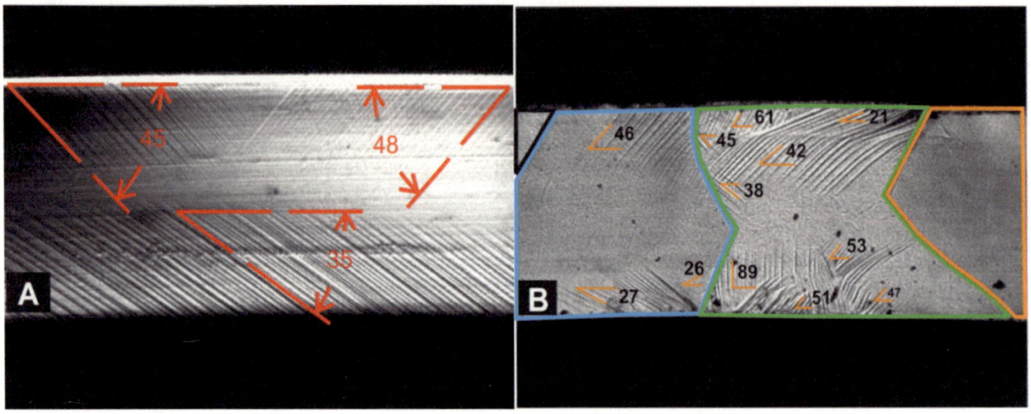

Figure 7. Stress-Induced martensitic variants: a) Monocrystal and b) Polycrystal

The Figure 7b shows the polycrystalline case. Here the central grains showed several martensitic variants; the first grain marked in blue showed one martensitic variant around 46º in tension but in compression the same grain showed two variants the first one was around 27º and the second one was around 26º that were different to those in tension; the compression variants are almost perpendicular each other; the grain marked in green is located at the center of the support span. So it is obvious that there are more martensitic plates in this region.

The martensitic plates are in several directions, all of them are close to -45º or 45º; nevertheless, there are a couple of variants that appeared in those directions that have less probability to exist according to the applied force direction and the direction of the early plates of martensite. These variants where located at 89º and 21º. It has to be pointed out that these variants appear in the same grain, which has a fixed crystalline orientation. So why do several variants of martensitic plates exist in the same grain? A possible answer is the granular interaction due to the martensitic transformation for the polycrystalline sample. It is clear that in the present work the crystalline orientation of the samples was not measured; however it is possible to infer the growth variants which have more chance to appear, if crystal orientations are guessed.

To identify these variants was used the list of habit planes and directions of Cu-Al-Be system reported by Kaouache et al.[21]. In addition to this list, it was used the procedure proposed by Bucheit et al. to get the surface transformation in single crystals [32]. Using the previous information, the plane stress transformational diagrams for typical crystalline orientations of the Cu-Al-Be in the austenitic phase were calculated in this work. The diagrams present irregular polygonal regions making evident the material anisotropy (Figure 8). This shows the existence of specific variants according to the state of stress (in typical cases like: tension-tension, tension-compression and compression- compression), habit plane and crystal orientation.

The plane stress transformational diagrams are located among [001], [011] and [-111] directions. It has to be noticed that these diagrams show significant differences between each other according to those guessed crystalline orientations; this means that there will be asymmetry and anisotropy between tension and compression in this material. This result agrees with the images that show the stress-induced martensitic transformation under 3-point bending (Figure 6).

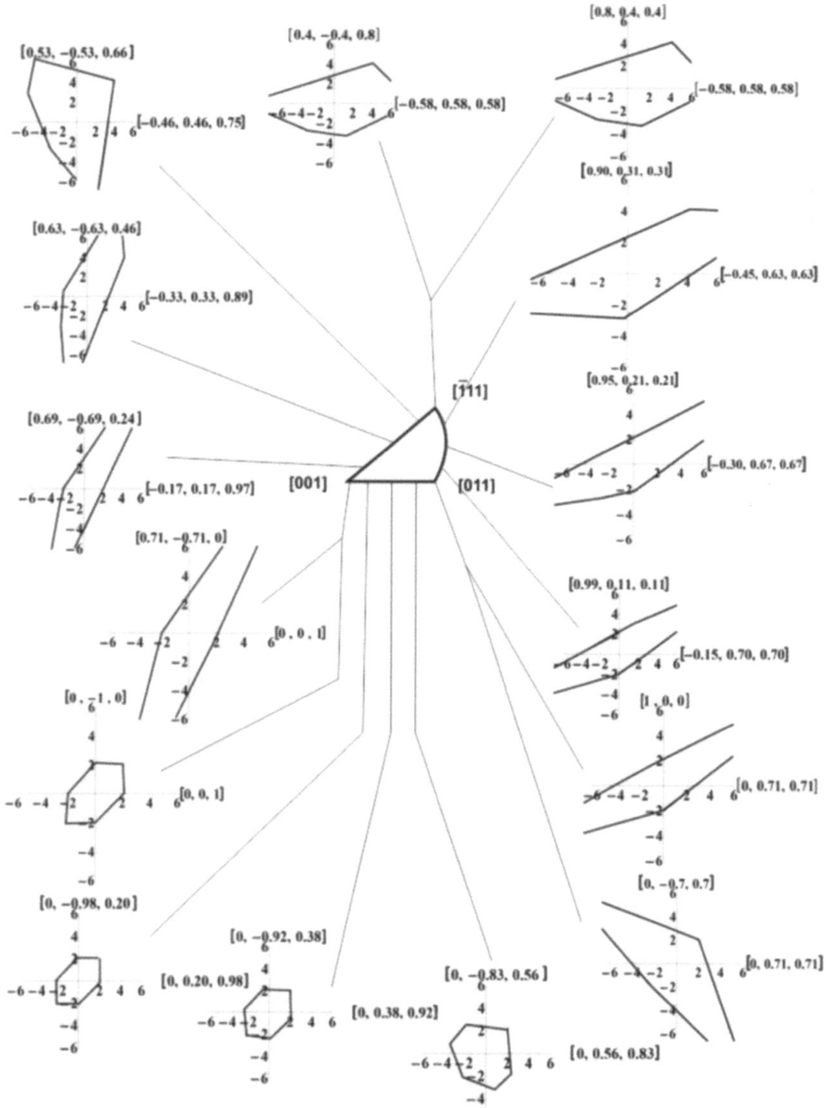

Figure 8. Plane stress transformational diagrams for typical crystalline directions of the Cu-Al-Be beta phase associated to different crystalline orientations represented in a stereographic projection.

From Figure 8 were selected four cases where several martensite variants can appear in tension and/or compression under the same applied force direction. The variants were identified according to the calculation proposed by Buccheit et al. for each martensitic variant taking in to account the guessed orientation. In general the diagrams present different variants with the same possibility to appear in accordance with the current state of stress; however, in this case the simple tension and compression cases are shown. These cases are shown in Figure 9; the mentioned variants present exactly the same trace in each circumstance. Here it is shown that several variants can appear because the Schmid Factor is equal for each variant; furthermore they have the same chance to grow up in the crystal. It is important to say that each variant present different mark on the observation surface as it is shown Figure 7.

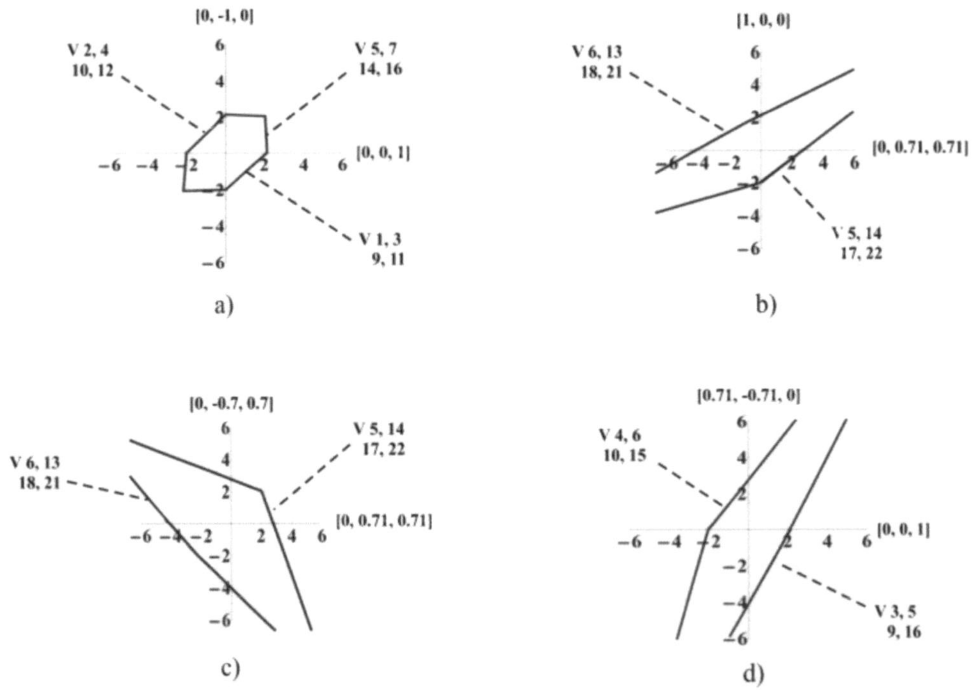

Figure 9. Plane stress transformational diagrams of CuAlBe and the Growth of several martensitic variants in a crystal with a fixed crystal orientation under the the same applied force direction. A) this figure shows the growth of 12 possible variants and figure b), c) and d) show four possible variants.

In order to observe the contribution of several martensitic variants to the micromechanical behavior of CuAlBe, digital image correlation was used get the displacement vector fields at different state of stress. The displacement vector fields are showed in Figure 10. Here it can be observed the curvature of the beam under bending.

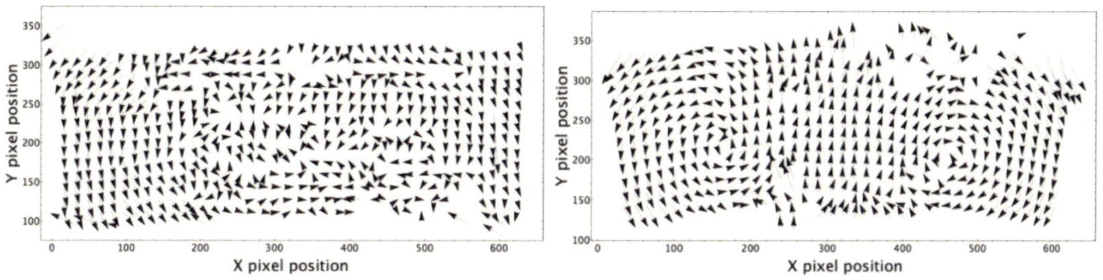

a) b)

Figure 10. Displacement vector fields of samples under 3-point bending: a) Monocrystal and b) Polycrystal.

From these displacement vector fields it is possible to observe that a moment *(M)* is acting in the both sides of the samples. In the polycrystalline case this effect is more evident. Isolating the rotation in "*xy*" plane and the "*y*" displacements was obtained the displacements in the "*x*" direction. These fields were completely overlapped on the corresponding images; they are showed in Figure 11. Here it is possible to observe that the samples are under tension and compression simultaneously.

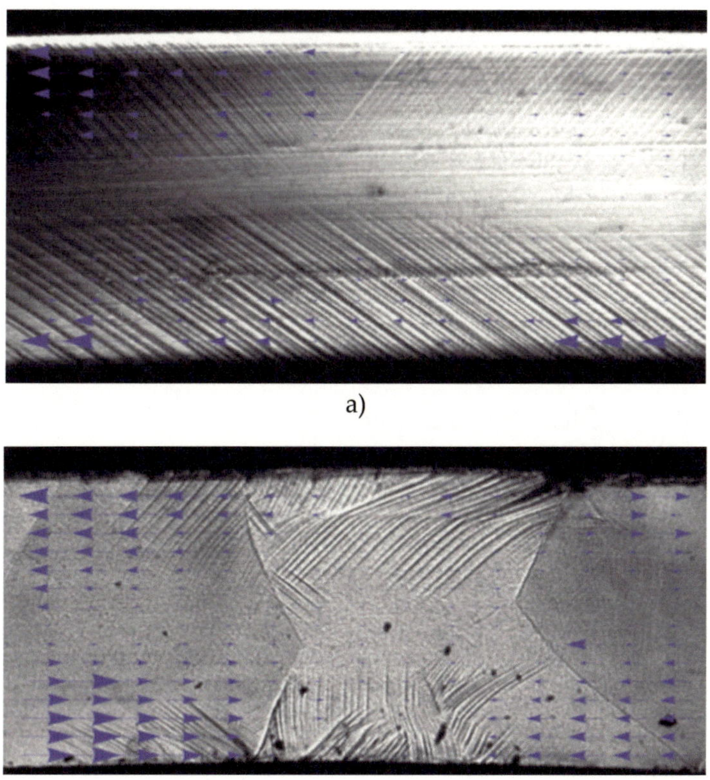

a)

b)

Figure 11. Overlapped displacement vector fields on CuAlBe Sample Undergoing 3 point bending: a) Martensitic variants in Monocrystal under tension and compression; b) Martensitic variants in a polycrystal. Martensitic reorientation phenomena is observed.

In both cases, the parts subjected to tension shows the expected *"x"* displacements; nevertheless, the region under compression for the monocrystal did not present the expected displacement vector field. This is due to the growth of a single variant. On the other hand, the region in compression of the polycrystal showed the expected displacement vector field showing compression. In all cases can be observed that the samples of CuAlBe present non-homogeneous behavior but this methodology was able to detect these small contributions in displacement caused by different martensitic variants.

4. Conclusions

This practical methodology was able to observe the micro and macromechanical behavior of Cu-Al 11.2 wt.%-Be 0.6 wt% polycrystal and Cu-Al 11.2 wt.%-Be 0.5 wt% monocrystal shape memory alloy undergoing a stress-induced martensitic transformation by a 3-point bending. The evolution of the martensitic transformation was registered by the CCD and was detected the non-symmetric behavior in tension and compression for both cases mono and polycrystalline samples. The overlapped displacement vector fields show the non-homogenous behavior of monocrystalline and polycrystalline samples of these alloys. This methodology was also able to detect the granular interaction in 2D confined grains; this may due to the interaction between the growth of martensite plates that modified the local state of stress in the grain and their neighbors. It is evident that there is a re-orientation effect of the martensitic phase while the load increases. This interaction provokes the apparition of several martensitic variants in different directions; this was in good agreement with the stress transformation diagrams for CuAlBe alloy.

Author details

R.J. Martínez-Fuentes, F.M. Sánchez-Arévalo and G.A. Lara-Rodríguez
Instituto de Investigaciones en Materiales, Universidad Nacional Autónoma de México, Cd. Universitaria, México

F.N. García-Castillo, J. Cortés-Pérez and A. Reyes-Solís
Centro Tecnológico Aragón, FES Aragón, Universidad Nacional Autónoma de México, Cd. Nezahualcoyotl Edo. De México, C.P, México

Acknowledgement

The authors wish to thank for the financial support to the DGAPA-UNAM through the IACOD program, project number I1101211 and PAPIIT project number IN111310. The authors are grateful to Esteban Fregoso Israel and Adriana Tejeda Cruz for their technical support.

5. References

[1] Olson, M.; Cohen, G. B.; Clapp, P.C. On the classification of displacive phase transformations. *Proceedings of the international conference on martensitic transformation. ICOMAT 79*, Cambridge-Massachussets U.S.A. 1979 pp. 1-11.

[2] Czichos, H. "Adolf Martens and the research of martensite." *Proceedings of the european conference on martensitic transformation in science and technology*. Bochum, Alemania, 1989; pp. 3-14.

[3] Patoor, E.; Berveiller, M. *Les alliages à mémoire de forme. Technologies de pointe*; Hermes, PARIS, 1990 ; pp 09-63.

[4] Wayman, C. M.; Duerig, T. M. *An introduction to martensite and shape memory, Engineering Aspects of Shape Memory Alloys*. Butterworth-Heinemann, London (UK), pp. 3-20.

[5] Otzuka, K.; Wayman, C. M.; Nakay, K.; Sakamoto, H,; Shumizu K. *Acta Metallurgica*, 1976. Vol.24, pp. 207-226.

[6] Yang, J. H.; Wayman, C. M. *Mater. Charact*. 1992. vol 28, pp. 23-35.

[7] Yang, J. H.; Wayman, C. M. *Mater. Charact*. 1992.vol. 28, pp. 37-47.

[8] Higuchi, A.; Suzuki, K.; Matsumoto, Y.; Sugimoto, K.; Komatsu, S.; Nakamura, N."Shape memory effect in Cu-Al-Be ternary alloy." In: *Proceedings of the international conference on martensitic transformations*, ICOMAT 1892. Leuven, Belgica. pp. 767-772.

[9] Higuchi, A.; Suzuki, K.; Sugimoto, K.; Nakamura, N. "Thermal stability of Cu-Al- Be shape memory alloy." In: *Proceedings of the international conference on martensitic transformations*, ICOMAT 1896. Nara, Japan. pp. 886-890.

[10] Belkahla S.; Flores H.; Guenin G. *Mater. Sci. Eng. A-Struct. Mater. Prop. Microstruct. Process*. 1993, A169. p. 119-124.

[11] Jurado, M.; Castan, T.; Mañosa, L.; Planes, A.; Bassas, J.; Alcobe, X; Morin, M. *Philos. Mag*. 1997, Vol. 5. pp. 1237-1250.

[12] Hautcoeur, A.; Eberhardy, A.; Patoor, E.; Berveiller M. "Thermomechanical behavior of monocrystalline Cu-Al-Be shape memory alloys and determination of the metastable phase diagram." *Journal de physique IV. Colloque C2, suplément au Journal de Physique III*.1995,Vol. 5. pp. c2-459 to c2-464.

[13] Siredey, N.; Eberhardt, A. *Mater. Sci. Eng. A-Struct. Mater. Prop. Microstruct. Process*.. 2000, A290, pp. 171-179.

[14] Balo, N.; Ceylan, M.; Aksoy, M. *Mater. Sci. Eng. A-Struct. Mater. Prop. Microstruct. Process*. 2001, A311, pp. 151-156.

[15] Balo, N.; Ceylan M. *J. Mater. Process. Technol*. 2002 Vol. 124. pp. 200-208.

[16] Pinh Zhang, Aibin Ma., Sheng Lu, Guanguo Liu, Pinghua Lin, Jinghua Jiang, Chenglin Chu. Effect of grain refinement on the mechanical properties of Cu–Al–Be–B shape memory alloy. Materials and Design. 2011. Vol. 32. Pp. 348-352.

[17] C de Albuquerque V. H. A de melo T. A. Ferreira D. Mediros R. Tavares J. M. Evaluation of grain refiners influence on the mechanical properties in a CuAlBe shape memory alloy by ultrasonic and mechanical tensile testing. Materials and Design. 2010. Vol.31. Pp.3275-3281.

[18] Bouvet C.; Calloch S.; Lexcellent, C. *Transactions of ASME*. 2002,Vol 124. pp.112-124.

[19] Chevalier, L.; Calloch, S.; Hild, F.; Marco, Y. Eur. *J. Mech. A/Solids*. 2001, vol. 20. pp. 169-187.

[20] Kaouache, B.; Berveiller, S.; Inal, K.; Eberhardt, A.; Patoor, E. *Mater. Sci. Eng. A-Struct. Mater. Prop. Microstruct. Process*. 2004, A378. pp. 232-237.

[21] Kaouache, B.; Inal, K.; Berveiller, S.; Eberhardt, A.; Patoor, E. *Mater. Sci. Eng. A-Struct. Mater. Prop. Microstruct. Process.* 2006, A 438-444. pp. 773-778.

[22] Siridey, N.; Patoor, E.; Berveiller, M.; Eberhardt, A. Int. J. Solids Struct. 1999, Vol. 36 pp. 4289-4315.

[23] Sánchez, F. M.; Pulos, G. *Materials science Forum*. 2006,Vol. 509. pp. 87-92.

[24] Sánchez-Arévalo, F. M.; Pulos, G. *Mater. Charact.* 2008,Vol. 59. Issue 11. pp. 1572-1579.

[25] Merzouki T., Collar C., Bourgeois N., Zineb T. B., Meraghni F. 2010. Mechanics of Materials Vol. 42 pp. 72-95.

[26] Bourgeois N., Meraghni F. , Zineb T. B. 2010. Physics Procedia. Vol. 10 pp. 4-10

[27] Willert C. E., Gharib M., 1991, Experiments in fluids. Vol. 10. pp. 181-193.

[28] Chu T. C., Ranson W. F., Sutton M.A., Peters W.H. 1985. "Aplications of digital correlation techniques to experimental mechanics". Experimental Mechanics. 25(3). pp. 232-244.

[29] Sánchez-Arévalo, F.M., García-Fernández, T., Pulos, G., Villagrán- Muniz, M., 2009. Use of digital speckle patter correlation for strain measurement in a CuAlBe shape memory alloy. Mater. Charact. 60, 775–782.

[30] Zhu J. J., Liew K.M. 2003. Description of deformation in shape memory alloys from DO3 austenite to 18R Martensite by group theory. Acta Materialia 51, 2443-2456.

[31] Patoor E., Lagoudas D. C. Entchev P. B, Brinson, C. Gao X. 2006. Shape memory alloy, Part 1: General properties and modeling of single Crystals. Mechanics of Materials, 38, 391-429

[32] Buchheit T. E. Wert J.A. 1994. Modeling the effects of stress state and crystal orientation on the stress-induced Transformation of NiTi single crystals. Metallurgical and Materials Transactions A. 25,283-238.

Section 4

Applications

Chapter 9

Applications of SMA Bundles in Practical Concrete Structures

Lei Li, Qingbin Li and Fan Zhang

Additional information is available at the end of the chapter

http://dx.doi.org/10.5772/51207

1. Introduction

Shape memory alloy (SMA for short) is a new kind of inter-metallic compound, which can "remember" its original size or shape. Since it was discovered that nickel-titanium showed shape memory effect (SME for short) in 1963, this metal had been being applied in many fields, such as aerospace technology, medical science, automotive industry, et al. At the same time, many constitutive models were developed to simulate its mechanical characteristics, yet there is not a satisfactory result because of its complicated microscopic mechanism.

In the recent twenty years, SMA composites were studied, and the matrix was almost polymer or fiber reinforced resin. During the recent ten years, researches on applying SMA in concrete structures were carried out, but yet limited in laboratory.

In this chapter, the constitutive characteristics of NiTi SMA and its application in practical concrete structures were investigated. Firstly, the transformation characters of the NiTi SMA adopted in this study were obtained through a differential scanning calorimeter (DSC) analysis technology, and then the properties of NiTi SMA during uncompleted transformation process and the effects of plastic deformation on the transformation were studied. The uniaxial tension, SME, and the constrained recovery process of NiTi SMA were examined through an improved 10KN universal material testing machine, effects of the constraint conditions, the maximum tension deformation and the initial phase of SMA on the constrained recovery process were compared and investigated.

To simulate the characteristics of NiTi SMA more effectively, a new constitutive model derived from the internal variable approach was constructed based on the DSC and the uniaxial tension experimental results, a new simple kinetics equation was presented, and the plastic deformation was considered in the constitutive equation.

As the application of SMA in practical concrete structures, a new concept for designing a smart concrete bridge in a freeway was presented against the overload problem in the transport field, and SMA was firstly applied in controlling the deformation of a concrete beam bridge, performances of the concrete beams and the concrete bridge with embedded SMA were examined through a sophisticated test program.

2. One-dimensional constitutive model of sma with an empirical kinetics equation

The characteristics of NiTi SMA were examined through sophisticated testing program, and the constitutive model of NiTi SMA with an empirical kinetics equation was investigated. Firstly, the transformation characters of the NiTi SMA were obtained through a differential scanning calorimetry (DSC) analysis technology, and the properties during incomplete and discontinuous transformation process and the effects of plastic deformation on the transformation were studied. The uniaxial tension, SME and constrained recovery process of NiTi SMA were examined through an improved 10KN universal material testing machine. Experimental results indicated that the phase transformation characters and the mechanical properties could be affected by the loading process considerably, and the plastic deformation should be taken into account. To simulate the characteristics of NiTi SMA more effectively, a one-dimensional constitutive model derived from the internal variable approach with the consideration of the plastic deformation was constructed based on the DSC and the uniaxial tension experimental results, in which a new simple empirical kinetics equation was presented, and the transformation temperature parameters were redefined according to the DSC experimental evidence. Comparison between the numerical and experimental results indicated that this constitutive model could simulate the phase transformation characters, the uniaxial tension, SME and the constrained recovery behavior of NiTi SMA well.

2.1. Background

Since the Shape Memory Effect (SME for short) was observed in Cu-Zn alloy (Greninger & Mooradian, 1938), and in NiTi alloy (Buehler, Gilfrich, & Wiley, 1963), SMA (especially NiTi alloy) has been widely used as intelligent material for its particular characteristics, such as large load capacity, high recovery strain up to 8%, excellent fatigue performance, variable elastic modulus with phase transformation, and especially the two main interesting properties, SME and Pseudo-elasticity (PE) due to the diffusionless martensitic transformation, as discussed elsewhere (Delaey et al., 1974; Krishnan et al., 1974; Warlimont et al., 1974; Funakubo, 1987; Otsuka & Wayman, 1998).

To simulate these specific properties, many constitutive laws have been proposed, such as the phenomenological models (Tanaka & Nagaki, 1982; Tanaka, 1986; Liang, 1990; Brinson, 1993; Zhu et al., 2002), micromechanics models (Sun & Hwang, 1993a, 1993b), 3D model for polycrystalline SMA based on microplane theory (Brocca, Brinson, & Bazant, 2002), etc.

Among them, phenomenological models based on the internal variable method were the most popular used in the practical engineering. The mechanical constitutive equation was derived from the principles of thermodynamics, martensite fraction as an internal variable was used to represent the stage of the transformation, and an empirical kinetics equation was proposed to describe this transformation governed by temperature and stress. The major difference of these models was its specific kinetics equation.

The first phenomenological model was derived by Tanaka and Nagaki (1982) from the first and second laws of thermodynamics and can be written as

$$\sigma = \rho_0 \frac{\partial \Phi}{\partial \varepsilon} = \sigma(\varepsilon, T, \xi) \tag{1}$$

where σ, ρ_0, Φ, ε, T and ξ represent the second Piola-Kirchhoff stress, the density, Helmholtz free energy, Green strain, temperature and the martensite fraction, respectively.

Equation (1) can be written by differential calculus as

$$\dot{\sigma} = \frac{\partial \sigma}{\partial \varepsilon}\dot{\varepsilon} + \frac{\partial \sigma}{\partial T}\dot{T} + \frac{\partial \sigma}{\partial \xi}\dot{\xi} = D\dot{\varepsilon} + \Theta\dot{T} + \Omega\dot{\xi} \tag{2}$$

where $D = \rho_0 \frac{\partial^2 \Phi}{\partial \varepsilon^2}$ is the modulus of the SMA materials, $\Theta = \rho_0 \frac{\partial^2 \Phi}{\partial \varepsilon \partial T}$ is related to the thermal coefficient of expansion, and $\Omega = \rho_0 \frac{\partial^2 \Phi}{\partial \varepsilon \partial \xi}$ can be regarded as the "transformation tensor".

The relationship between the martensite fraction, the temperature and the stress is expressed by an exponential kinetics equation as

$$\xi_{M \to A} = \exp\left[a_A(T - A_s) + b_A \sigma\right] \tag{3}$$

for the transformation from martensite to austenite, and

$$\xi_{A \to M} = 1 - \exp\left[a_M(T - M_s) + b_M \sigma\right] \tag{4}$$

for the transformation from austenite to martensite, where a_A, a_M, b_A, b_M are the material constants related to the transformation temperature, A_s and M_s are the start temperatures of austenite transformation and the martensite transformation, respectively.

Liang (1990) simplified the constitutive model based on Equation (2), material functions were assumed to be constants, and the constitutive relation can be written as

$$\sigma - \sigma_0 = D(\varepsilon - \varepsilon_0) + \Omega(\xi - \xi_0) + \Theta(T - T_0) \tag{5}$$

where the subscript "0" indicates the initial conditions of the materials. Martensite fraction, as a function of stress and temperature during transformation, is represented by an empirically based cosine models as

$$\xi_{M\to A} = \frac{\xi_0}{2}\cos[a_A(T - A_s) + b_A\sigma] + \frac{\xi_0}{2} \tag{6}$$

for the transformation from martensite to austenite, and

$$\xi_{A\to M} = \frac{1-\xi_0}{2}\cos[a_M(T - M_f) + b_M\sigma] + \frac{1+\xi_0}{2} \tag{7}$$

for the transformation from austenite to martensite, where ξ_0 is the initial martensite fraction, a_A, a_M, b_A, b_M are the material constants, A_s is the start temperature of austenite transformation, M_f is the finish temperature of martensite transformation.

Brinson (1993) redefined the martensite fraction based on the micromechanics of SMA material as

$$\xi = \xi_S + \xi_T \tag{8}$$

where ξ_s represents the fraction of the stress-induced martensite (or single-crystal martensite) and ξ_T denotes the fraction of the temperature-induced martensite with multiple variants.

Young's modulus was assumed to be non-constant from the experimental evidence as

$$D(\varepsilon,\xi,T) = D(\xi) = D_A + \xi(D_M - D_A) \tag{9}$$

where D_A and D_M are the Young's moduli of austenite and martensite.

The transformation tensor was also redefined as

$$\Omega(\xi) = \Omega(\xi_0) + (\xi - \xi_0)\Omega'(\xi_0) \tag{10}$$

Because of these improvements, phenomenological models based on the internal variable method could simulate the characteristics of SMA well, such as SME and PE, and have been used in some commercial program for its simple form. Whereas, DSC experiment indicated that there were no apparent transformation start and finish temperatures. And additionally, for the SMA materials used as actuators in practical engineering, not only the stress but also the thermo load determined the transformation process simultaneously, especially the latter. In this section, constitutive model was presented to simulate the characteristics of SMA materials for its practical utilization based on the former effort and the thermal and mechanical experiments. Transformation temperature parameters were redefined, a new simple empirical kinetics equation was presented, and a mechanical equation was developed to describe its mechanical character more directly.

2.2. Experiments and results

2.2.1. Thermodynamics experiment

2.2.1.1. Complete transformation

The Ti-49.5 wt % Ni in this research (binary, straight annealed) was one-way SMA. Its transformation property can be achieved through a DSC apparatus (type: Mettler Toledo DSC821 e), as shown in Fig. 1, the four transformation temperatures can be determined as: martensite transformation start and finish temperatures denoted by M_s and M_f are 33.8 °C and 23.3 °C, austenite transformation start and finish ones (A_s and A_f) are 41.9 °C and 59.6 °C, respectively. It should be noticed that these transformation start and finish temperatures were defined through the DSC diagram artificially, actually the transformation could take place slowly earlier than the transformation start temperature and remain unfinished after the transformation finish temperature, as shown in the experimental curve, thus the recovery forces induced by the transformation might increase or decrease beyond the transformation start and finish temperatures. Accordingly, a new series of parameters were adopted to describe this transformation character more realistically, where T_A and T_M denotes the temperatures corresponding to the peak points of the austenite and martensite transformations, A and M indicate the widths of the transformation peaks, $A \approx \left(A_f - A_s\right)/2e$ and $M \approx \left(M_s - M_f\right)/2e$, and e is the natural constant and equals to 2.71828.

Subtracting the exothermic or the endothermic part of non phase transformation process from the DSC curve (as seen in Fig. 1), normalizing the heat absorbing and releasing capacities, and taking the absolute value, martensite quality fraction ratio versus temperature can be achieved, as shown in Fig. 2.

Calculating the integral of the martensite quality fraction ratio, the relationship of martensite quality fraction versus temperature during the martensite and austenite transformation process can be obtained, as shown in Fig. 3.

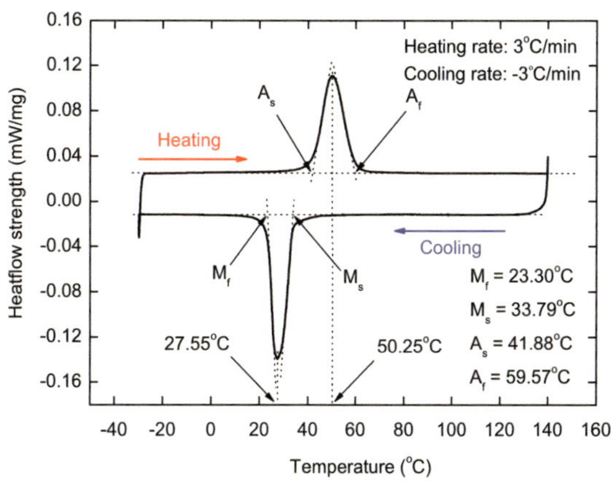

Figure 1. DSC diagram of SMA

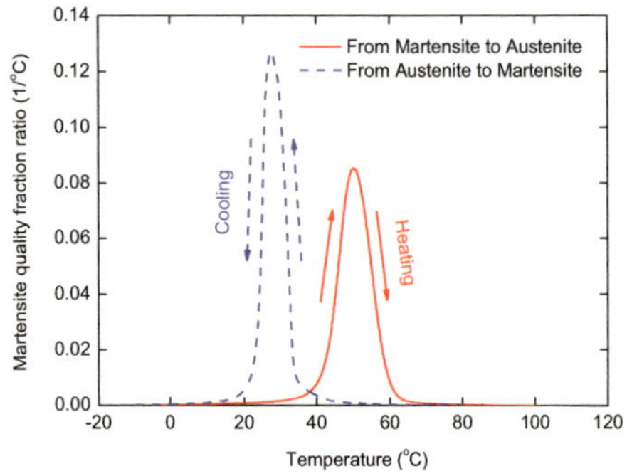

Figure 2. Variation rate of martensite quality fraction versus temperature

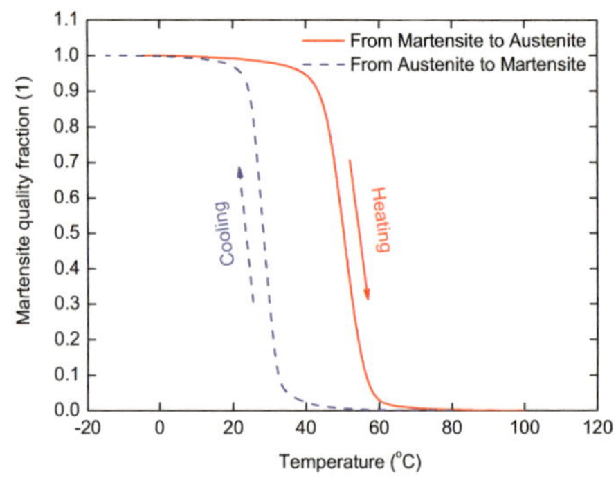

Figure 3. Variation of martensite quality fraction versus temperature

2.2.1.2. Incomplete and discontinuous transformation

If the initial state is not the complete martensite or austenite, the phase transformation will be different.

At the temperature of 140 °C, the initial state of SMA is pure austenite, cooling will induce the transformation from austenite to martensite, while at half of this transformation, heating will lead to the reverse transformation from the new generated martensite to austenite, the start and finish temperature and the shape of the peak are similar to the ones of the complete austenite transformation, but the peak value is smaller than the complete transformation one, as shown in Fig. 4.

Similarly, at the temperature of -30 °C, the initial state is pure martensite, heating will induce the transformation from martensite to austenite, while at half of the transformation, cooling will lead to the transformation from the new generated austenite to martensite, the start and finish temperatures and the shape of the peak are similar to those of the complete martensite transformation, but the peak value becomes smaller than the complete transformation one, as shown in Fig. 5.

These phenomena indicate that the martensite transformation intensity is related to the initial austenite quality fraction and the austenite transformation intensity is related to the initial martensite quality fraction. Accordingly, it is assumed that the martensite transformation intensity is proportional to the initial austenite fraction, and the austenite transformation intensity is proportional to the initial martensite fraction.

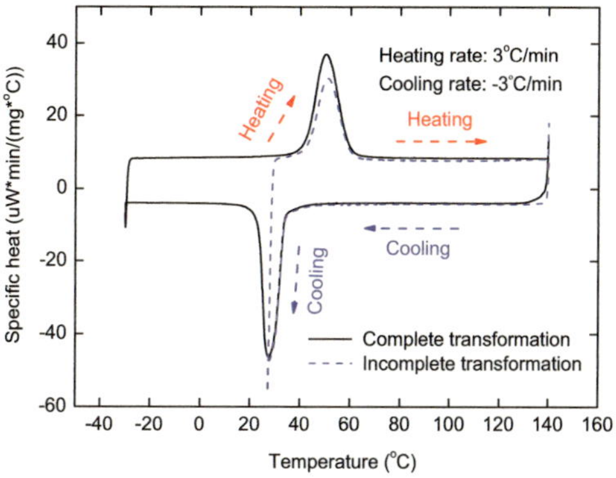

Figure 4. Incomplete transformation from martensite to austenite

During the transformation from austenite to martensite, i.e., the cooling process, if the small heating does not induce the transformation from martensite to austenite, the martensite transformation will continue during the subsequent cooling process, as shown in Fig. 6.

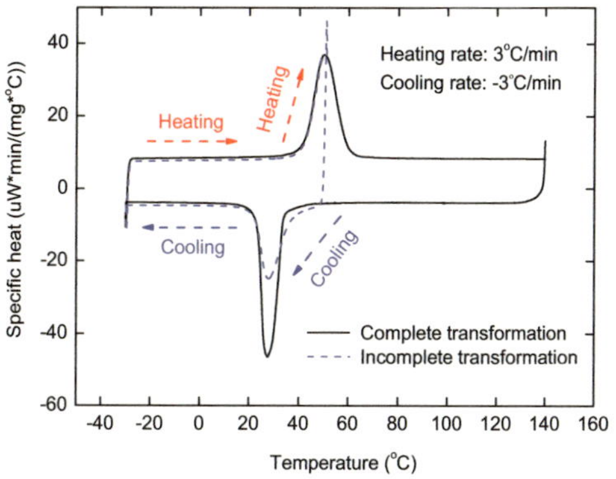

Figure 5. Incomplete transformation from austenite to martensite

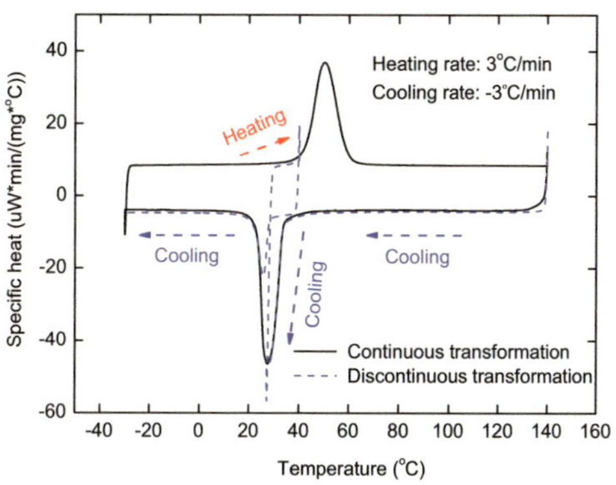

Figure 6. Discontinuous transformation from austenite to martensite

Similarly, during the transformation from martensite to austenite, i.e., the heating process, if the small cooling does not induce the transformation from austenite to martensite, the austenite transformation will continue during the subsequent heating process, as shown in Fig. 7.

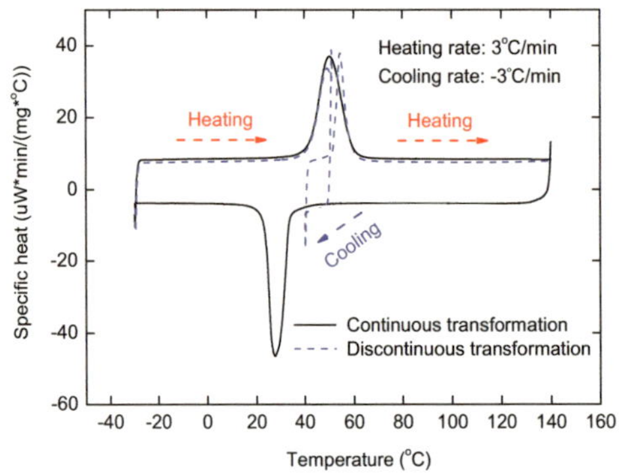

Figure 7. Discontinuous transformation from martensite to austenite

2.2.1.3. *Influence of plastic deformation*

Plastic deformation can also change the transformation properties of SMA materials. Stretch SMA to its strain limit, great plastic deformation will be induced, and the DSC experimental result can be achieved as shown in Fig. 8. The temperatures of the austenite transformation and martensite transformations become higher, the width of the peak become wider, and the peak value become smaller than these before tension. This phenomenon comes from the variation of the crystal structure and the internal stress induced by the plastic deformation. Actually, the transformation properties of the SMA crystal cell did not change, just the micro structure of SMA material varied.

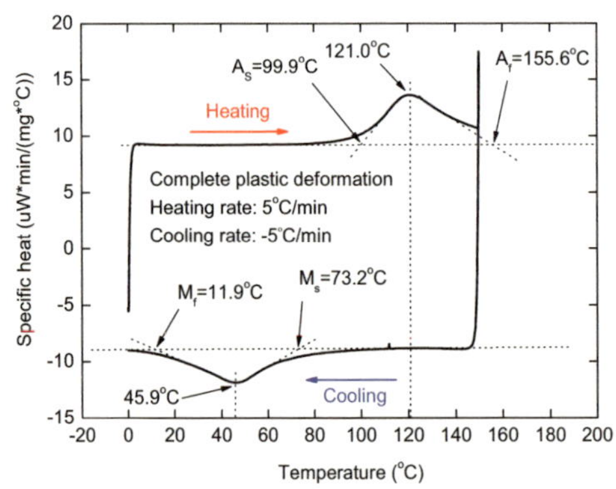

Figure 8. DSC analysis of plastic deformed SMA

2.2.2. *Mechanics experiment*

Mechanical experiment was carried out to determine the material characteristics and parameters through a WDW-10 universal material test machine. The experimental equipment is shown in Fig. 9. Pt100 platinum electric resistance temperature sensors were plastered on the SMA wire with silica gel to monitor its temperature, strain gauge was used to measure its strain. Load applied on the SMA wire, environment temperature, current intensity, and the voltage of SMA wire between the two points of the strain gauge were also monitored at the same time. All these signals were acquired through a 16-channel dynamic data acquisition system.

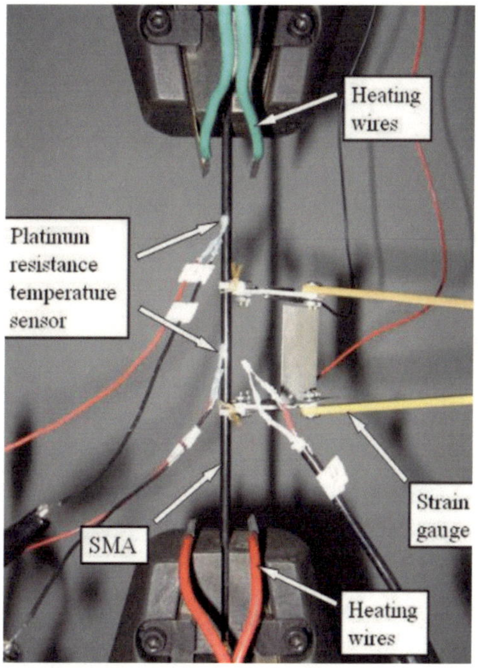

Figure 9. Uniaxial tension test equipment

2.2.2.1. Uniaxial tension test

The first test was made with constant ambient temperature. SMA wire samples were heated by boiled water firstly, and then cooled by ice water, thus the initial state of the samples was pure twinned martensite. Experimental results are shown in Fig. 10. There are two horizontal "plastic" stages during the tension process of 1mm diameter SMA wire. The first stage was derived from the detwinning process of the twinned martensite, i.e. from twinned martensite to single martensite, which can be named "pseudo-plasticity" that could be recovered by heating, and the second stage was induced by plastic deformation, which was the "real plasticity".

The second uniaxial tension test was carried out with different SMA temperatures. Similarly, SMA wire samples were also pretreated to obtain different initial state. For example, if the sample was heated by boiled water, then cooled by ice water, and then heated to the test temperature of 44.5 °C by electric current, the initial state will be twinned martensite. However, if the sample was heated by boiled water, and then cooled to the test temperature of 44.5 °C, the initial state will be austenite. During the experimental process, testing temperatures were kept constant by adjusting the current intensity.

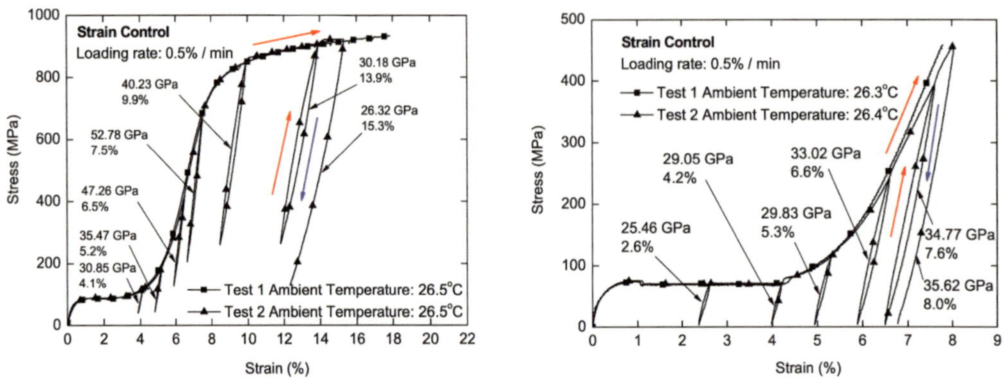

(a). SMA diameter: 1mm (b). SMA diameter: 5mm

Figure 10. Uniaxial tension test

Elastic moduli of austenite, twinned martensite and detwinned martensite can be determined as 85.5 GPa, 28.24 GPa and 25.75 GPa, respectively based on the experimental result, as shown in Fig. 11, where "M" means the initial state of the sample as twinned martensite, and "A" as austenite. At different ambient temperatures, the initial elastic moduli of austenite are the same. In the ambient temperature of 44.5 °C, no matter what the initial state is, the critical stress that leads to the stress-induced martensite transformation or the detwinning process of twinned martensite was almost equal.

During the heating recovery process under stress free conditions, the critical temperatures of the thermo-induced austenite transformation were almost the same. Otherwise, the unrecoverable stain of the specimen stretched from martensite initial state was a little bit bigger than that from austenite, as shown in Fig. 12.

For the sample starting from complete austenite, the critical stress inducing martensite transformation increased with the temperature. The statistic result is shown in Fig. 13.

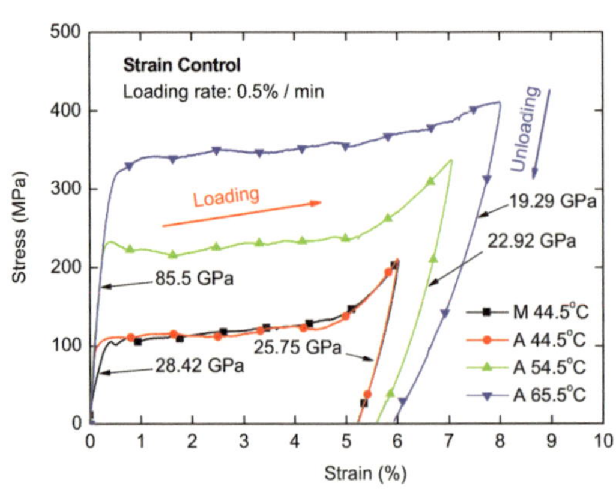

Figure 11. Uniaxial tension processes under different ambient temperatures

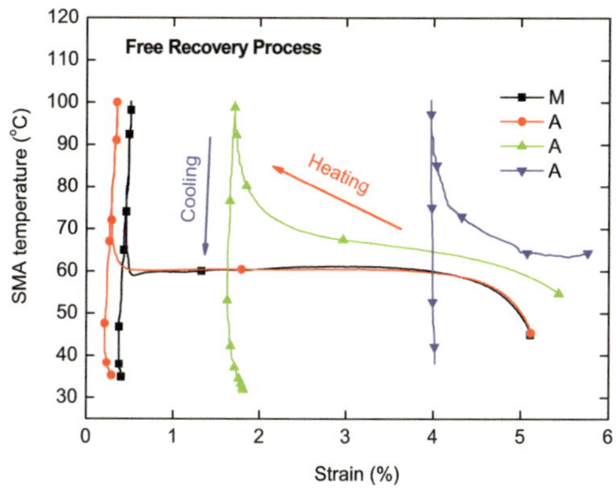

Figure 12. Free recovery processes of SMA after loading with different temperatures

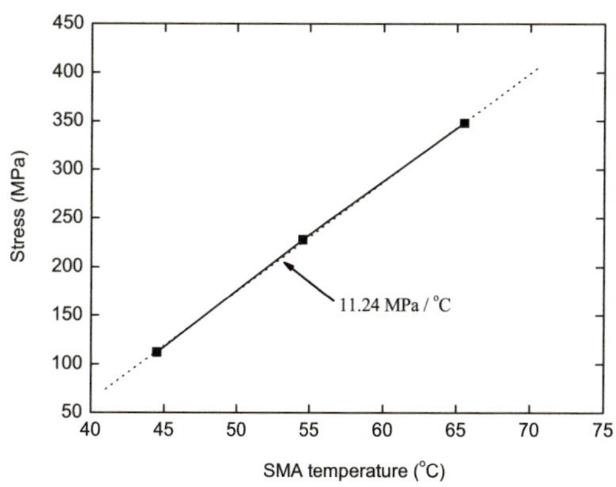

Figure 13. Variation of critical stress with stress induced martensite transformation temperature

2.2.2.2. SME test

SMA samples with 5mm in diameter were firstly stretched to 4.2%, 5.3%, 6.6%, 7.6%, respectively, and then heated in strain free conditions, this loading cycle was repeated for three times.

For the sample with the loading strain of 5.3%, variation of stress with strain during the uniaxial tension process is shown in Fig. 14, and the variation of strain with temperature during the free heating process is shown in Fig. 15. Stress-strain curve of SMA with other maximum loading strains are similar to Fig 14 and Fig. 15.

Experimental results indicated that the plastic deformation increased with the loading times, and during the free recovery process, critical temperature inducing the reverse martensite transformation decreased with loading times, and the recoverable strain became smaller and smaller.

Comparing the samples with different maximum tension strain, variations of stress with strain and that of strain with temperature during the second loading times and free recovery process are shown in Fig.s 16 and 17, respectively. Test results during the first and the third loading cycles are similar to each other.

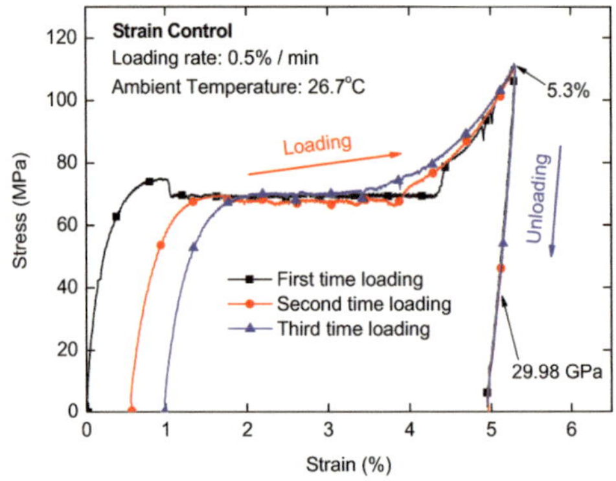

Figure 14. Stress-strain curve of SMA in different loading times

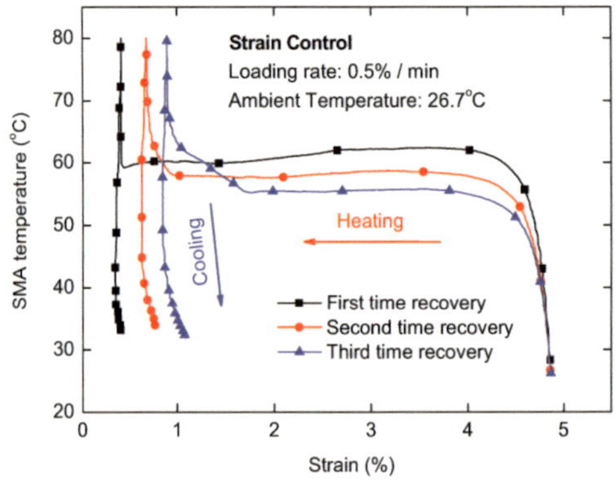

Figure 15. Temperature-strain curve of SMA in different loading times

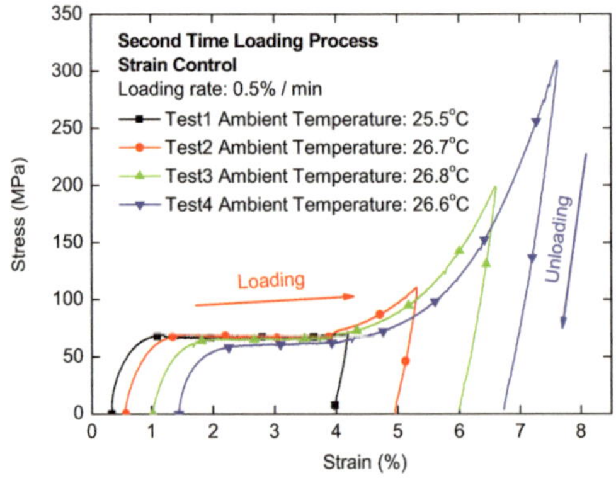

Figure 16. Stress-strain curve of SMA in different maximum loading deformations

Additionally, statistical results indicated that the critical temperature inducing the reverse martensite transformation increased with the maximum deformation, decreased with the loading/unloading times, as shown in Fig. 18. At the same time, the plastic deformation increased with the maximum deformation and the loading/unloading times, as shown in Fig. 19.

According to the experimental curve, the plastic deformation increased with the initial tension strain during the loading and unloading process, and the critical temperature inducing the austenite transformation increased with the initial tension strain, also increased with the plastic deformation.

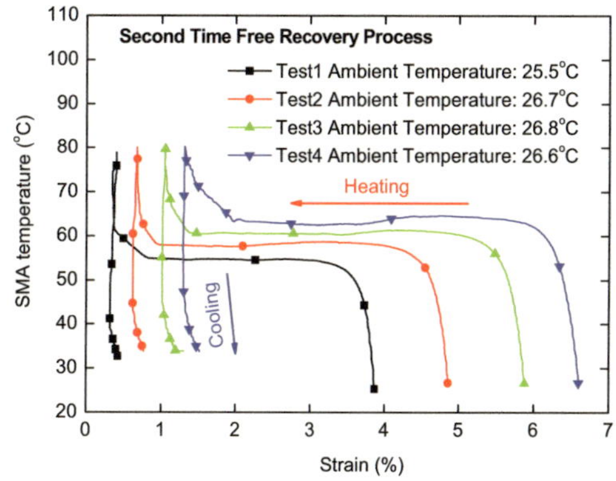

Figure 17. Temperature-strain curve of SMA in different maximum loading deformations

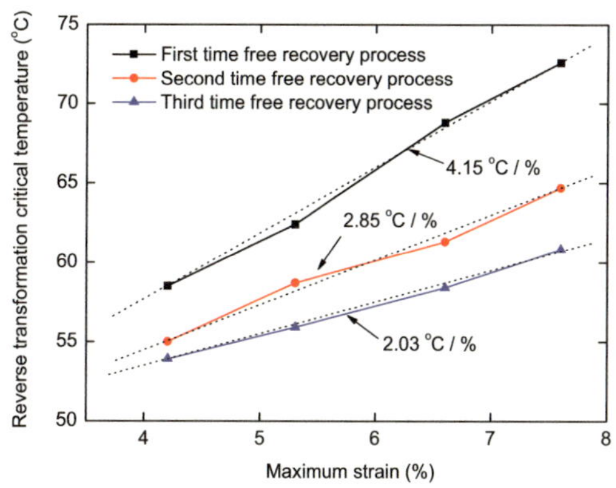

Figure 18. Variation property of transformation critical temperature with maximum loading strain

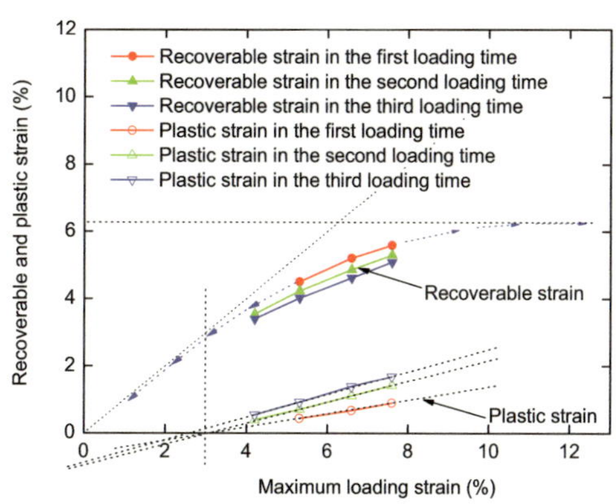

Figure 19. Variation property of recoverable and plastic strain with maximum loading strain

2.2.2.3. Recovery test under constrained condition

The recovery test under constrained condition contained five loading paths: (1) strain-control loading to 4.2% under normal temperature; (2) heating to 90 °C and then cooling to 30 °C under elastic constrained condition in the first time; (3) heating to 90 °C and then cooling to 30 °C under elastic constrained condition in the second, third and fourth times; (4) unloading to free condition; (5) heating to 100 °C and then cooling to 30 °C under free condition.

In the initial state of twinned martensite, the elastic constrained recovery test of SMA is shown in Fig. 20. Test results under different constrained conditions, initial state and maximum strain are similar in form.

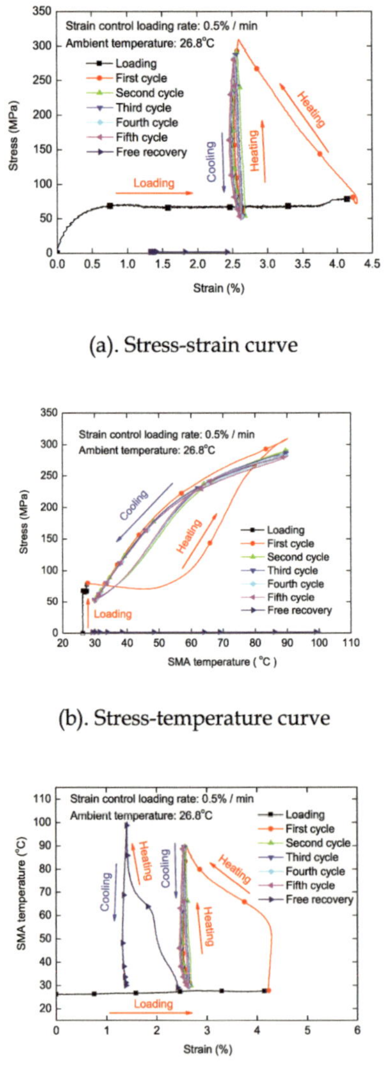

(a). Stress-strain curve

(b). Stress-temperature curve

(c). Temperature-strain curve

Figure 20. Constrained recovery test

Because constraint elastic modulus during the first heating process was smaller than that of the subsequent second to fifth heating/cooling process, relationships between stress and strain, stress and temperature, and temperature and strain, were different correspondingly. In the first heating process, the stress increased gradually, the strain decreased and some deformation of SMA restored. In the following heating and cooling process, the strain almost remained, the stress increased with temperature rising and decreased with temperature dropping, forming a hysteretic circle, and the variation rate of stress with temperature changed in the range of 2.4 MPa ~ 8 MPa / °C. During the heating process under stress free state, the strain recovery course of SMA with temperature showed two obvious stages, as shown in Fig. 20 (c).

2.3. Constitutive model

2.3.1. *Physical equation*

According to the macro-phenomenon, deformation of SMA during thermo and stress loading process can be divided into four components: (1) macro recoverable deformation caused by martensite transformation, mainly determined by single crystal martensite percentage; (2) elastic deformation; (3) Thermal expansion deformation; (4) Plastic deformation. Thus, the macro physical equation of SMA in strain form during the temperature and stress loading process can be expressed as

$$\begin{aligned}\varepsilon &= \varepsilon^R + \varepsilon^E + \varepsilon^T + \varepsilon^P \\ &= \varepsilon_{res} \cdot \xi_{SM} + \frac{\sigma}{\left(\xi_{SM} \cdot E_{SM} + \xi_{TM} \cdot E_{TM} + \xi_A \cdot E_A\right)} + \left(\xi_{SM} \cdot \eta_{SM} + \xi_{TM} \cdot \eta_{TM} + \xi_A \cdot \eta_A\right) \cdot \left(T - T_0\right) + \varepsilon^P \end{aligned} \tag{11}$$

where ε is the dimensionless macro total strain, positive for stretch; ε^R, ε^E, ε^T, ε^P denote the dimensionless recoverable transformation, elastic, thermal expansion and plastic strains, respectively; ε_{res} is the maximum recoverable strain; σ is the macro stress, positive for stretch (unit: MPa). E_{SM}, E_{TM}, E_A denote the elastic moduli of single crystal martensite, twinned martensite and austenite, respectively (units: GPa). η_{SM}, η_{TM}, η_A denote the thermal expansion coefficients of single crystal martensite, twinned martensite and austenite, respectively (units: $1/^{O}C$). ξ_{SM}, ξ_{TM}, ξ_A denote the quality fractions of single crystal martensite, twinned martensite and the austenite, respectively, with $\xi_{SM} + \xi_{TM} + \xi_A$ equals to 1. *T* is the temperature of SMA (unit: °C). T_0 is the initial temperature of SMA (unit: °C).

2.3.2. *Kinetics equation*

DSC experiments showed that, there were no obvious transformation start and finish temperatures during the transformation process. In this section, the single transformation characteristic temperatures, as well as the parameters that reflect the width of the endothermic and exothermic peaks, are adopted to simulate the phase transformation of SMA, and the corresponding kinetics equation is established at the same time.

During the transformation process, there are three crystal structures in SMA material, twinned martensite, single martensite, and austenite, respectively, thus there are six phase transformation processes theoretically. In this section, each phase transition process is analyzed based on the DSC experiments, and the corresponding transformation principles and conditions are presented.

2.3.2.1. Transformation from twinned martensite to austenite

Transformation condition: Assuming in one-dimensional terms, only heating can lead to this phase transition from twinned martensite to austenite.

Transformation principle: For the transformation from twinned martensite to austenite, if the reverse transformation from austenite to twinned martensite did not occur during the process, a small cooling will not affect its subsequent transformation process, as shown in Fig. 7.

However, if the reverse transformation process happened and induced a lot of austenitic to twinned martensite, then the extent of the following transformation from twinned martensite to austenite required reduction in accordance with the quality fraction of the current twinned martensite, as shown in Fig. 4.

Thus, the kinetics equation from twinned martensite to austenite can be acquired as

$$\xi_{TM} = \frac{\xi_{TM0}}{1 + e^{\frac{T - T_A}{A}}} \tag{12}$$

where ξ_{TM0} is the initial quality fraction of twinned martensite; T_A is the characteristic temperature for austenite transformation, and numerically equivalent to the temperature where the complete austenite transformation goes into the half, and is equivalent to the peak temperature of the DSC endothermic curve approximately; A is the transformation parameter reflecting the width of the endothermic peak, and numerically equivalent to $\frac{A_f - A_s}{2e}$, here A_s, A_f denote the start and finish temperatures of austenite transformation defined by the traditional method. e is the natural constant and equals to 2.71828.

2.3.2.2. Transformation from austenite to twinned martensite

Transformation condition: Assuming in one-dimensional terms, only cooling can lead to this phase transition from austenite to twinned martensite.

Transformation principle: For the transformation from austenite to twinned martensite, if the reverse transformation from twinned martensite to austenite did not occur during the process, a small heating will not affect its subsequent transformation process, as shown in Fig. 6.

However, if the reverse transformation process happened and induced a lot of twinned martensite to austenitic, the extent of the following transformation from austenite to twinned martensite required reduction in accordance with the quality fraction of the current austenite, as shown in Fig. 5.

Thus, the kinetics equation from austenite to twinned martensite can be acquired as

$$\xi_{TM} = \frac{\left(1-\xi_{TM0}\right)}{1+e^{\frac{T-T_M}{M}}} + \xi_{TM0} \tag{13}$$

where T_M is the characteristic temperature for martensite transformation, and numerically equals to the temperature where the complete martensite transformation goes into the half, and is equivalent to the peak temperature of the DSC exothermic curve approximately; M is the transformation parameter reflecting the width of the exothermic peak, and numerically equivalent to $\frac{M_s - M_f}{2e}$, here M_s, M_f denote the start and finish temperatures of martensite transformation defined by the traditional method.

2.3.2.3. Transformation from twinned martensite to single martensite

Transformation condition: Assuming in one-dimensional terms, only the increase of stress can lead to this phase transition from twinned martensite to single martensite.

Transformation principle: During the uniaxial tension process, for the transformation from twinned martensite to single martensite, if the none macro press stress occur during unloading process, this unloading will not affect the transformation of the subsequent reloading process.

During the phase transition process, taking the twinned martensite as austenitic, this transformation will be similar to the transformation from austenite to single martensite, only with the different phase transition peak temperature, and the transformation parameters and impact coefficient of the stress on the temperature still adopted the values of transformation from austenite to twinned martensite.

Therefore, based on the kinetics of the transformation from austenite to twinned martensite, kinetics of the transformation from twinned martensite to single one can be achieved as

$$\xi_{SM} = \frac{1}{1+e^{\frac{T-\frac{\sigma^i}{C_{M'}}-T_{M'}}{M'}}} \tag{14}$$

or

$$\xi_{TM} = \frac{1}{1+e^{\frac{\sigma^i-\sigma^i_{crit}}{C_{M'}\bullet M'}}} \tag{15}$$

where σ^i denotes the average internal stress between the crystals (unit: MPa); $T_{M'}$ is the transformation characteristic temperature; M' is the transformation parameter; $C_{M'}$ denotes the impact coefficient of stress on transformation temperature; σ^i_{crit} denotes the average internal stress between the crystals where the full de-twinned process occurred on the half (unit: MPa).

2.3.2.4. Transformation from single martensite to twinned one

Transformation condition: For uniaxial tension process, macro-stress and the temperature will not induce the transformation from single martensite to twinned one. However, under repeated stretch-press loading, the martensites with different directions will change its direction repeatedly, and this process corresponds to the SMA rubber-like plastic phenomenon.

2.3.2.5. Transformation from single martensite to austenite

Transformation condition: Assuming in one-dimensional terms, heating and the decrease of stress can lead to this phase transition from single martensite to austenite.

Transformation principle: As similar to the transformation from twinned martensite to austenite, for the transformation from single martensite to austenite, if the reverse transformation from austenite to single martensite did not occur during the process, a small cooling or increase of stress will not affect its subsequent transformation process.

However, if the reverse transformation process happened and induced a lot of austenitic to single martensite, then the extent of the following transformation from single martensite to austenite required reduction in accordance with the quality percentage of the current single martensite.

Thus, the kinetics equation from single martensite to austenite can be acquired as

$$\xi_{SM} = \frac{\xi_{SM0}}{1 + e^{\frac{T - \frac{\sigma^i}{C_A} - T_A}{A}}} \tag{16}$$

where ξ_{SM0} is the initial quality fraction of single martensite.

2.3.2.6. Transformation from austenite to single martensite

Transformation condition: Assuming in one-dimensional terms, cooling and increase of stress can lead to this phase transition from austenite to single martensite.

Transformation principle: As similar to the transformation from austenite to twinned martensite, for the transformation from austenite to single martensite, if the reverse transformation from single martensite to austenite did not occur during the process, a small heating or decrease of stress will not affect its subsequent transformation process.

However, if the reverse transformation process happened and induced a lot of single martensite to austenitic, then the extent of the following transformation from austenite to single martensite required reduction in accordance with the quality percentage of the current austenite.

Thus, the kinetics equation from austenite to single martensite can be acquired as

$$\xi_{SM} = \frac{\left(1-\xi_{SM0}\right)}{1+e^{\frac{T-\frac{\sigma^i}{C_M}-T_M}{M}}} + \xi_{SM0} \tag{17}$$

2.3.2.7. Parameter variation properties of the kinetics

At the same time, DSC experiments showed that, even for the same material components and the same diameter of the SMA, the transformation of the material would be very different after experiencing loading and plastic deformation, as shown in Fig. 8, and these changes must be considered in the constitutive model.

For the SMA of 1mm in diameter, in austenite transformation process, after undergoing the plastic deformation, the transformation peak temperature changes from 50.25 °C to 121.0 °C, and the transformation parameter changes from 3.25 to 10.25. Similarly, in the martensite transformation process, the transformation peak temperature changes from 27.55 oC to 45.9 °C, and the transformation parameter changes from 1.93 to 11.28.

It must be pointed out that these variations are apt for the SMA with 1 mm in diameter. For the SMA with 5 mm in diameter, the change scope of these parameters will be different and can be determined by experiment.

2.3.3. Consideration of the plastic deformation

In the first loading process, after stretching to different deformations and heating under free condition, plastic deformations for different initial deformations are shown in Fig. 19.

Experimental results indicated that, when the maximum tensile strain exceeded 3%, the plastic deformation increment would be linearly related to the maximum tensile strain increment, thus, the experience equation of the plastic evolution could be achieved as follows

$$\varepsilon^P = 0.198 \times \left(\varepsilon - 0.03\right) \text{ and } \varepsilon^P \geq 0 \tag{18}$$

It must be pointed out that, this equation is the fitted results of the uniaxial tensile test at room temperature. For another SMA material under different temperature, variation of the plastic deformation with the strain will be very different. Additionally, this equation is only applicable to simulate plastic evolution characteristics during the first loading process, but not the second and the third loading processes.

2.4. Numerical simulation and comparison with the experimental results

Two parts are included in this section, one is the determination of the parameters in the kinetics equation based on the DSC test results and the calculation of the phase transformation, uniaxial tension, SME and the constrained recovery curves, the other is the comparison with the experimental results.

2.4.1. Determination of the parameters in constitutive equation

The transformation characteristics was calculated through the kinetics equation directly and compared with the experimental results. The SMA uniaxial tension, SME and the constrained recovery stress-strain-temperature curves were calculated through the incremental form of the constitutive model. The Parameters used in the calculating process were determined through the following principles:

1. Transformation characteristic temperature and parameters of the kinetics were determined through the DSC test results;
2. The macro physical parameters, such as the elastic modulus, thermal expansion coefficient adopted the experimental result;
3. Because the characteristics temperature and transformation parameters change with the loading process, therefore, in the calculating program, these parameters were adjusted according to the test results of SMA with 1 mm in diameter;
4. The critical value of the internal stress was adjusted based on the experimental critical stress in uniaxial tension.

2.4.2. Verification of SMA transformation property

According to the variation curve of martensite quality percentage with the temperature of martensite transformation and reverse one (Fig. 3), the parameters of the kinetics equation can be achieved through the numerical fitting of the test results, as shown in Table 1.

	Transformation characteristics temperature (°C)	Transformation parameter (°C)
Martensite transformation process	28.2	1.947
Reverse martensite transformation process	50.14	2.999

Table 1. Transformation parameters determined by DSC test

Using the parameters calibrated by the numerical fitting, variation curve of martensite quality percentage with temperature can be calculated. The results with comparison of the test ones are shown in Fig. 21. Additionally, variation curve of martensite quality percentage changing rate with temperature can be calculated as shown in Fig. 22, with comparison of the test ones.

These comparisons between the numerical and the experimental results indicate that the kinetics equation proposed can simulate the transformation properties of SMA more practically, there are not obvious start and finish temperatures in the calculated curve during the transformation process, the calculated exothermic and endothermic peaks are consistent well with the experimental ones.

2.4.3. Comparison of the uniaxial tension, SME and controlled recovery curve

Through the constitutive model with differential form, the uniaxial tension, SME and controlled recovery curve of SMA with 5 mm in diameter can be calculated by program, and compared with the test results, as shown in Fig.s 23-25.

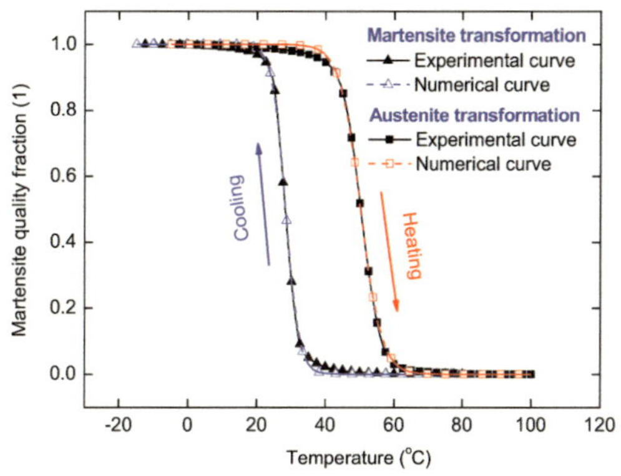

Figure 21. Comparison between experimental and calculated curve of martensite quality fraction versus temperature

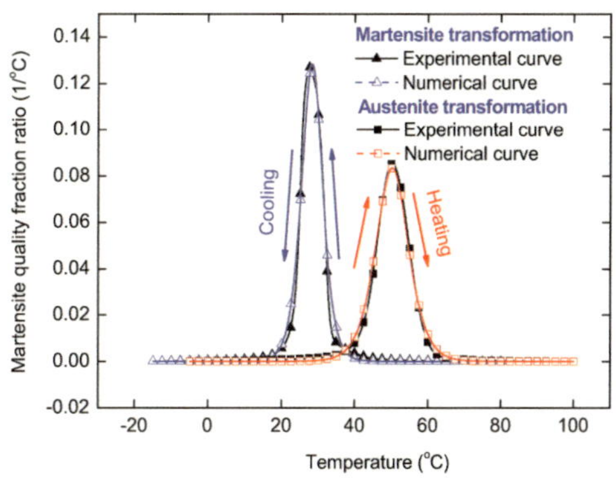

Figure 22. Comparison between experimental and calculated curve of martensite quality fraction ratio versus temperature

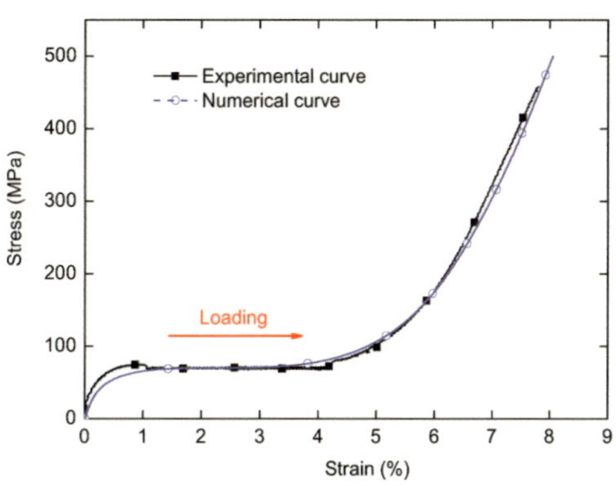

Figure 23. Comparison between experimental and calculated curve of uniaxial tension process

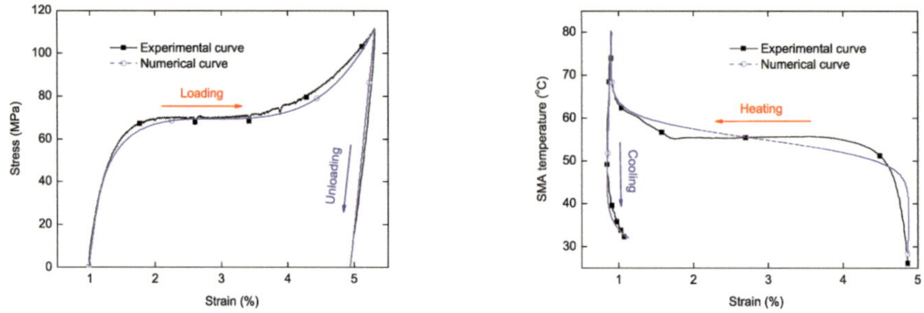

(a). Stress-strain curve (b). Temperature-strain curve

Figure 24. Comparison between experimental and calculated curve of the third times SME process

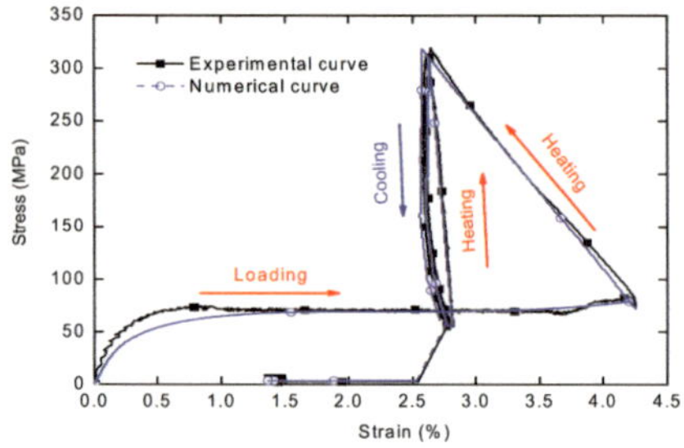

(a). Stress-strain curve

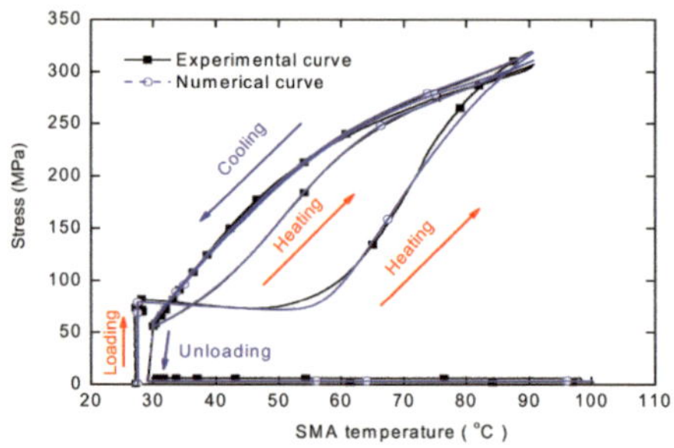

(b). Stress-temperature curve

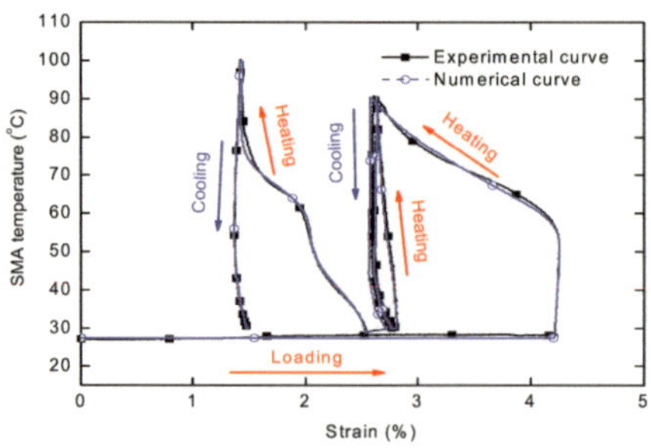

(c). Temperature-strain curve

Figure 25. Comparison between experimental and calculated curve of constrained recovery process

Numerical results indicate that, based on the well forecast for the phase transformation properties, the constitutive model can simulate the uniaxial tension, SME and controlled recovery process of SMA more accurately.

2.5. Conclusions

A constitutive model in strain form with a new simple empirical kinetics equation is presented based on the DSC and uniaxial tension test, and the plastic deformation is considered. Transformation temperature parameters that denote the peak temperature and peak width of the endothermic and exothermic process were adopted to simulate the transformation characteristics of SMA more practically. Transformation conditions and properties of the six transformation processes were analyzed.

At the same time, physical and kinetics parameters were identified through the DSC and the uniaxial test analysis for SMA, programs were adopted for calculating the phase transformation and uniaxial tension process. Comparison between the simulating and experimental results shows that the established constitutive model can simulate the martensite and the reverse transformation, as well as the uniaxial tension, SME and constraint recovery process, more accurately.

3. Behavior of smart concrete beams with embedded SMA bundles

The behavior of smart concrete beams with embedded SMA bundles was investigated. Two beams measuring 1996cm×99cm×85cm, which will be integrated into a smart bridge in a freeway, were manufactured and examined. Each beam contained six trusses of SMA bundles used as actuators to achieve recovery force. The SMA bundles were connected with pre-stressing steel strands and separated from the concrete matrix, so that the temperature interchange between SMA bundles and the matrix could be decreased as small as possible. Some temperature sensors, reinforcement meters and displacement sensors were used to monitor the active control effect of SMA bundles, all the data were acquired through a 16-channel dynamic data-acquisition system, and each beam was examined several times with different activating current intensity. Experimental results indicated that the recovery force induced by SMA bundles was significant and controllable, the deflection generated by the SMA bundles at the middle span of the beam was about 0.44mm, and the capability of resisting overload of each beam was about 2.98 KN (average). A relationship between SMA temperature and activating / inactivating time was also formulated. The conclusion is that SMA could be used in civil engineering structures either from technological or economic aspect.

3.1. Background

With the developments of materials science, engineering technology, data acquisition and computer technology, 'Intelligent' materials, such as Electrorheological fluid, piezoelectric ceramic, shape memory alloys, etc, have opened the door for many useful applications, e.g.

civil engineering, automotive industry, aerospace technology, medical science (Crawley, 1994; Matsuzaki, 1997; Srinivasan & Mcfarland, 2001). Implementation of smart structures has become feasible, and much effort has been made in these fields.

3.1.1. SMA smart structure

Because of the merits of SMA, many researchers have been trying their efforts to apply it in smart structures. These applications were mainly based on the two properties of SMA: one is the shape change with temperature, where SMA fibers are pre-strained during the curing process and prevented from a transformation to austenite, if the temperature rises during the operational cycle, the fibers transform into austenite and tend to contract, and the recovery tensile stresses develop due to the constraint of the matrix; The other is the mechanical characters variation, where unstrained SMA fibers are embedded within the composite, increase or decrease of the temperature alters the elastic modulus or the damping property of SMA, consequently changing the property of the composite. Due to these properties, SMA is usually used to generate bending (Chaudhry & Rogers, 1991; Lagoudas & Tadjbakhsh, 1992), control bucking and postbucking (Baz & Tampa, 1989; Choi et al., 1999, 2000), induce or depress vibration (Baz, Imam, & Mccoy, 1990; Anders, Rogers & Fuller, 1990; Baz, Poh, & Gilheany, 1995; Lau, Zhou, & Tao, 2002), isolate seism (Graesser & Cozzarelli, 1991).

Jonnalagadda, Kline, & Sottos (1997) investigated the interaction between SMA wires and a host polymer matrix by correlating local displacements and stress fields induced by the embedded wires with SMA/polymer adhesion, interfacial bond strength was measured for four different SMA surface treatments: untreated, acid etched, hand sanded and sandblasted. Song, Kelly, & Agrawal (2000) presented the design and experimental results of the active position control of a SMA wire actuated composite beam, which has a honeycomb structure with SMA wire embedded in one of its face sheets, a robust controller was designed and implemented to actively control the tip position of the composite beam.

3.1.2. SMA-based concrete structures

In the recent ten years, many researches have been done for the application of these intelligent materials in civil engineering. Maji & Negret (1998) carried out their laboratory studies on concrete (mortar) beams (30.5cm×2.5cm×1.3cm), where SME in NiTi was utilized as a way of inducing additional pre-stressing in concrete. Strands made with NiTi SMA ($4\Phi 0.64mm$) were elongated beyond their plastic limit and subsequently embedded in concrete beams, upon electrical heating, the SMA strands shrunk and induced deflection and failure in concrete. In this study, the temperature rise of the beam was formulated based on heat transfer theory, assuming that the temperature of SMA was the average temperature of the beam.

Deng et al. (2003) studied the behavior of concrete specimens uniaxially embedded with shape memory alloy wires actuated by electrical current. Two kinds of specimens with different number of SMA wires were studied, the diameter of the SMA employed in the matrix was 3.5mm and the pre-strains were 8% and 6%. Many factors affecting the behavior of concrete specimens were examined through experiments, such as actuating electrical current mode, initial pre-strain of SMA, actuation times, and initial ambient temperature. The experimental results indicated that the axial strain could be adjusted easily by changing the value of electrical current intensity or the actuating time of the SMA actuator.

However, the application of smart concrete beams with embedded shape memory alloy bundles has not been reported to date. The above researches were mostly carried out in laboratory, but a potential practical use of smart concrete beams needs an adequate study on them, because the characteristics of small specimens, as well as the manufacture process, may be very different with those of the engineering structures, thus it is necessary to make a practical engineering investigation in civil structures. Accordingly, two concrete beams measuring 1996cm×99cm×85cm were manufactured and investigated in this section.

3.2. Experimental program

3.2.1. Materials

The materials used for the host concrete were type I Portland cement, crushed limestone, and river sand. A high-rang water reducer was used to achieve good workability. The mix proportion by weight was cement: sand: aggregate: water: admixture = 1: 1.187: 2.412: 0.329: 0.008. The compressive strength of the concrete (cubic specimen) for 28 days was 53.47 MPa.

The seven-wire high-strength steel strands with the diameter of 15.24mm were used to apply conventional pre-stressing force in this experiment, its tensile strength is 1860 MPa, the working stress was controlled to 1395 MPa, and the modulus is 180 GPa. As ordinary reinforced concrete beam, steel reinforcement was used to construct the reinforcing cage.

The Ti-49.5 wt % Ni for our research (binary, straight annealed) was one-way SMA. Its transformation temperatures were determined as 25.3 °C (Mf), 36.3 °C (Ms), 44.8 °C (As) and 66.4 °C (Af) through a DSC (differential scanning calorimetry) analysis (Fig. 26).

The specific stress-strain behavior of this SMA bar under different temperatures and constrained condition were, as shown in Fig. 27, achieved through a sophisticated test method with a 10KN universal material testing machine.

The maximum recovery strain was larger than 5%, and the tensile strength was about 940 MPa (corresponding to the ambient temperature: 22.3 °C) with the ultimate strain of 19.1% (Fig. 27a). When the SMA is free to deform at a relative lower temperature, large plastic strain (about 4%) is induced, but upon heating, this plastic strain disappears, with only a small unrecovery strain (smaller than 0.2%) in the end (Fig. 27b). Whereas, if the SMA recovery is constrained, large stresses should be induced (Fig. 27c), this behavior is the principle of SMA used as actuators.

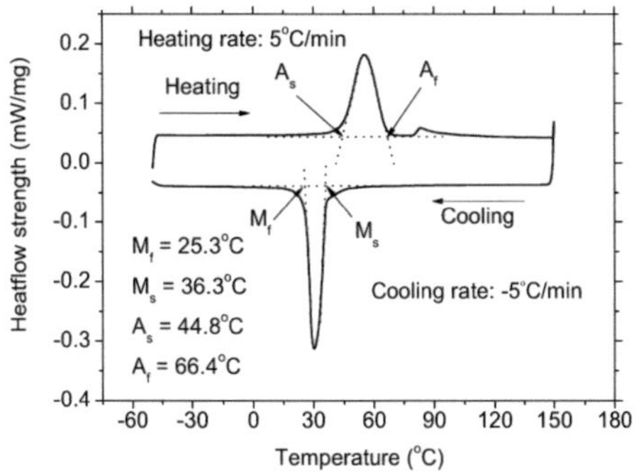

Figure 26. DSC thermogram for NiTi SMA

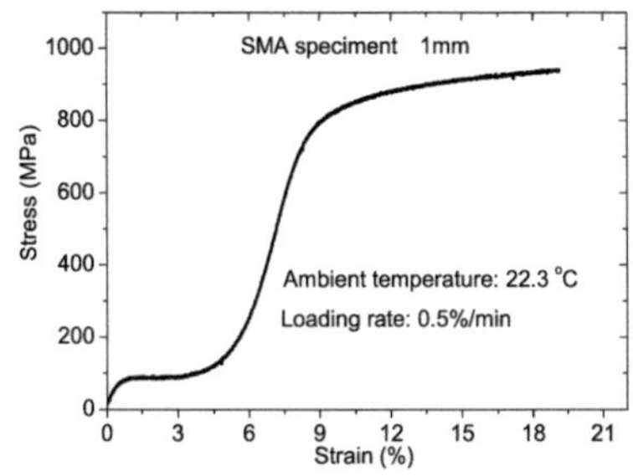

(a). Complete uniaxial tension stress-strain curve of SMA

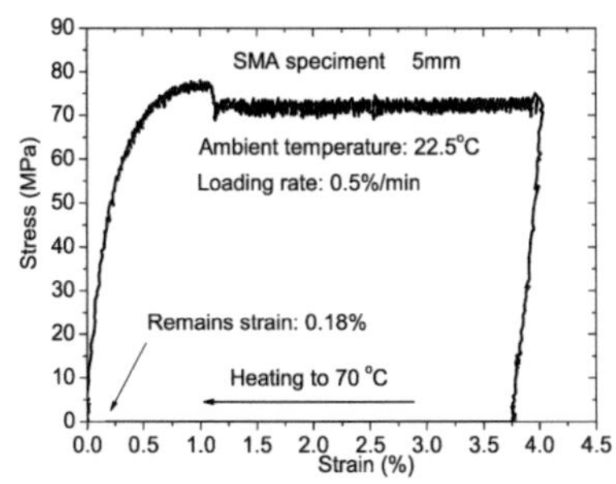

(b). SME of SMA

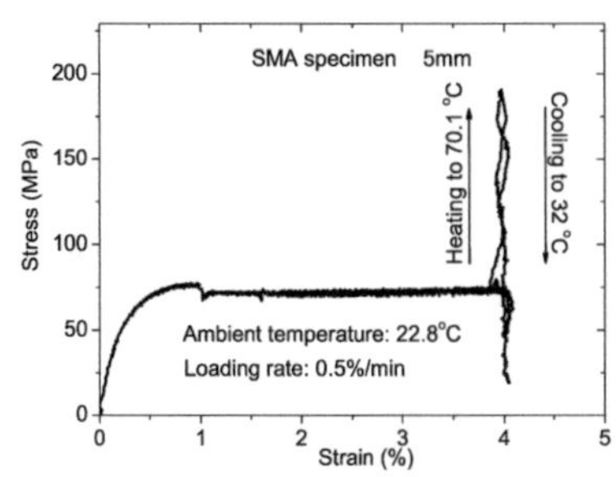

(c). Constrained curve of SMA

Figure 27. Stress-strain behavior of NiTi SMA

As depicted in Fig. 28, a seven-wire SMA bundle with dimension of $\Phi 15.3mm$ was presented to obtain large recovery force, where one SMA wire with diameter of 5.3mm was surrounded by six SMA wires with diameter of 5.0 mm to obtain good clamping capability.

3.2.2. *Smart concrete beam manufacture*

The smart concrete beam with dimensions 1996cm×99cm×85cm was manufactured, and the cross section is shown in Fig. 29. Sixteen bounds of pre-stressed high-strength steel strands were mounted on the bottom of the beam to resist the normal loads; six bounds of SMA bundle-steel strand union body, numbered 1 to 6 from left to right, respectively, were constructed upside the steel strands to create recovery forces.

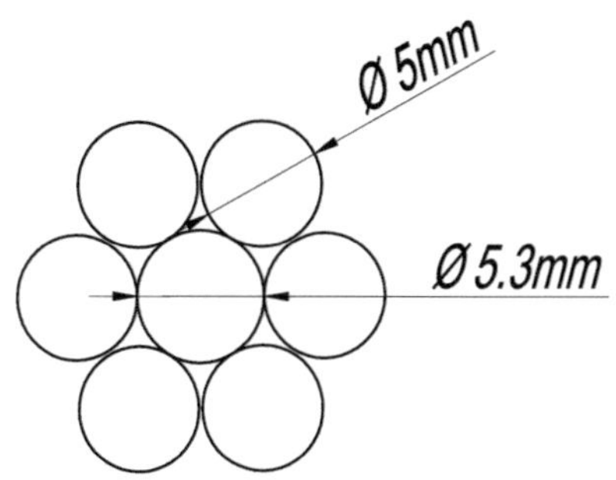

Figure 28. Diagram of SMA bundle

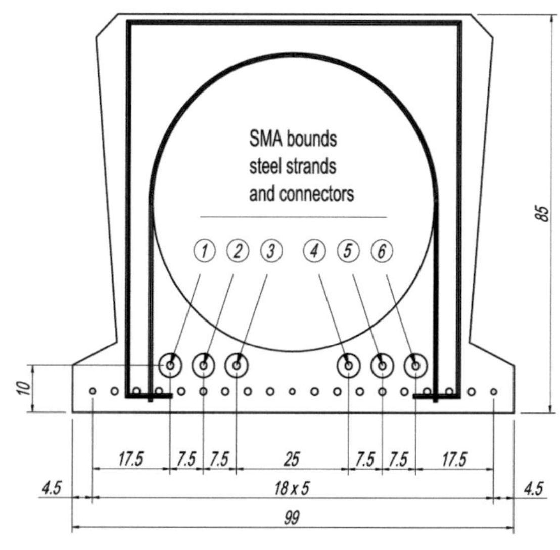

Figure 29. Distribution of cross-section of the smart concrete beam

A new SMA embedding method in concrete was presented, which can be seen in Fig. 30. The SMA bundle measuring 3.5 m long was mounted in the middle of the beam to create recovery forces, and connector was used to connect the SMA bundle with the steel strands. The recovery forces generated by the SMA bundles were transferred to the concrete matrix by the steel strands on the bonding zone measuring 3 m at both ends of the beam. When the

SMA bundles were activated by electricity, additional bending force was generated and applied to the beam, In the unbonded zone, SMA bundles and steel strands were separated from concrete by PVC pipes, therefore the SMA bundles and the connectors could shift freely, and the temperature influence of SMA bundles to the matrix could be decreased as small as possible.

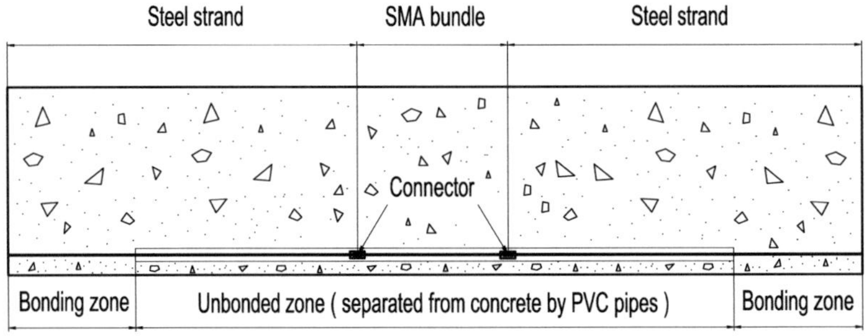

Figure 30. Distribution of longitudinal-section of the smart concrete beam

Sensing and power system of these smart beams can be seen in Fig. 31. Temperature sensors and reinforcement meters were mounted on the SMA bundle, six trusses of SMA bundles were connected in series by voltaic wires to receive the electric power, and copper wiring terminals were used to connect the conductor wires with the SMA bundles. All the sensor wires were laid along the beam at a lower level position and led out at the end of the beam. Two voltaic wires were laid along the beam at a higher level position and led out at the same end of the beam. Considering the effect of steel reinforcement cage, the electric circuit must be constructed as a self-consistent system to avoid short circuit, and all sensor wires and electric cable should be protected from the damage of the construction machinery.

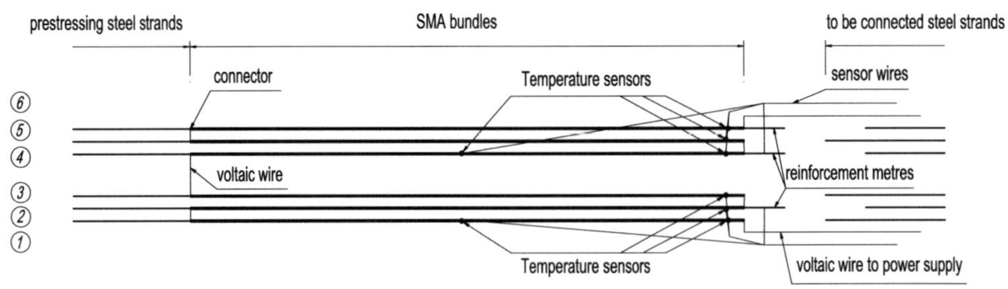

Figure 31. Layout of the sensors and wires

Initial tension of the SMA bundles was carried out after the initial elongation of the steel strands, and the initial strain of the SMA bundles reached 4%. After finishing the stretching of the SMA bundles and steel strands (Fig. 32), concrete was deposited into the steel model and steam was used during the curing period of the concrete.

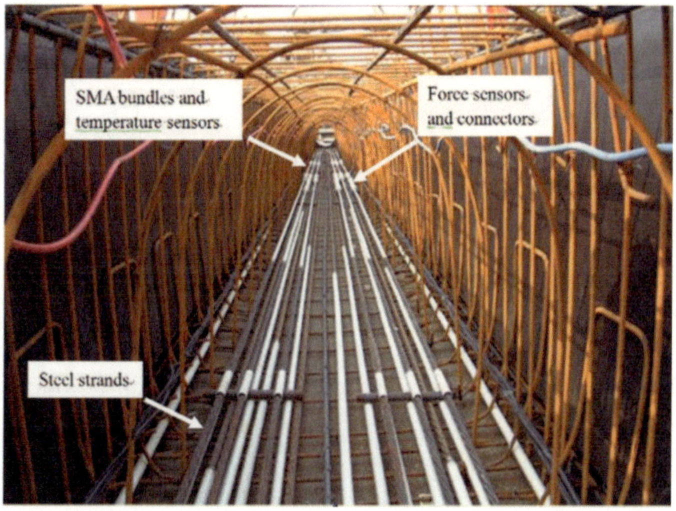

Figure 32. Framework of steel reinforcement with stretched SMA bundles

After about three days, when the concrete strength reached 80 percent of its 28–day strength, SMA bundles and steel strands were relaxed and the concrete beams were lifted out from the model, and tests of these beams were carried out subsequently.

3.2.3. Testing apparatus and procedure

The beam to be monitored was laid on an experimental bench for freely-supported condition, and the testing site is shown in Fig. 33. When the SMA bundles were activated, additional bending forces were generated, and the beam was inflected at the same time. Contrarily, when the SMA bundles were inactivated, the recovery forces disappeared and the beam recurred to the original state, as shown in Fig. 34.

Figure 33. Experiment site

Figure 34. Behavior of the beam with SMA activated and inactivated

The SMA bundles were heated with electricity. Because of the low electrical resistance of the SMA bundles, the necessary voltage was low, but the required current capacity was large, therefore, a 12.5V, 125A variable capacity voltage source was employed to each beam. Experimental results indicated that the resistance of the SMA bundles in each beam was about 0.166 Ω.

Three displacement sensors were mounted at three quartiles of the beam. Deflection of the beam, the temperatures of the SMA bundle, and the recovery forces generated by the SMA bundles, were monitored synchronously during the progress of the experiment. All the data were acquired and analyzed preliminarily through a 16-channel dynamic data-acquisition system. The sampling frequency was 10 Hz, and the maximum sampling frequency could reach 10000Hz.

3.3. Results and discussions

Two beams were manufactured and each was examined for several times. For convenience, beams, SMA bundles and temperature sensors were all numbered by the following principle:

1. Two beams were numbered 1 to 2.
2. Serial number of the SMA bundles of each beam was composed of the number of the beam and relative position label in the same beam, i.e. 2-4. The first number indicated that the SMA bundles belonged to the No. 2 beam and the second number indicated that this is the No. 4 SMA bundle, as depicted in Fig. 29.
3. Temperature sensors were marked by the number of the SMA bundle and the position label, i.e. 1-3-T1. The first number of the temperature sensor label indicates the number of the beam, and the second denotes the number of the SMA bundle where the temperature sensors are mounted on. The symbol T1 denotes that the temperature sensors are mounted in the middle of the SMA bundles, and T2 denotes that the temperature sensors are mounted at the end of the SMA bundles. Thus the flag 1-3-T1 indicates the temperature sensor mounted in the middle of 1-3 SMA bundle.

3.3.1. Deflection of the smart concrete beam

The recovery forces generated by the SMA bundles would induce a bending force on the beam, and cause a deformation of the beam, as depicted in Fig. 34. The deformation of the beam was indicated by the displacement of the three quartiles on the beam. The effect of the environmental temperature, which was discussed in the following content, was eliminated during this analytic process. Each beam was activated and monitored for several times, and the results were almost similar. The representative results of beam 1 and beam 2 were shown in Fig. 35.

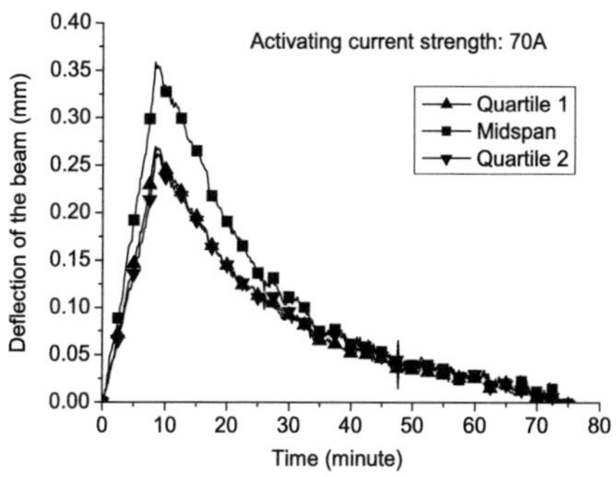

(a). Beam 1 in No.3 activating times

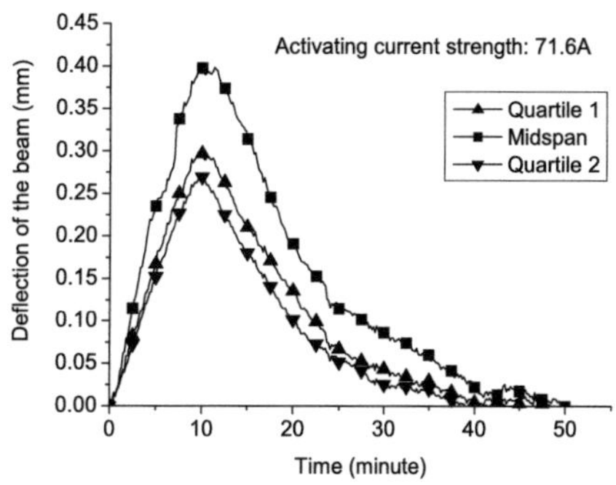

(b). Beam 2 in No.9 activating times

Figure 35. Deflection of the beam versus time

3.3.2. *Temperature of SMA bundles*

When the beam was activated with voltage, temperature of the SMA bundles would rise, and when the current source was shut down, the temperature would drop. The maximum temperature rising speed was at the initial point, as time went on, it decreased and approached to zero, then the thermo-genesis was equal to the heat dissipation and temperature kept up a constant maximum value. Different temperature sensors mounted on the different SMA bundles in a beam detected different temperatures, as shown in Fig. 36. This phenomenon came from the different heat conducting boundary condition of the different SMA bundles. But the results of the same temperature sensor under different activating times were little different, as depicted in Fig. 37, only the temperature dropping section had a little difference yielding from different temperature environment in different activating times.

3.3.3. *Recovery force generated by SMA bundles with temperature*

When the SMA bundles were activated by voltage, the temperature of these bundles raised and the recovery forces generated and increased, but when the current was cut off, the temperature dropped and the forces decreased. The different force generated by different SMA bundles, this difference yield from the different initial force of the SMA bundles. Force variation with time of some bundles was shown in Fig. 38, and each symbol indicated 1.5 minutes. The recovery forces almost increased proportionally with the rising of the temperature, just as depicted in Fig. 39. Hysteresis cycles were observed in the curve of force versus temperature of SMA bundle 2-4, this phenomenon will be investigated later.

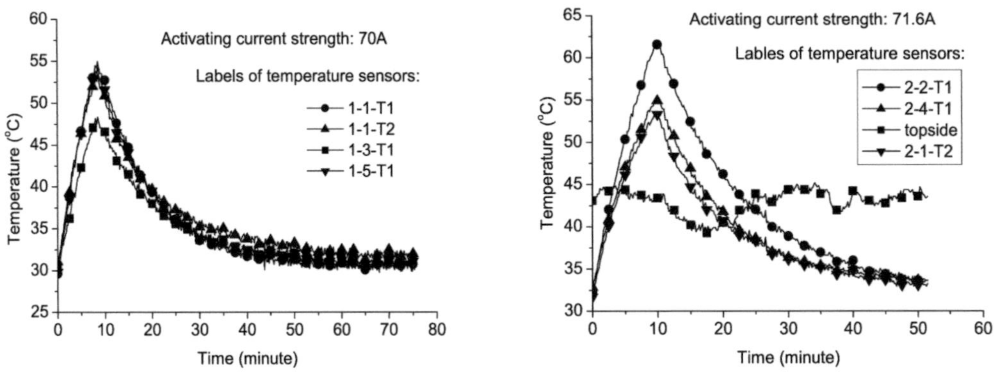

(a). Temperature sensors of beam 1 in No.3 activating times

(b). Temperature sensors of beam 2 in No.9 activating times

Figure 36. Temperature versus time for different temperature sensors in the same beam

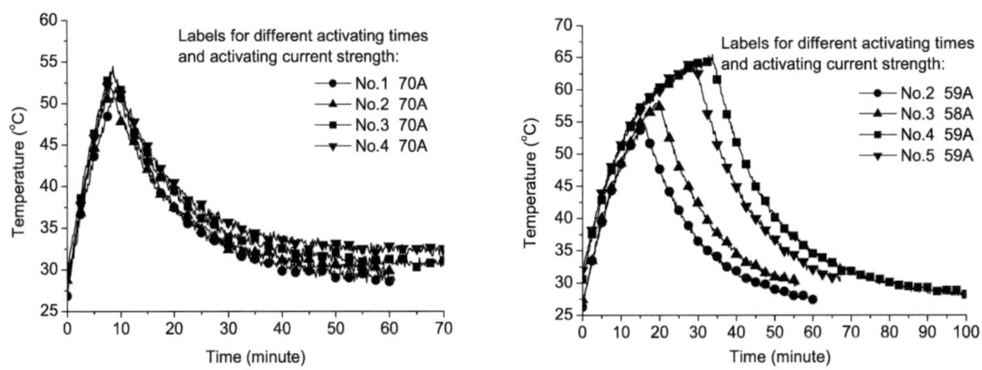

(a). Temperature sensor 1-5-T1

(b). Temperature sensor 2-2-T1

Figure 37. Temperature versus time for different activating times of the same temperature sensor

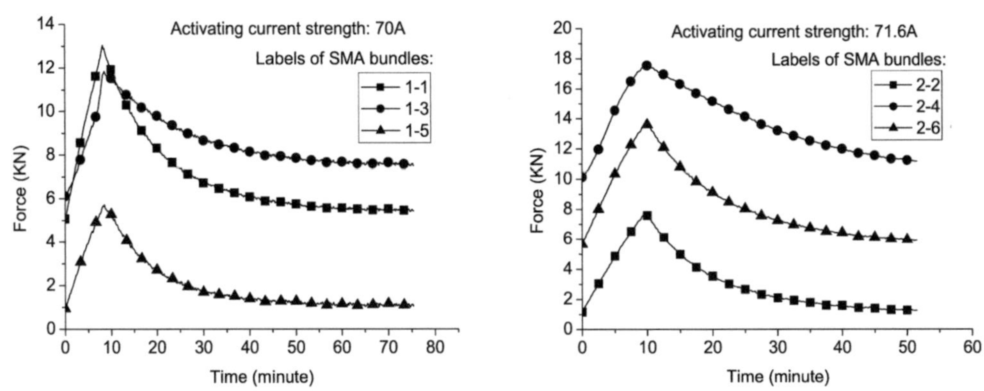

(a). SMA bundles of beam 1 in No.3 activating times

(b). SMA bundles of beam 2 in No. 9 activating times

Figure 38. Force versus time for different SMA bundles of the same beam

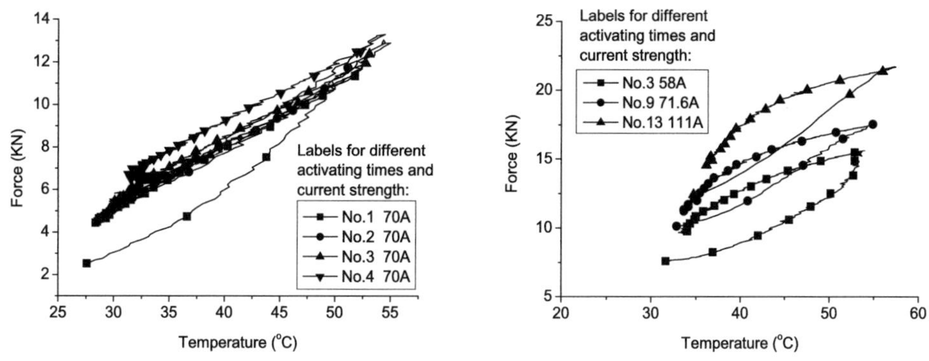

(a). SMA bundle 1-1 (b). SMA bundle 2-4

Figure 39. Force versus temperature for different activating times of the same SMA bundles

3.3.4. Effect of activating current

Effect of the activating current was investigated, current with the strength of 58A, 70A and 110A were applied to the SMA bundles, higher current would induce a faster rising of the temperature of the SMA bundles, thus, the recovery forces increased faster, and the beam reacted with a higher speed, but the recover processes were almost similar, just as depicted in Fig. 40, 41 and 42.

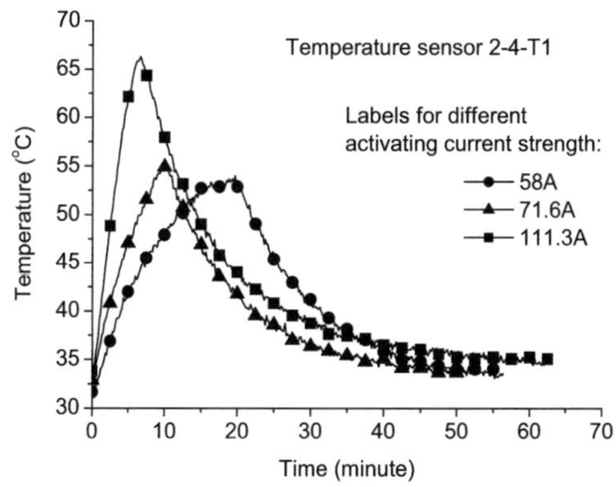

Figure 40. Temperature versus time of temperature sensor 2-4-T1 for different activating current

3.3.5. Effect of environment temperature

During the experiment, the upside temperature of the beam would increase because of the sunshine, while that of the lower side varied rarely. Therefore, this uneven temperature distribution would induce the deflection of the beam. Deformation and the variation of the temperature of the beam also monitored to explain this effect on the deflection of the beam, and the result of this effect was depicted in Fig. 43. Monitoring of No.1 beam was carried out at 9:00 AM and finished at 8:00 PM. The maximum temperature on the upside of the beam (T0) occurred at 3:00 PM, but the maximum deflection the beam at 5:40 PM, this time lag showed that it needed heat conductivity from the surface to the inner. Temperature on the lower side of the beam varied rarely. In the morning, right-hand of the beam faced the sunshine, and in the afternoon, the left-hand of the beam, therefore, temperature at the right-hand of the beam (T2 and T3) increased faster than that of the left-hand of the beam (T1) in the forenoon, and slower in the afternoon, just as depicted in Fig. 43. During this monitoring course, all the SMA bundles were inactivated, thus the temperatures of the SMA bundles were equal to these of the concrete matrix. From this result, it can be seen that the effect of the environment temperature is considerable.

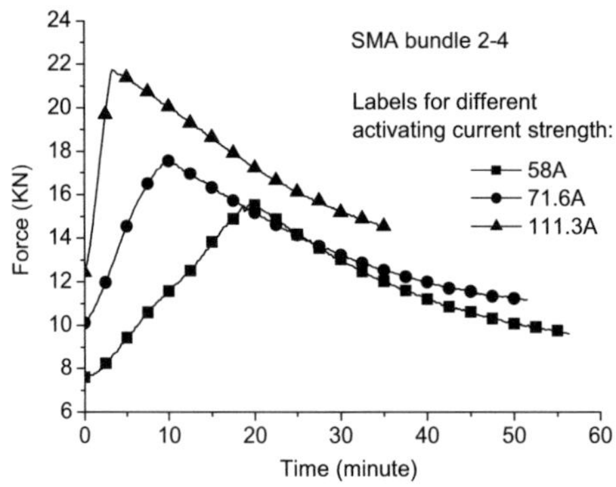

Figure 41. Force versus time of SMA bundle 2-4 of different activating current

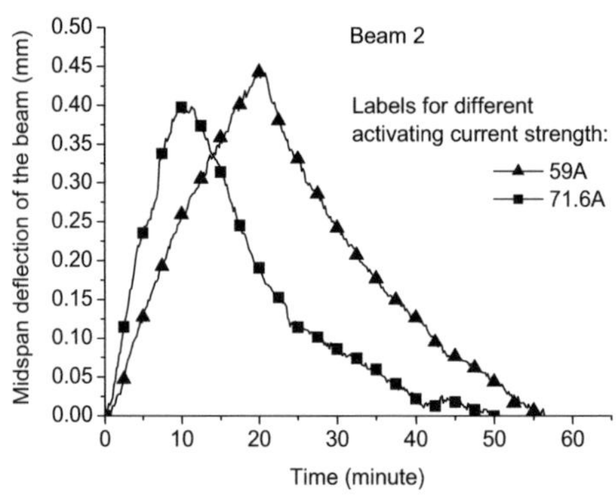

Figure 42. Middle span deflection versus time of beam 2 for different activating current

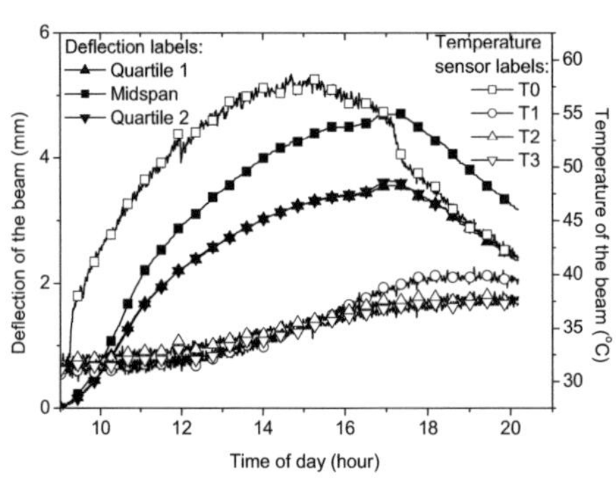

Figure 43. Fig. 43 Effect of environment temperature on deflection and temperature of the beam

3.4. Theoretical analysis

3.4.1. Resisting overload capability

Assuming that a short-term concentrated overload P was applied to the beam, as depicted in Fig. 44, then the overload inducing 1 mm displacement at the middle span of the beam could be calculated.

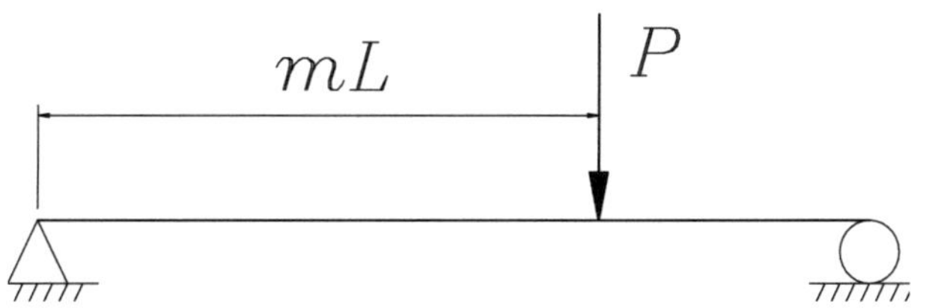

Figure 44. Schematic of load application model

When the overload was applied to the middle span of the beam, the maximum flexural torque at the middle span of the beam generated by the overload was:

$$M_{\max} = \frac{PL}{4} \tag{19}$$

Where P is the overload, and L is the span of the beam.

Thus, the deflection at the middle span of the beam could be given as:

$$f_{\max} = \frac{1}{12} \frac{M_{\max} L^2}{B_{s1}} \tag{20}$$

Where $f_{\max}$ is the maximum deflection of the beam (at the middle span); L is the span of the beam measuring 1996cm; B_{s1} is the flexural rigidity of the beam for short-term load; $M_{\max}$ is the maximum flexural torque at the middle span of the beam.

Thus, the maximum concentrated overload inducing 1 mm displacement at the middle span of the beam could be calculated as:

$$P = \frac{48 f_{\max} B_{s1}}{L^3} = 6.73 \quad \text{KN} \tag{21}$$

Assuming that value of the overload inducing a deflection of 1 mm at the middle span of the beam is equal to that of the resisting force in the negative direction at the middle span of the beam, thus the statistic result of the deflections at the middle span of the beam and the resisting force generated by the SMA bundles for each beam could be seen in table 2.

Beam serial number	The maximum deflection at the middle span of the beam (unit: mm)	Equivalent resisting force (unit: KN)	Activating current strength (unit: A)	Activating time (unit: minute)
No. 1	0.402	2.7055	70	9.3
No. 2	0.482	3.2439	71	16.3

Table 2. Statistic result of deflections at the middle span of the beam and equivalent resisting force of the beams

3.4.2. *Curve fitting of experimental results*

Considering a section of SMA bundle, if T is defined as the temperature of the SMA bundle, then the increase of T is given as

$$\rho c V_{\mathrm{SMA}} \frac{dT}{d\tau} = q_v V_{\mathrm{SMA}} - a\mu F_{\mathrm{SMA}} \left(T - t_f\right) \tag{22}$$

Where ρ, c, τ and V_{SMA} denote density, specific heat, time and capacity of the SMA bundle; q_v is the rate of heat production; a, μ, F_{SMA} and t_f denote the coefficient of heat convection, quotient of contact face form, area of the heat-exchange surface and ambient temperature.

The rate of heat production $q_v V_{\mathrm{SMA}}$ can also be expressed as follows:

$$q_v V_{\mathrm{SMA}} = I^2 R_{\mathrm{SMA}} \tag{23}$$

Where I and R_{SMA} denote the current strength and electric resistance of SMA.

Thus, equation (4) could be written as:

$$\frac{dT}{d\tau} + \frac{a\mu F_{\mathrm{SMA}}}{\rho c V_{\mathrm{SMA}}} t = \frac{I^2 R_{\mathrm{SMA}}}{\rho c V_{\mathrm{SMA}}} + \frac{a\mu F_{\mathrm{SMA}}}{\rho c V_{\mathrm{SMA}}} t_f \tag{24}$$

Defining

$$\mathrm{X} = \frac{I^2 R_{\mathrm{SMA}}}{\rho c V_{\mathrm{SMA}}} \tag{25}$$

$$\mathrm{Y} = \frac{a\mu F_{\mathrm{SMA}}}{\rho c V_{\mathrm{SMA}}} \tag{26}$$

Then the solution to equation (24) could be given as:

$$T(\tau) = Ae^{-Y\tau} + \frac{\mathrm{X}}{\mathrm{Y}} + t_f \tag{27}$$

Where A is related to the initial condition during temperature rising period.

In the same manner, the decrease of T could be given as:

$$\frac{dT}{d\tau}+\frac{a\mu F_{\text{SMA}}}{\rho cV_{\text{SMA}}}T=\frac{a\mu F_{\text{SMA}}}{\rho cV_{\text{SMA}}}t_f \tag{28}$$

Thus from equations (26) and (28), we have:

$$T(\tau)=Be^{-Y\tau}+t_f \qquad (\text{When } I=0) \tag{29}$$

Where B is related to the initial condition during temperature dropping period.

The experimental curve could be fitted well with this theoretical curve, just as depicted in Fig. 45.

Considering equation (22), temperature rising rate of the SMA bundles $dT/d\tau$ could be given as:

$$\frac{dT}{d\tau}=\frac{I^2R_{\text{SMA}}}{\rho cV_{\text{SMA}}}+\frac{a\mu F_{\text{SMA}}}{\rho cV_{\text{SMA}}}(t_f-T) \tag{30}$$

Assuming that the temperature T was equal to the environment temperature t_f at the beginning, then the initial temperature rate of the SMA bundles $dT/d\tau$ can be given as

$$\frac{dT}{d\tau}=\frac{R_{\text{SMA}}}{\rho cV_{\text{SMA}}}I^2 \tag{31}$$

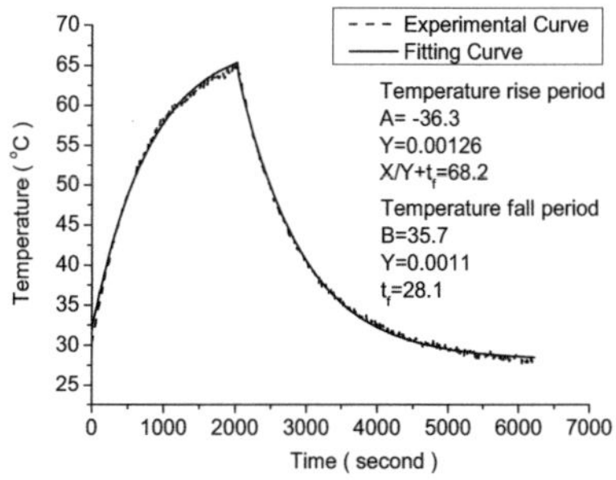

Figure 45. Fitting curve of temperature versus time of temperature sensor 2-2-T1 in NO.4 activating times

According to the above theory, statistic values of the initial temperature rising rate were fitted with the numerical curve, just as depicted in Fig. 46. From it, one can see that the activating current strength had significant effects on the behavior of the smart beam, and higher activating current could be used to achieve a faster reacting speed of the beam. According to the above analyses, the action speed was the proportion square of the current strength.

Theoretically, the temperature of SMA is proportional to the square of the applied current because of this electric heating method. As a matter of fact, this relationship may be affected by many factors, e.g. the heat transfer from the SMA to the concrete matrix. Here, the well accord between the experimental data and the fitting curve indicated that influences of heat transfer between the SMA and the concrete matrix could be decreased effectively by separating SMA bundles from concrete matrix by PVC pipes.

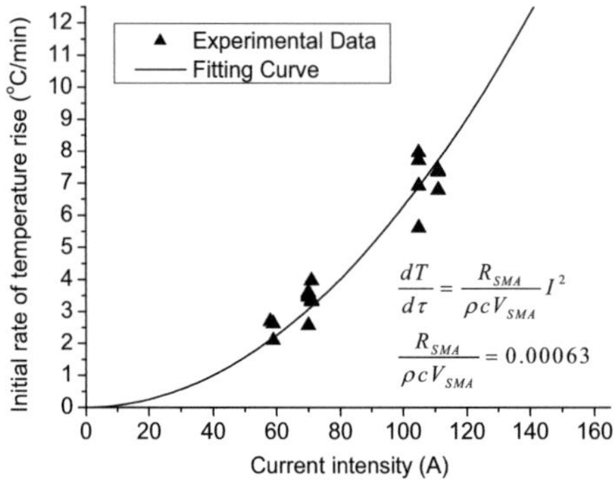

Figure 46. Fitting curve of temperature rising rate versus current intensity

3.5. Conclusions

In this section, the manufacture process and behavior of the smart concrete beams with embedded SMA bundles were investigated through an extensive experimental program. The main parameters, such as temperature and recovery force of SMA bundles and the deflection of the concrete beam, were monitored through this program. Major factors influencing the behavior of the beams were examined. From these results, the following conclusions can be drawn.

1. SMA can be used in practical civil engineering structures to resist overload, the effect is considerable. SMA bundles can change the mechanical performance of the smart concrete beam as needed. In this section, a new SMA embedding method in concrete was presented and applied to this experiment -- only a short section of SMA bundle could generate large force, which was mainly due to the large recoverable deformation of SMA.
2. The recovery force created by the SMA bundles was almost proportional to the temperature of the SMA bundles, and hysteresis cycles were observed in the recovery force-temperature curve of some SMA bundles. The recovery force generated by each SMA bundle was different, which came from the different initial condition for different bundles. But the increments of the recovery force of each SMA bundle in each activating time were almost the same.
3. The temperature of different SMA bundles was different, and temperature of a same SMA bundle in different activating times was also little different. These differences yielded from the different heat conducting condition of different SMA bundles and the different environment boundary condition in different activating times. Due to these differences, more force sensors and temperature sensors would be obligatory for more comprehensive monitoring of the beam.

4. The effect of the environment was remarkable, and displacement at the middle span of the beam induced by the environment temperature was about twelve as times as that caused by the recovery force of SMA bundles. Thus, this effect must be considered during the experimental and analytic process.

Some other factors, such as the activating times, the length of the unbonded section, the length of the SMA bundles, the pre-strain of the SMA bundles, and the curing condition of the concrete, etc., can also affect the behavior of the concrete beam. These effects deserve more investigation.

4. Behavior of a smart concrete bridge with embedded SMA bundles

Since there are many differences between the laboratory researches and the practical applications of using this smart material to structural active controlling; thus, it would be worthful and necessary to study how to apply SMA on a practical concrete bridge and how to use the "smart" forces. Accordingly, the objective of this section is to investigate the behavior of a smart concrete bridge with embedded SMA bundles.

The bridge was a simply-supported pre-stressed concrete one with 20 meters long and 12.5 meters wide, and composed of ten ordinary concrete beams and two smart concrete ones measuring 1996cm×99cm×85cm. When the SMA bundles were activated, additional bending forces were generated, and the beam was inflected to resist excessive load at the same time. Otherwise when the SMA bundles were inactivated, the smart forces disappeared and the beam recovered to the initial state. In order to monitor the active control effect of the SMA bundles, some temperature sensors, reinforcement meters and displacement sensors were employed. All the data were acquired through a 16-channel dynamic data-acquisition system. The bridge was examined several times with different activating current intensity, and some static load tests were also carried out to evaluate the capability of resisting overload of the bridge. The experimental results indicate that the smart forces induced by SMA bundles were significant and controllable, the deflection generated by the SMA bundles at the middle span of the bridge was about 0.03 mm, and the capability of resisting excessive load of the smart beams was 5.6 ~ 7.5 KN. It is also shown that SMA could be used as active controlling actuator in civil engineering structures.

4.1. Experimental program

4.1.1. Design concept of the smart concrete bridge

A complete smart concrete bridge in a freeway should contain the following three parts: load and speed sensors mounted at several miles beyond the bridge; control unit and power supply; actuators, temperature and force sensors, and the bridge matrix, the design concept of the smart bridge was demonstrated in Fig. 47.

The two smart beams will be fixed on the smart bridge. In the session of the freeway, load and speed sensors will be amounted at several miles away from the bridge. Therefore, SMA bundles embedded in the smart beams can be activated before the load exerts on it and recovery forces generate to resist the external load. Additionally, the recovery force generated by the SMA bundles is only used to resist the excessive load when needed, normal load is resisted by ordinary structures, such as steel reinforcements and steel strands. After the load passes the bridge, SMA bundles will be inactivated.

4.1.2. Manufacturing the beams and the bridge

The bridge was composed of twelve concrete beams, and the two smart concrete beams were fixed on the No. 5 and No. 6 from outer side (Fig. 48). Gaps between the beams were filled up with concrete, and then the surface concrete and the pavement were constructed subsequently.

4.1.3. Testing apparatus and procedure

The two smart concrete beams were electrically activated several times to measure the effect of the recovery forces. To evaluate the performance of the bridge, the static load tests were also carried out by parking a 202KN-weight truck on the nine positions of the bridge (Fig. 49).

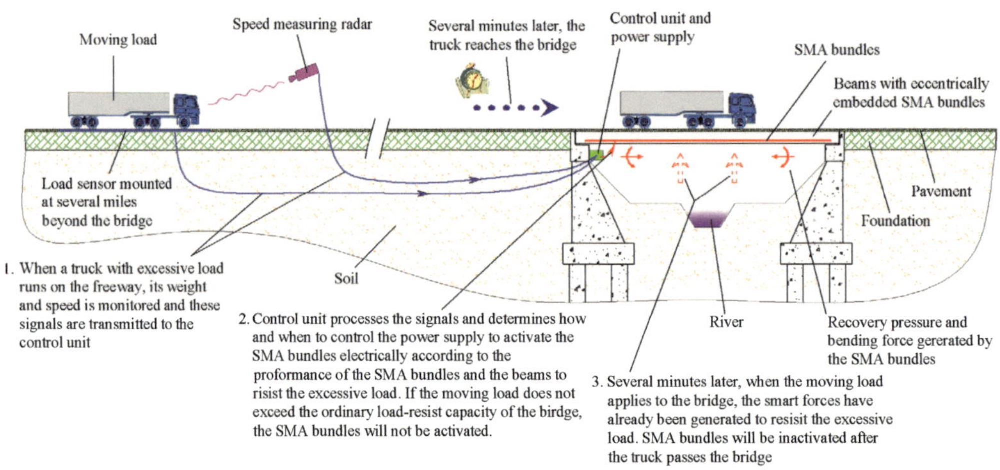

Figure 47. Diagram of the design concept about a smart bridge of a freeway

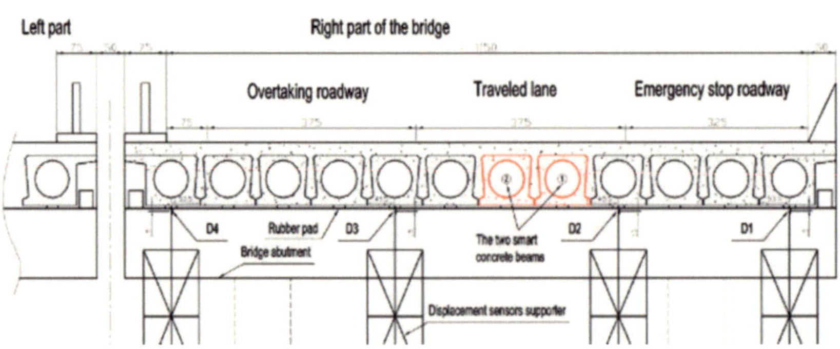

Figure 48. Schematic of mounting deflection sensors

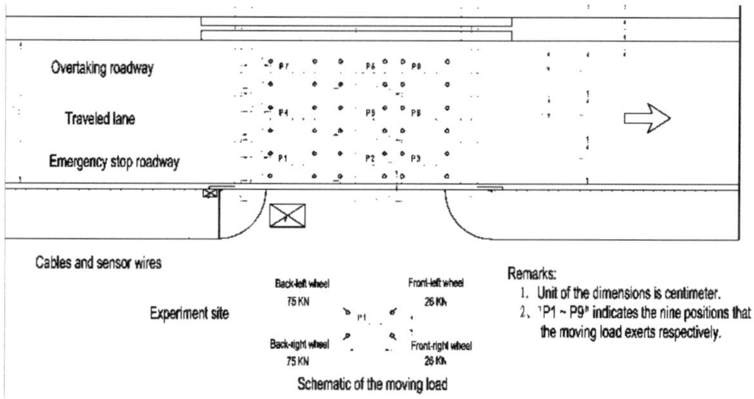

Figure 49. Schematic of static load test

4.2. Results and discussions

In order to record easily, the SMA bundles, temperature and force sensors were numbered serially. Two beams were numbered 1 and 2, and SMA bundles, 1 to 6. Thus SMA bundles and the force sensors can be indicated with the number of the beam and the relative position label in the same beam. The temperature sensor was denoted via the number of the SMA bundle and its position label (i.e. T1 denotes the temperatures mounted at the middle of the SMA bundle, T2 the end).

4.2.1. Activating test results of the bridge

To evaluate the effect of the recovery force, the SMA bundles of the two smart concrete beams were activated two times without applying any external load on the bridge. Displacement sensors were mounted at the bottom of No 1, 4, 8, 12 beams of the bridge (denoted by D1, D2, D3, D4 respectively) to monitor its deformation, as shown in Fig. 48.The unit of the dimensions is centimeter, and D1, D2, D3, D4 indicates the four positions that the displacement sensors mounted on respectively. Temperatures, as well as recovery forces, were also measured synchronously during the experiment.

When the SMA bundles were activated, the bridge was deformed and upward deflections were induced. Deflections at the middle span of the bridge increased and decreased during the activating and inactivating, as shown in Fig. 50. The maximum deflection of the bridge was about 0.03mm. Displacement sensor mounted at the D1 position did not work normally, so the result will be not considered in the analysis.

During the heating and cooling process, temperatures of SMA bundles also increased and decreased, as shown in Fig. 51, and recovery forces of SMA bundles increased and decreased proportionally with these temperatures, as shown in Fig. 52. The recovery force created by the SMA bundles was almost proportional to the temperature of the SMA bundles, and hysteretic cycles were observed in the recovery force-temperature curve of the SMA bundles.

4.2.2. Static test results of the bridge

The static load was applied on the nine positions of the bridge, as shown in Fig. 49, deflections at the four representative points of the bridge (indicated as D1, D2, D3, D4) were demonstrated in Fig. 53.

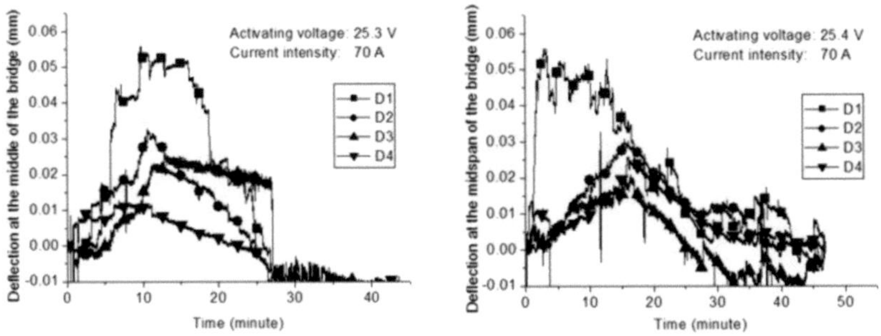

(a) No. 1 activating times **(b) No. 2 activating times**

Figure 50. Deflection at the middle of the bridge versus time

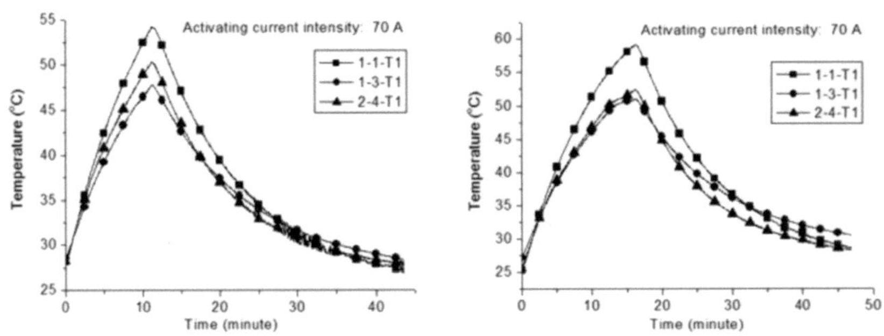

(a) No. 1 activating times **(b) No. 2 activating times**

Figure 51. Temperature of SMA bundles versus time

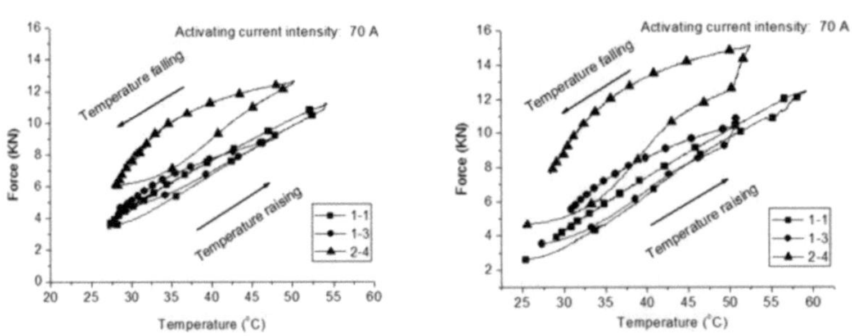

(a) No. 1 activating times **(b) No. 2 activating times**

Figure 52. Recovery force versus temperature

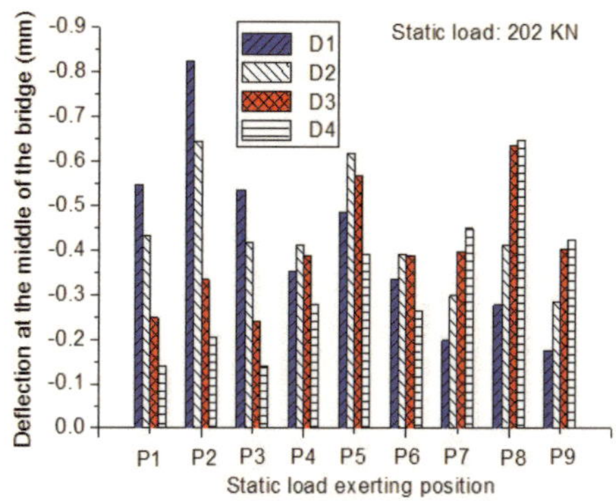

Figure 53. Deflection of the bridge under different load exerting positions

4.2.3. *Effect analysis of the recovery forces*

The capability of resisting excessive load of recovery forces was estimated through the following principle: all the deflections at the D1, D2, D3, D4 induced by the recovery forces are larger than or equal to these reverse ones induced by the external load. Therefore, by comparing the average maximum deflections of the activating test and deflections of the static load test, the maximum resisting load capability could be obtained. Different position the excessive load exerted on is, different the resisting load of the SMA bundles.

When the load ran on the middle of the traveled lane, deflection of the beam No.8 induced by the recovery forces reached the reverse one by the external load firstly, thus the resisting load of the recovery forces was about 7.53 KN shown in Fig. 54 (a), and on the overtaking roadway, 5.55 KN in Fig. 54 (b).

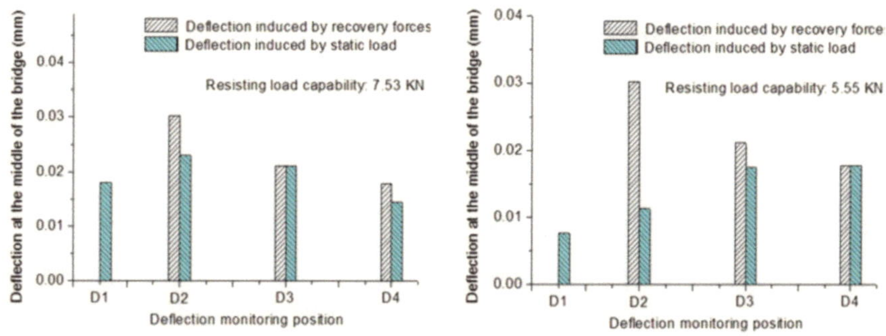

(a) middle of the traveled lane **(b) middle of the overtaking roadway**

Figure 54. Resisting load analysis

4.3. Conclusions

In this section, the manufacturing process and the behavior of the smart concrete bridge with embedded SMA bundles were investigated. From these results, the following conclusions can be drawn: (1) SMA can be used in practical concrete bridges to resist external load with considerable effect. SMA bundles can change the mechanical performance of the smart concrete beam as needed; (2) The capability of resisting excessive load of the recovery forces can be varied with the different exerting position of the external load, thus the optimum design of the smart concrete structures will be necessary to obtain the best control effect.

5. Remarks

In this chapter, the constitutive characteristics of NiTi SMA and its application in practical concrete structures were investigated. The experimental results indicated that the mechanical properties and the phase transformation characters of NiTi SMA could be affected by the loading process considerably. Comparison between the numerical and experimental results indicated that the proposed model could simulate the phase transformation characters, the uniaxial tension and the constrained recovery behavior of NiTi SMA more effectively. Experimental results indicated that the additional recovery forces generated by the SMA could be changed by altering its temperature, and can be used to adjust the deflection of the concrete bridge and to enhance the load-bearing capability of the concrete bridge.

Author details

Lei Li
State Key Laboratory of Hydroscience and Engineering, Tsinghua Univ., Beijing, P. R. China
China Three Gorges Corporation, Beijing, P. R. China

Qingbin Li*
State Key Laboratory of Hydroscience and Engineering, Tsinghua Univ., Beijing, P. R. China

Fan Zhang
Dept. of Transportation of Henan Province, Zhengzhou, P. R. China

6. References

Anders, W. S., Rogers, C. A., & Fuller, C.R. 1990. "Vibration and low frequency acoustic analysis of piecewise-activated adaptive composite panels," *Journal of Composite Materials*, Vol.26, No.1, pp.103-120.

Baz, A., Imam, K., & Mccoy, J. 1990. "Active vibration control of flexible beams using shape memory actuators," *Journal of Sound and Vibration*, Vol.140, No.3, pp.437-456.

Baz, A., Poh, S., & Gilheany, J. 1995. "Control of the natural frequencies of Nitinol-reinforced composite beams," *Journal of Sound and Vibration*, Vol.185, No.1, pp.171-183.

Baz, A., & Tampa, L. 1989. "Active control of buckling of flexible beams," *Failure Prevention & Reliability*, ASME, pp.211-218.

---

* Corresponding Author

Brinson, L. C. 1993. "One-dimensional constitutive behavior of shape memory alloys: thermomechanical derivation with non-constant material functions and redefined martensite internal variable," *Journal of Intelligent Material Systems and Structures*, Vol.4, No.2, pp.229-242.

Brocca, M., Brinson, L. C., & Bazant, Z. P. 2002. "Three-dimensional constitutive model for shape memory alloys based on microplane model," *Journal of the Mechanics and Physics of Solids*, Vol.50, No.5, pp.1051-1077.

Buehler, W. J., Gilfrich, J. V., & Wiley, R. C. 1963. "Effect of low-temperature phase changes on the mechanical properties of alloys near composition TiNi," *J. Appl. Phys.*, Vol.34, No.5, pp.1475-1477.

Chaudhry, Z., & Rogers, C. A. 1991. "Bending and shape control of beams using SMA actuators," *J. Intelligent Mat. Syst. Struct.*, Vol.2, No.4, pp.581-602.

Choi Sup, Lee Jung Ju, Seo Dae Cheol, & Choi Sun Woo. 1999. "The active buckling control of laminated composites beams with embedded shape memory alloy wires," *Composite Structures*, Vol.47, No.1-4, pp.679-686.

Choi Sup, Lee Jung Ju, Seo Dae Cheol, & Dong Kusong. 2000. "A study on the bucking and postbuckling control of composite beams with embedded NiTi actuators," *Journal of Composite Materials*, Vol.34, No.17, pp.1494-1510.

Crawley Edward F. 1994. "Intelligent structures for aerospace: a technology overview and assessment," *AIAA Journal*, Vol.32, No.8, pp.1689-1699.

Deng Zongcai, Li Qingbin, Jiu Anquan, & Li Lei. 2003. "Behavior of Concrete Driven by Uniaxially Embedded Shape Memory Alloy Actuators," *Journal of Engineering Mechanics*, Vol.129, No.6, pp.697-703.

Delaey, L., Krishnan, R. V., Tas, H., & Warlimont, H. 1974. "Thermoelasticity, pesudoelasticity and the memory effects associated with martensitic transformations (part 1 structural and microstructrual changes associated with the transformations)," *Journal of Materials Science*, Vol.9, No.9, pp.1521-1535.

Funakubo Hiroyasu. (Translated from Japanese by Kennedy J.B.) 1987. "Shape Memory Alloys," *Gordon and Breach Science Publishers*, New York.

Greninger, A.B., & Mooradian, V.G. 1938. "Strain transformation in metastable beta copper-zinc and beta copper-tin alloys," *Trans. Met. Soc. AIME*, Vol.128, pp.337-368.

Graesser, E. J., & Cozzarelli, F. A. 1991. "Shape-memory alloys as new materials for aseismic isolation," *ASCE Journal of Engineering Mechanics*, Vol.117, No.11, pp.2590-2608.

Greninger, A. B., & Mooradian, V. G. 1938. "Strain transformation in metastable beta copper-zinc and beta copper-tin alloys," *Trans. Met. Soc. AIME*. Vol.128, pp.337-368.

Jonnalagadda, K., Kline, G. E., & Sottos, N. R. 1997. "Local displacements and load transfer in shape memory alloy composites," *Experimental Mechanics*, Vol.37, No.1, pp.78-86.

Krishnan, R. V., Delaey, L., Tas, H., & Warlimont, H. 1974. "Thermoelasticity, pesudoelasticity and the memory effects associated with martensitic transformations (part 2 the macroscopic mechanical behaviour)," *Journal of Materials Science*, Vol.9, No.9, pp.1536-1544.

Liang Chen. 1990. "Constitutive modeling of shape memory alloys," PhD Thesis, VPI, Blacksburg, VA.

Lagoudas Dimitris C., & Tadjbakhsh Iradj G. 1992. "Active flexible rods with embedded SMA fibers," *Smart Mater. Struct.*, Vol.1, No.2, pp.162-167.

Lau Kin-tak, Zhou Limin, & Tao Xiaoming. 2002. "Control of natural frequencies of a clamped-clamped composite beam with embedded shape memory alloy wires," *Composites Structures*, Vol.58, No.1, pp.39-47.

Maji Arup K., & Negret Ihosvany. 1998. "Smart prestressing with shape-memory alloy," *Journal of Engineering Mechanics*, Vol.124, No.10, pp.1121-1128.

Matsuzaki Yuji. 1997. "Smart strutures research in Japan," *Smart Mater. Struct.*, Vol.6, No.4, pp.1-10.

Otsuka, K., & Wayman, C. M. 1998. "Shape Memory Materials," Cambridge University Press, Cambridge, UK; New York, USA.

Song Gangbing, Kelly Brian, & Agrawal Brij N. 2000. "Active position control of a shape memory alloy wire actuated composite beam," *Smart Mater. Struct.*, Vol.9, No.5, pp.711-716.

Srinivasan, A.V., & Mcfarland D. Michael. 2001. "Smart Structures: Analysis and Design," Cambridge University Press, Cambridge, UK; New York, USA.

Sun Qingping, & Hwang Keh Chin. 1993a. "Micromechanics modeling for the constitutive behavior of polycrystalline shape memory alloy – I. Derivation of general relations," *J. Mech. Phys. Solids*, Vol.41, No.1, pp.1-17.

Sun Qingping, & Hwang Keh Chin. 1993b. "Micromechanics modeling for the constitutive behavior of polycrystalline shape memory alloy – II. Study of the individual phenomena," *J. Mech. Phys. Solids*, Vol.41, No.1, pp.19-33.

Tanaka K. 1986. "A thermomechanical sketch of shape memory effect: one-dimensional tensile behavior," *Res Mechanica*, Vol.18, pp.251~263

Tanaka, K., & Nagaki, S. 1982. "A thermomechanical description of material with internal variable in the process of phase transitions," *Ingenieur-Archiv*, Vol.51, No.5, pp.287-299.

Warlimont, H., Delaey, L., Krishnan, R. V., & Tas, H. 1974. "Thermoelasticity, pesudoelasticity and the memory effects associated with martensitic transformations (part 3 thermodynamics and kinetics)," *Journal of Materials Science*, Vol.9, No.9, pp.1545-1555.

Zhu Jiujiang, Liang Naigang, Huang Weimin, & Liu Zhihong. 2002. "A thermodynamic constitutive model for stress induced phase transformation in shape memory alloys," *International Journal of Solids and Structures*, Vol.39, No.3, pp.741-763.

Chapter 10

NiTi Shape Memory Alloys, Promising Materials in Orthopedic Applications

Marjan Bahraminasab and Barkawi Bin Sahari

Additional information is available at the end of the chapter

http://dx.doi.org/10.5772/52228

1. Introduction

A large number of materials are continuously being developed to meet the requirements for different engineering applications including biomedical area. However, development of a material in this field is a challenging issue especially for those devices that are implanted in the human body, because the material must fulfill an array of fundamental biological and mechanical requirements. Among these, orthopedic applications require careful attention as a result of ageing population worldwide, large number of injuries and the demand for higher quality of life. A wide range of materials including metals, alloys, ceramics, polymers and composites are currently used in this area, but unfortunately, some have shown tendencies to cause device failure after long term use in the body since they cannot fulfill some vital requirements (Geetha, Singh et al. 2008; Bahraminasab, Hassan et al. 2010). Developing or applying an optimal material, therefore, can cause the implant to last longer and to avoid the huge cost related to the inappropriate or unsuitable choice of materials. Shape memory alloys (SMA) have provided new insights into biomedical area for cardiovascular, orthopedic and dental applications, and for making advanced surgical instruments. The biomedical success of these materials is due to their unusual properties, which makes them superior to conventional materials. Among many SMAs, NiTi alloy is considered to be the best because of its superb characteristics. NiTi alloy possesses most of the necessities for orthopedic implantation and is used in a large number of applications. Therefore, it is worth to highlight the orthopedic applications of this material. The reminder of this chapter is organized as follows; section 2 discusses on biocompatibility of NiTi shape memory alloy in both bulk and porous forms, followed by a brief review of some current NiTi applications in section 3. After that section 4 introduces one of the potential applications of this material in orthopedics, which is the femoral component of knee prosthesis. The chapter ends with discussion and conclusion in in sections 5 and 6, respectively.

2. Biocompatibility of NiTi

Enthusiasm to apply NiTi in implants has been tempered by the concern related to its biocompatibility. Biocompatibility is related to the capability of material to exist in contact with human body tissues without causing an unacceptable degree of harm to the body. Furthermore, the materials should have the ability to interact with the biological environment to improve the biological response, the tissue-surface bonding, and to enhance the ability to undergo a progressive degradation when new tissues grow and heal. Of the two metal elements in NiTi alloys, titanium is well recognized to be biocompatible with superb long-term corrosion resistance. In contrast to Ti, Ni release from the surface of NiTi implants has been a concerning issue because the dissolution and concentration of Ni ions or wear debris above a certain amount causes some allergic reaction and biocompatibility problems including pneumonia, chronic sinusitis and rhinitis, nostril and lung cancer for patients (Mantovani 2000; Kapanen, Ryhanen et al. 2001; Machado and Savi 2003). However, most of in vivo studies of NiTi implantation and in vitro experiments exhibited good biocompatibility of this material (Castleman, Motzkin et al. 1976; Ryhaenen, Kallioinen et al. 1998; Kapanen, Ryhanen et al. 2001). The good corrosion resistance of this material can be attributed to its crystal structure stability, which impedes the separation and release of Ni ions. Therefore, it seems that the actual risk of large Ni leaching from wear and corrosion phenomena may potentially be overexaggerated (Es-Souni and Fischer-Brandies 2005). The titanium content of these alloys is readily oxidized and creates an outmost protective titanium oxide layer which act as a barrier to chemical attack and corrosion and confines the diffusion of Ni ions. This oxide layer, however, is not permanent and can be depleted by wear, corrosion, and fatigue. In this situation, NiTi repassivates and regenerate the oxide layer. The integrity of the protective titanium oxide layer is influenced by surface roughness, inhomogeneities, residues, porosity, and geometry. A certain toxicity usually observed in vitro studies, is most likely due to the higher amount of Ni concentrations in vitro that are not possible to achieve in vivo (Shabalovskaya 1996). The biocompatibility of NiTi alloys has been reported to be equal or better than that of titanium, Co–Cr alloys and stainless steels (Shabalovskaya 1996; Ryhanen 2000; Es-Souni and Fischer-Brandies 2005). Treatments such as surface oxidation, plasma immersion ion implantation (PIII), and laser surface modification reduce the amount of Ni leaching from the surface of NiTi implant to negligible amounts. For example, a treatment like dual electropolishing (EP) and photoelectrocatalytically oxidation (PEO) makes NiTi suitable for hard tissue replacements (Chu, Guo et al. 2009). A problem associated with this material, which seems to be important for orthopedic implants, is slow osteogenesis process and growth of a fibrous layer at interface of bone-implant (Berger-Gorbet, Broxup et al. 1996; Chen, Yang et al. 2004) generating weak anchorage between the implant and adjacent bone, and finally leading to micro-motion and loosening of the implant. Nevertheless, creation of a thin apatite layer on NiTi components in situ showed a large amount of new bone directly contacting with the host bone (Chen, Yang et al. 2004). In addition to this, since porous biomaterials for implants (either as porous coating or integral porous body) have attracted researchers' interest, it has been tried to produce porous NiTi alloys with different fabrication techniques such as self-propagating high temperature synthesis process, capsule-free hot isostatic pressing and metal injection molding (Figure 1). The amount of porosity, pore size, microstructure and mechanical properties of the porous NiTi may depend upon the fabrication method. Porous NiTi has been reported to have good biocompatibility, comparable to the conventional porous stainless steel and titanium implant materials (Thierry, Merhi et al. 2002). However, the large exposed surface area directly interfacing the adjacent bone and tissue makes the issue of Ni release more serious than dense NiTi. Minimal Ni release can be obtained by surface engineering and treatments such as thermal annealing, oxygen plasma immersion ion implantation, hydroxyapatite (HAP) coatings, pre-soaking in simulated body fluid (SBF) solution, TiN and TiO_2-PVD coatings, and combinations thereof.

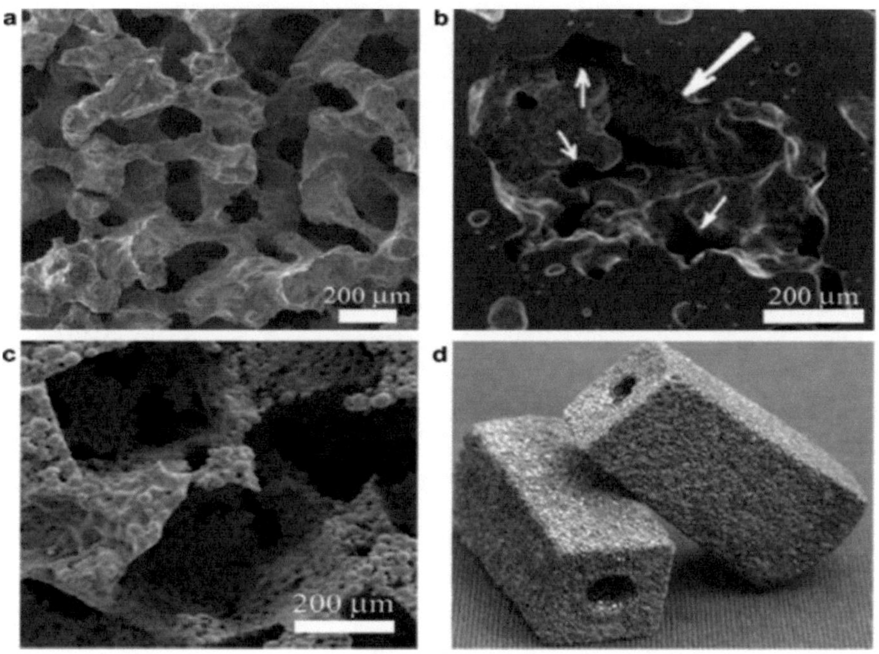

Figure 1. Scanning electron microscope Photographs of porous NiTi fabricated by three different techniques: (a) self-propagating high temperature synthesis process (about 65% porosity, 100–360 μm) (b) capsule-free hot isostatic pressing with argon expansion (42% porosity, 50–400 μm) (c) metal injection molding (70% porosity, 355–500 μm). (d) Image of commercial porous NiTi implant (Actipore™, Biorthex, Canada) for spinal fusion made by self-propagating high temperature synthesis process (Bansiddhi, Sargeant et al. 2008)

3. A brief review of NiTi SMAs in orthopedic applications

Shape memory alloys are a group of metallic materials with some unusual properties such as one-way and two-way shape memory effects, superelastic effect, high damping property and rubber-like effect. These characteristics make the material suitable for different orthopedic applications such as load-bearings, plates for bone fracture repair, internal fixators for long bone shafts, spinal correctors, vertebral spacers and bone distraction devices. Some of these applications are explained in the following subsections.

3.1. Spinal vertebral spacer

The spinal vertebral spacer is one of the applications of this material in orthopedics. The insertion of the spacer (disc) between two vertebrae provides the local reinforcement of the spinal column, avoiding any traumatic motion during the healing process. The employ of a shape memory spacer enables the use of a constant load regardless of the patient position with some degree of motion. This device is used to treat scoliosis. Figure 2 presents spinal vertebrae spacer in the in the original shape (right) and martensitic state (left).

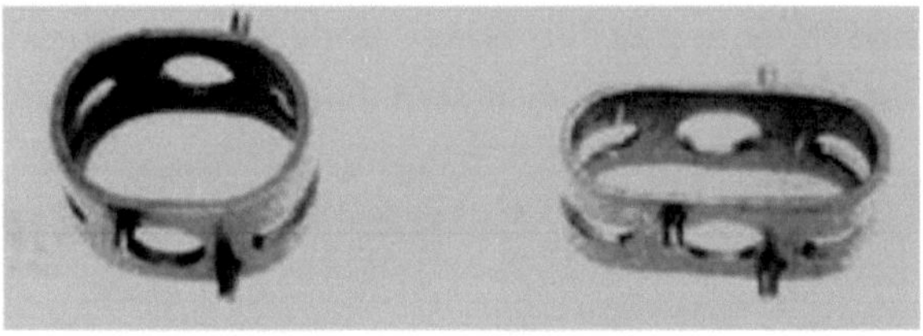

Figure 2. Spinal vertebrae spacer (Duerig, Melton et al. 1990)

3.2. Spinal rod

Shape memory rod has been applied as a tool to help the scoliosis correction (Figure 3). NiTi is used in this application due to its ability to return to some predefined shape when subjected to a thermal treatment. It is expected that the spinal rod has the ability to keep the spine force loaded postoperatively, and it appears that this will take the advantage of spine viscous behavior to obtain extra correction. Furthermore, a postoperative fusion may prevent the long term failure of the system. The additional postoperative correction is expected to be obtained before the occurrence of this vertebral fusion.

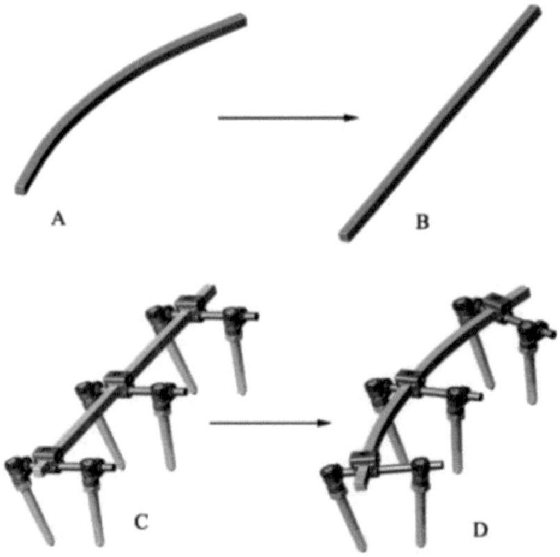

Figure 3. Spinal rod; (A) The original rod shape in the cold condition, (B) The straightened rod before insertion and heat treatment, (C) The implantation of straightened rod with anchorage system, (D) The recovered original curve of the rod with anchorage system after heat treatment (Wever, Elstrodt et al. 2002)

3.3. Medical staples

Another application of NiTi is related to the healing process of broken and fractured bones, using the shape memory effect. The shape memory orthopedic staples (Figure 4) are placed directly into the region of the break to compress the two parts of the bone. These staples, in their opened shape, are implanted to the fractured site of the bone, while through heating the staples tend to close, compressing the separated part of fracture. In this application, the heating is performed by an external device, and not due to the body temperature.

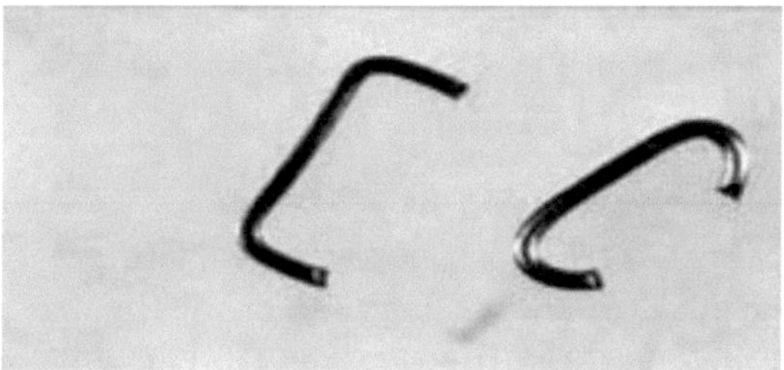

Figure 4. Medical staple before and after distraction (Laster, MacBean et al. 2001)

3.4. Plates for fractured bone

Shape memory plates also have been used to heal and recover the fractured bones, in the injured area where it is not possible to apply cast such as facial areas, nose, jaw, and eye socket. They are inserted to the fracture and fixed with intermediate screws (Figure 5). This maintains the original alignment of the bone and enables cellular regeneration. When these plates are heated, they tend to recover their previous shape (because of the shape memory effect) and exert a constant and uniform force on the two broken sections, which causes to join separated parts of fractures and helps in the healing process.

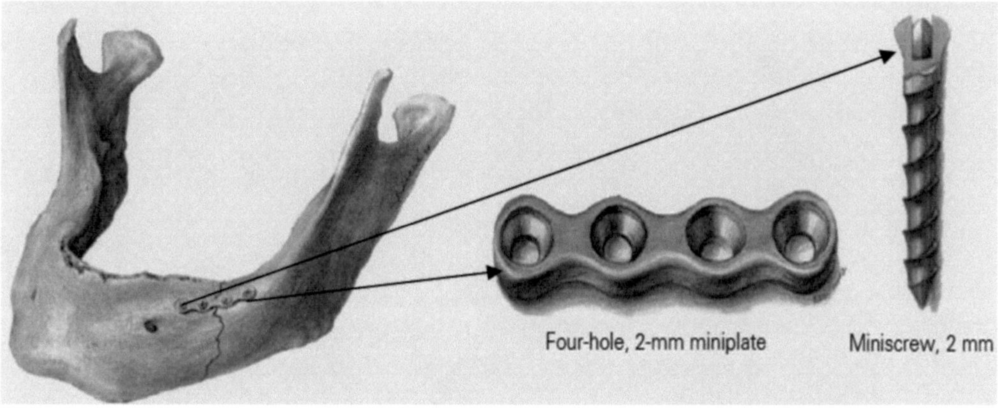

Figure 5. SMA plate for fractured human jaw bone and details of the plate and the screw (Machado and Savi 2003)

3.5. Nails for marrow cavity

These implants are applied where long bones, such as the femur, are broken. Treatment of this type of broken bones involves the hollowing out the bone marrow cavity of the two bone sections followed by the reconnection and introduction of a nail to enable the healing of the break. SMAs are used for nail material to apply controlled force to bone. The SMA nail can be manufactured to the desired shape for forced diaphyseal bone bending. Cooling down to the martensitic phase allows insertion of the shaped nail into the cavity, while at body temperature, the nail returns to its original shape, leading to a bending force.

3.6. Cruciate ligament

Considering the resistance to rupture and the maximum elastic deformation, the NiTi SMA can approach the behavior of a natural knee ligament. However, SMA could answer the demand for severe mechanical conditions imposed on the ligament prosthesis; the lack of fatigue resistance is a limiting factor for the prosthesis that is subjected to the repeated cyclic movements of the knee (Hagemeister, Yahia et al. 1995; Mantovani 2000). It should be pointed out that a new design of prosthetic ligament may lead to a solution for this weakness. NiTi alloy designed (in 1992) to reconstruct the anterior cruciate ligament in which the Nitinol strips were used with filaments, utilizing the unique property of Nitinol to change its shape by heating and cooling. The Nitinol alloy chosen for this application had an austenitic finish temperature of about 35°C, below the temperature of body. The contraction of the Nitinol wires, therefore, occurred (due to body temperature) on warming; caused the tibia and femur to be pulled together. On the other hand, Nitinol strips (used for anchorage to the tibia and femur) were also deflected by body heat after passing through the tibia and femur. The deflection of the strips, combined with the contraction of the wires, acted as a spring and counterbalanced the prosthesis loading during knee motion (Hedayat, Rechtien et al. 1992).

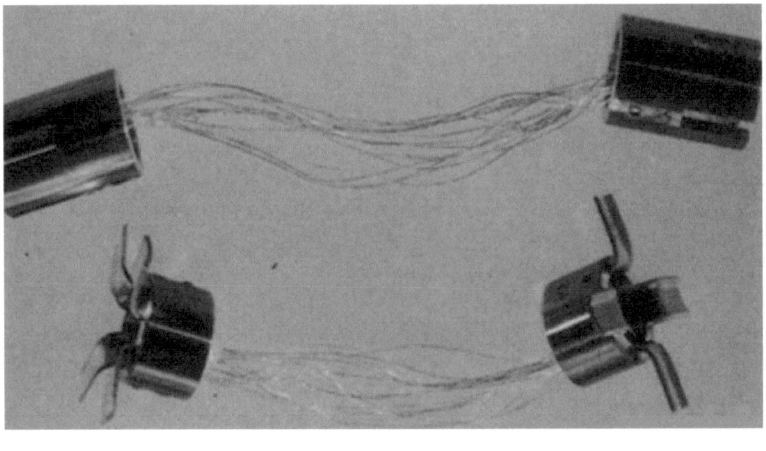

Figure 6. A carbon-coated NiTi prosthesis for reconstruction of the anterior cruciate ligament

3.7. Gloves in physiotherapy

NiTi SMAs are utilized in orthopedic treatment for physiotherapy of partially atrophied muscles. Physiotherapy gloves, which can reproduce the original movements of hands, were developed by locating the SMA wires on the region of fingers (Figure 7). These wires possess the two-way shape memory effect, so that heating the glove shortens the length of the wires and close the hand while cooling returns the wires back to their former shape and open the hand. In fact, the wires made of NiTi can withstand a large number of heating and cooling cycles, over time, without a decrease in performance (Gobert, Hoang et al. 2004).

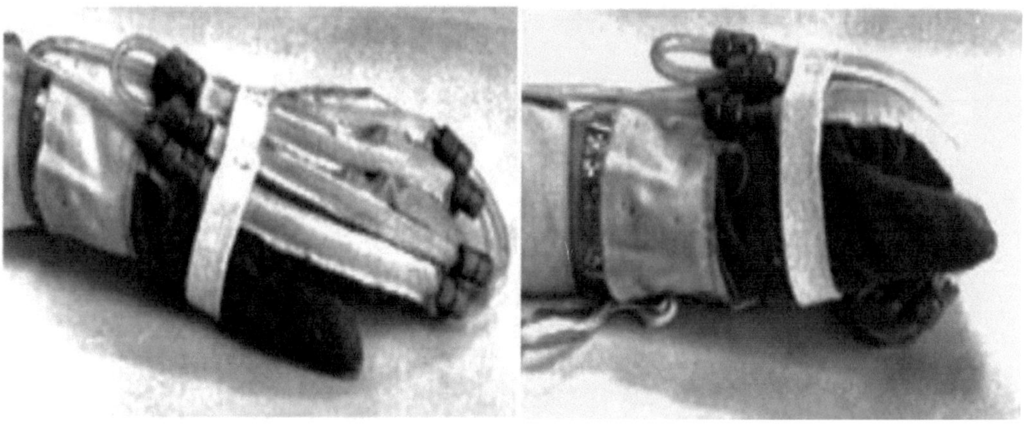

(a) (b)

Figure 7. Gloves with SMA wires: (a) position at low temperature; (b) position at high temperature (Machado and Savi 2003)

4. Case study: Potential advantages of NiTi SMAs in knee implants

One of the most important current debates in orthopedic is the total joint replacements particularly hip and knee because of the simultaneous increasing number of both replacement and revision surgeries (Kurtz, Ong et al. 2007; Carr and Goswami 2009). One of the most serious concerns associated with revision surgery is the aseptic loosening of the components. Excessive wear between articular surfaces, stress shielding of the bone by prosthesis, and development of a soft tissue at the bone-implant interface are the main leading causes for aseptic loosening. Applying the best material for the implant components can reduce the wear debris, improve the load transfer system, and provide anchorage between the bone and the component interfacing the bone. Therefore, the optimal material can reduce the risk of prosthesis loosening.

4.1. Knee prosthesis components

Total knee replacement (TKR) typically has three main components: femoral component, tibial component (consisting of tibial tray and tibial insert), and the patellar component (Figure 8). The tibial insert and the patellar component are plastic parts such as ultra high molecular weight polyethylene (UHMWPE) or cross-linked polyethylene (XLPE). The femoral component and the tibial tray are usually made of metals and alloys including titanium alloys, stainless steels or cobalt chromium alloys. Femoral component replaces the distal femur, hence; it interfaces the bone from the upper side and articulates against the polyethylene insert from the lower side. As a result, aseptic loosening of this component is involved with all the three main causes and it appears to be a more challenging issue. Therefore, there is a need to apply an optimal material for this component to reduce or avoid the loosening problem and provide longer lasting knee prosthesis.

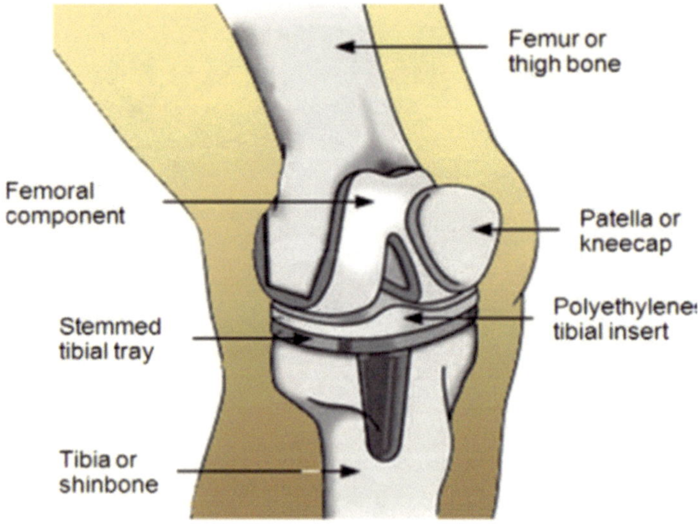

Figure 8. Total knee replacement components (Bahraminasab and Jahan 2011)

4.2. Biomaterial requirements for femoral component

Implant materials must balance some requirements which are essential for prosthesis to have well performance with no rejection after long term use in the body. These requirements vary from one application to another. The required properties for femoral component of TKR include strength, elastic modulus, ductility, density, corrosion and wear resistance, biocompatibility and osseointegration (Bahraminasab and Jahan 2011). High strength is required to avoid fracture of the component, and ductility is, also, needed to avoid brittle failure of the implant under loading conditions. Furthermore, the weight and density of the biomaterial must be comparable to that of bone. Low elastic modulus plays an important role for heavily loaded joint such as TKR. Large difference between Young's modulus of implant component biomaterial and the adjacent bone can contribute to stress shielding effect, which causes the subsequent bone resorption and aseptic loosening. In addition to this, lower Young's modulus means higher damping capacity and resilience leading to more absorption of impact energy and dampening of the maximum stress between the bone and the articular implant. High corrosion resistance is, also, desirable due to the corrosive body fluid. The metallic implants release unfavorable (non-biocompatible) ions that either can accumulate in tissues, near the implant, or they might be transported to the other parts of the body (Okazaki and Gotoh 2005) causing toxic effects. Low wear resistance or high coefficient of friction between the articular surfaces can result in aseptic loosening of the component. Furthermore, wear debris is biologically active and provides adverse inflammatory response, which destruct the healthy bone supporting the actual prosthesis. Osseointegration refers to the bone healing process. The incapability of the implant surface to bond with the adjoining bone and tissues leads to micro-motions and formation of a fibrous tissue around the implant promoting aseptic loosening process (Viceconti, Muccini et al. 2000; Geetha, Singh et al. 2008). The biocompatibility for total joint replacements has been defined as; optimizing the rate and quality of bone apposition to the material, minimizing the release rate of corrosion and the tissue response to the released particles, minimizing the release rate of wear debris and the tissue reaction to this debris, and optimizing the biomechanical environment to minimize disturbance to homeostasis in the bone and its surrounding soft tissue (Williams 2008). Among the above described properties, resistance to wear, low elastic modulus and acceptable osseointegration are the necessities avoiding aseptic loosening problem.

4.3. Superiority of NiTi compared to the current materials

NiTi alloys combine high strength, unique fatigue resistance, and good ductility (Bahraminasab, Hassan et al. 2010). These materials exhibit high resistance to wear even more than Co-Cr-Mo alloy (the most accepted current material for femoral component). The wear resistance in conventional materials depends upon some mechanical factors such as hardness, toughness, and work-hardening. However, for NiTi shape memory alloys, other parameters are believed to be the reasons for high wear resistance and low coefficient of friction such as the recovery of the superelastic deformation (Yan 2006), pseudoelasticity effect and strength (Abedini, Ghasemi et al. 2009). Further, the low Young's modulus of this alloy also decreases the maximum contact pressure and accordingly the wear rate (Yan 2006). High wear resistance of this material makes it a potential for the applications such as femoral component of TKR or components of other joint replacements in which wear is of crucial importance. Most of wear tests carried out on NiTi were pin-on-disk tests with NiTi disk and steel bearing pin. However, in the knee joint prosthesis femoral condyles (made of metals) articulate against the tibial tray (made of UHMWPE), thus; it is required to obtain the coefficient of friction between NiTi and UHMWPE. To the best of the authors' knowledge, the values of friction coefficient and wear rate between these two materials have not been reported in the literature. Knee joints function as dynamically loaded bearings that are subjected to 10^8 loading cycles in 70 year lifetime. The average friction coefficient of the load bearing synovial joints including hip and knee is around 0.02 and the wear factor is approximately 10^6 mm^3/N. However, the coefficient of friction for materials implanted in the body varies between 0.05 to 0.16, which depends on the mate material and the type of test lubricant (Geetha, Singh et al. 2008). Therefore, it would be interesting to test this material against UHMWPE (or XLPE) to obtain the friction coefficient and wear rate in the conditions mimicking the natural knee joint situations and to have a precise comparison with the existing materials. On the other hand, low stiffness or low Young's modulus of NiTi, which is much lower than those of Co-Cr based alloys and stainless steels and much closer to that of bone (less than 30 GPa), minimizes the stress shielding effect and the subsequent aseptic loosening. It also means higher damping capacity and resilience, which can highly affect the absorption of impact energy and reduce the peak stress between the bone and the articular prosthesis (Bahraminasab and Jahan 2011). Another issue that is worth to be highlighted is the superelastic behavior. Human body especially the skeletal part is subjected to stresses during daily activities such as walking, stair climbing and lifting objects. The stresses experienced by the bone vary from one activity to another, and also vary at different time during an activity. These stresses will cause deformation and change the shape, thus it is important that once an activity is completed, the implant return to its designated shape. The elastic property ensures this but over a long period of use, superelastic behavior of NiTi shape memory alloy may help for longer implant life (no unrecoverable or residual strain). In addition to the dense or bulk NiTi, porous form of these alloy have very high potentials to be used in orthopedics because the interconnected open pores and large surface area enables the body fluids transportation and accelerates the healing process. Furthermore, it also allows tissue and bone cells in-growth, which makes strong anchorage between the prosthesis and the adjoining bone, provides long-term fixation, and reduces knee implant aseptic loosening. Porous NiTi alloy, therefore, presents better osteoconductivity and osteointegration than bulk NiTi alloy. Appropriate amount of porosity and pore size, and suitable fabrication technique make it possible to obtain a combination of good contact between bone and implant, high strength, low elastic modulus, high toughness and high energy absorption (Ryhanen, Niemi et al. 1997; Zhu, Yang et al. 2008). It is possible to achieve low stiffness in the range of bone elastic modulus to minimize stress shielding effect, and provide high damping capacity. This material may offer other advantages such as super-elasticity after tissue in-growth, superb mechanical stability within the host tissue because of shape-recovery characteristic, and morphology similar to that of bone. Figure 9 shows a comparison of NiTi and trabecular bone porosity.

To the best of authors' knowledge, NiTi has not been used for prosthetic femur, however, based on finite element analysis (FEA) study and multi-criteria decision-making (MCDM) method, these materials are superior to the currently used metallic materials for this application (Bahraminasab and Jahan 2011; Bahraminasab, Sahari et al. 2011). MCDM is usually used for contemporary materials selection problems in which the suitability of candidate materials is assessed against multiple criteria rather than considering one single factor (Jahan, Ismail et al. 2010) to avoid misuse of materials and the respective huge cost. The MCDMs approaches have been developed for biomedical applications (Jahan, Bahraminasab et al. 2011; Jahan, Mustapha et al. 2011) among which comprehensive VIKOR method (Jahan, Mustapha et al. 2011) was used to select the best material for femoral component of TKR. Based on the results of the evaluation, porous NiTi (SMA) was the optimal metal alternative with the confidence level of 100% and dense NiTi (SMA) was ranked second with the confidence level of 73 %. The compositions of the current and promising metals considered for MCDM material selection and comparison of their properties are given in Table 1 and Table 2 respectively. FEA allows changing material properties of the mechanical components and predicts the performance before manufacturing any prototypes. A finite element analysis study on the knee joint under static

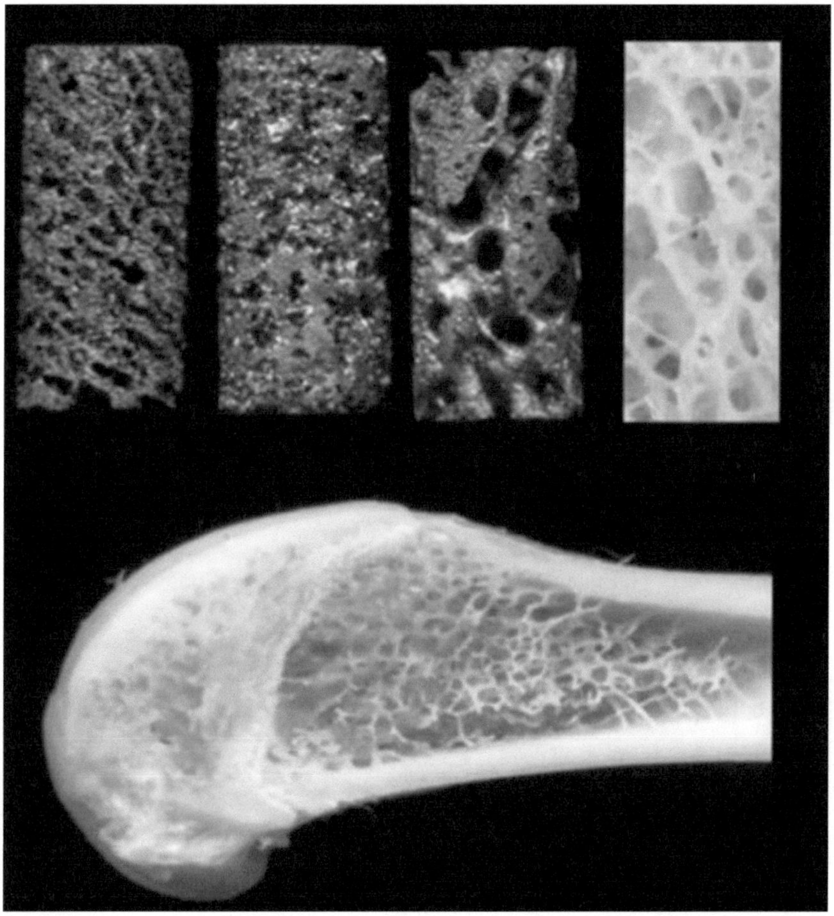

Figure 9. Porous NiTi compared with trabecular bone (taken from http://herkules.oulu.fi/isbn9514271246/html/c721.html)

load of 800 N at 0° of flexion angle demonstrated that NiTi (SMA) provided higher Von Mises stresses on the femoral bone than Co-Cr and Ti-6Al-4V alloys, as shown in Figure 10 and Figure 11, meaning that this material can reduce the stress shielding, and consequently aseptic loosening of the implant (Bahraminasab, Sahari et al. 2011). FEA and MCDM partly covered the theoretical aspects related to the use of NiTi as femoral component. However, there is a long distance from theoretical development of biomaterials to the practical applications including in vivo and vitro tests.

Materials number	Materials' names	Compositions
1	Stainless steel L316 (annealed)	Fe balancing, 17-20% Cr,10-14%Ni, 2-4% Mo, 0.03-0.08% C, 2% Mn and 0.75% Si
2	Stainless steel L316 (cold worked)	
3	Co-Cr alloys (Wrought Co-Ni-Cr-Mo)	Co balancing, 19-21% Cr, 9-11% Ni, 14.6-16%W, 0.13% Mo, 0.05-0.15 %C, 0.48%Si and maximum 2%Mn &3%Fe
4	Co-Cr alloys (Cast able Co-Cr-Mo)	Co balancing, 27-30% Cr, 2.5% Ni, 5-7% Mo, 0.75%Fe, 0.36%C and maximum 1%Mn &Si
5	Ti alloys (Pure Ti)	0.3% Fe, 0.08% C, 0.13% O2, 0.07% N2
6	Ti alloys (Ti-6 Al-4V)	Ti balancing, 5.5-6.5% Al, 3.5-4.5% V, 0.25% Fe and 0.08% C
7	Ti-6Al-7Nb (IMI-367 wrought)	Ti balancing, 5.50 - 6.50 % Al, <= 0.080 % C, <= 0.0090 % H, <= 0.25 % Fe, 6.50 - 7.50 % Nb, <= 0.050 % N, <= 0.20 % O, <= 0.50 % Ta
8	Ti-6Al-7Nb (Protasul-100 hot-forged)	
9	NiTi shape memory alloy	Ni 55.0 - 56.0 %, Ti 43.835 - 45.0 %, C <= 0.050 %, Fe <= 0.050 %, O <= 0.050 %, H <= 0.0050 %, Other <= 0.010 %
10	Porous NiTi shape memory alloy	Ni–49.0at.%Ti, 16% porosity

Table 1. Candidate materials for femoral component and their compositions

Material number	Density (g/cc)	Tensile Strength (MPa)	Modulus of Elasticity (GPa)	Elongation (%)	Corrosion resistance	Wear resistance	Osseointegration
1	8	517	200	40	high	Above average	Above average
2	8	862	200	12	high	Very high	Above average
3	9.13	896	240	10-30	Very high	Extremely high	High
4	8.3	655	240	10-30	Very high	Extremely high	High
5	4.5	550	100	54	Exceptionally high	Above average	very high
6	4.43	985	112	12	Exceptionally high	High	very high
7	4.52	≥ 900	105 – 120	≥ 10	Exceptionally high	High	very high
8	4.52	1000-1100	110	10-15	Exceptionally high	High	very high
9	6.50	≥1240	≥48	12	Extremely high	Exceptionally high	Average
10	4.3<	1000	15	12	Very high	Exceptionally high	Exceptionally high

Table 2. Properties of candidate materials

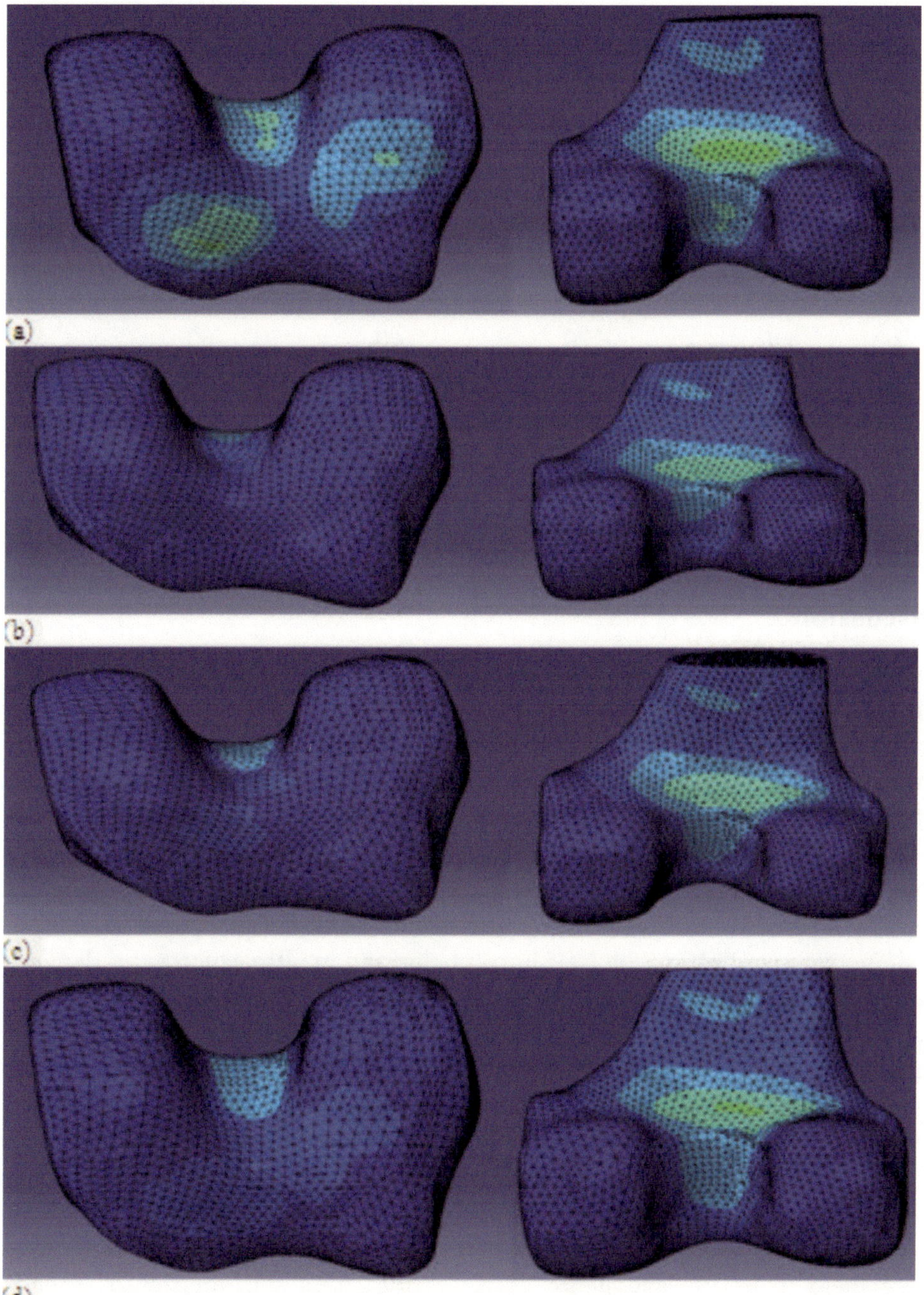

(a)

(b)

(c)

(d)

Figure 10. Stress pattern on the femur for (a) natural knee (b) Cr-Co alloy (c) Ti-6Al-4V (d) NiTi (SMA)

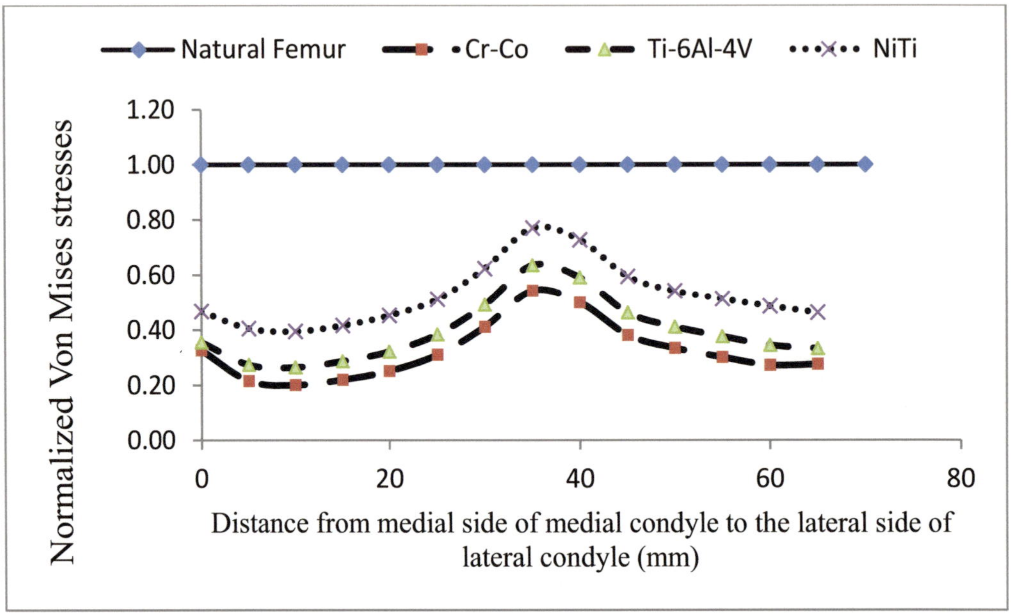

Figure 11. Comparison of stress values for natural femur, Cr-Co alloy, Ti-6Al-4V and NiTi at 30 mm posterior to anterior

5. Discussion

Development and selection of optimal biomaterials for orthopedic applications is a challenging task due to dealing with complex biological and mechanical requirements on one hand, and growing number of damaged organs replacements with synthetic materials on the other hand. Shape memory alloys, particularly NiTi alloys have been considered for many orthopedic applications either in practice or as promising material for future use. There are several main aspects of NiTi that favor its use as implant materials. These include biocompatibility (especially corrosion resistance), the ability to return to its original shape at a specific temperature, and material property values comparable to those of natural bone. Most of in vivo investigations on NiTi implantation and in vitro experiments demonstrated good biocompatibility of this material. Furthermore, higher level of biocompatibility also can be achieved by surface modification of NiTi alloys. The ability of NiTi to restore its original shape (after deformation) at a specific temperature makes the material suitable for a wide range of applications; we recall the staples, plates for fractured bone and gloves for physiotherapy. Additionally, the superelastic behavior of this material may provide benefits for human skeletons, especially for joint replacements where it is subjected to cyclical stresses during daily activities. The closeness of the implant and the natural bone material properties (elastic modulus), particularly at the interface reduces the stress difference between these two parts and hence decreases the stress shielding effect. Therefore, it appears that NiTi SMA is a very suitable candidate for orthopedic implant applications, among which several have been well established but some other suggested applications, such as femoral component of total knee replacement, require more investigations and preclinical tests. Co-Cr based alloy, which currently is the most accepted material for prosthetic femur, and other metals including Ti and its alloys and stainless steel cannot completely fulfill the desired requirements for long term use in the body, and the revision surgery is usually performed 10-15 years after the implantation of knee prosthesis. Therefore, there is still a need for material solution in knee joint replacement. However, applying a new biomaterial for a specific application may impose either redesign of the component or material tailoring for the existing design. Both bulk and porous NiTi are high potential materials in orthopedic applications such as this. Bulk NiTi has high wear resistance and fairly low elastic modulus, which are the two important properties to avoid aseptic loosening of femoral component, but unfortunately, it lacks osseointegration and bioactivity. To overcome this deficiency, surface engineering are being conducted on this material. Generation of porosity in material structure is also a way to obtain bioactivity and well integration with the bony bed, which simultaneously can further reduce the elastic modulus as well. Porous NiTi, therefore, is a quite high potential material for this application. However, when the porosity increases, corrosion behavior of NiTi requires careful consideration. Porous NiTi has many orthopedic applications including spinal fixation, acetabular hip prostheses, and permanent osteosynthesis plates.

For using a biomaterial for knee joint prosthesis, several medical and engineering skills are needed. Knowledge on the anatomy of the natural knee joint, the histology of cortical and cancellous bone, cartilage, meniscus, ligament, tendon and synovial fluid, the physiology of circulation and of the bone growth and bone loss, the biomechanics of the knee joint during daily activities, orthopedic implantation methods, pathology (e.g. osteoarthritis and osteoporous) and biocompatibility are all the essential medical skills. The engineering skills include tribology, kinematics, fracture mechanics, fatigue and elasticity/plasticity theory. Therefore, a multidisciplinary team of surgeons and materials scientists and engineers are needed for new material applications in total knee replacement. Generally, the requirements for femoral component may necessitate for the components of other joint replacement with similar mechanical and biological conditions. However, it should be pointed out that the required properties are site specific in human body, and in selection process of materials for different joints, the alterations must be carefully taken into account. NiTi as a promising material can be widely investigated theoretically and clinically, as a future research, in biomedical engineering especially for orthopedic applications.

6. Conclusion

From the above discussion, the following conclusions can be made:

1. NiTi shape memory alloy is a biocompatible material with low elastic modulus and high wear resistance which makes it suitable for use in orthopedic applications, particularly joint replacements such as knee and hip.
2. NiTi SMA has a unique property to "remember" its shape at a specific temperature and return to that shape when the specific temperature is reached. This property provides

advantages for use in human body as an artificial organ since the human body experiences stress changes during activities.

3. NiTi SMA has relatively low Young's modulus, and it can be reduced by generation of porosity in the structure to be comparable with that of bone which results in lower stress difference between the implant and bone or stress shielding effect, and hence, increase the implant life.
4. Porous NiTi not only offers elastic modulus in the range of human bone modulus, but also promotes the growth and penetration of bone cells and tissues into the implant and therefore provides strong anchorage, avoiding loosening of the implant.

Author details

Marjan Bahraminasab and Barkawi Bin Sahari
Department of Mechanical and Manufacturing Engineering, Universiti Putra Malaysia, Malaysia

Barkawi Bin Sahari
Institute of Advanced Technology, ITMA, Universiti Putra Malaysia, Malaysia

7. References

Abedini, M., H. M. Ghasemi, et al. (2009). "Tribological behavior of NiTi alloy in martensitic and austenitic states." Materials and Design 30(10): 4493-4497.

Bahraminasab, M., M. R. Hassan, et al. (2010). "Metallic biomaterials of knee and hip - A review." Trends in Biomaterials and Artificial Organs 24(2): 69-82.

Bahraminasab, M. and A. Jahan (2011). "Material selection for femoral component of total knee replacement using comprehensive VIKOR." Materials and Design 32: 4471-4477.

Bahraminasab, M., B. B. Sahari, et al. (2011). "Finite element analysis of the effect of shape memory alloy on the stress distribution and contact pressure in total knee replacement." Trends in Biomaterials and Artificial Organs 25(3): 95-100.

Bansiddhi, A., T. D. Sargeant, et al. (2008). "Porous NiTi for bone implants: A review." Acta Biomaterialia 4: 773–782.

Berger-Gorbet, M., B. Broxup, et al. (1996). "Biocompatibility testing of NiTi screws using immunohistochemistry on sections containing." Journal of Biomedical Materials Research 32: 243-248.

Carr, B. C. and T. Goswami (2009). "Knee implants–Review of models and biomechanics." Materials and Design 30(2): 398-413.

Castleman, L. S., S. M. Motzkin, et al. (1976). "Biocompatibility of nitinol alloy as an implant material." Journal of Biomedical Materials Research 10(5): 695-731.

Chen, M. F., X. J. Yang, et al. (2004). "Bioactive NiTi shape memory alloy used as bone bonding implants." Materials Science and Engineering: C 24(4): 497-502.

Chu, C. L., C. Guo, et al. (2009). "Microstructure, nickel suppression and mechanical characteristics of electropolished and photoelectrocatalytically oxidized biomedical nickel titanium shape memory alloy." Acta Biomaterialia 5(6): 2238-2245.

Duerig, T., K. Melton, et al. (1990). Engineering aspects of shape memory alloys, Butterworth-Heinemann.

Es-Souni, M. and H. Fischer-Brandies (2005). "Assessing the biocompatibility of NiTi shape memory alloys used for medical applications." Analytical and bioanalytical chemistry 381(3): 557-567.

Geetha, M., A. K. Singh, et al. (2008). "Ti based biomaterials, the ultimate choice for orthopaedic implants–A review." Progress in Materials Science 54(3): 397-425.

Gobert, A. N., C. Hoang, et al. (2004). The Design and Testing of a Therapeutic Glove for Patients Suffering from Arthritis Pains in the Hands.

Hagemeister, N., L. H. Yahia, et al. (1995). "Development of a NiTi anterior cruciate ligament prosthesis." Journal of de physique IV 5: C2-403-408.

Hedayat, A., J. Rechtien, et al. (1992). "Phase transformation in carbon-coated nitinol, with application to the design of a prosthesis for the reconstruction of the anterior cruciate ligament." Journal of Materials Science: Materials in Medicine 3(1): 65-74.

Jahan, A., M. Bahraminasab, et al. (2011). "A target-based normalization technique for materials selection." Materials and Design 35: 647–654.

Jahan, A., M. Y. Ismail, et al. (2010). "Material screening and choosing methods - A review." Materials and Design 31(2): 696-705.

Jahan, A., F. Mustapha, et al. (2011). "A comprehensive VIKOR method for material selection." Materials and Design 32(3): 1215-1221.

Kapanen, A., J. Ryhanen, et al. (2001). "Effect of nickel–titanium shape memory metal alloy on bone formation." Biomaterials 22(18): 2475-2480.

Kurtz, S., K. Ong, et al. (2007). "Projections of primary and revision hip and knee arthroplasty in the United States from 2005 to 2030." Journal of Bone and Joint Surgery - Series A 89(4): 780-785.

Laster, Z., A. D. MacBean, et al. (2001). "Fixation of a frontozygomatic fracture with a shape-memory staple." British Journal of Oral and Maxillofacial Surgery 39(4): 324-325.

Machado, L. G. and M. A. Savi (2003). "Medical applications of shape memory alloys." Brazilian Journal of Medical and Biological Research 36: 683-691.

Mantovani, D. (2000). "Shape memory alloys: Properties and biomedical applications." JOM Journal of the Minerals, Metals and Materials Society 52(10): 36-44.

Okazaki, Y. and E. Gotoh (2005). "Comparison of metal release from various metallic biomaterials in vitro." Biomaterials 26(1): 11-21.

Ryhaenen, J., M. Kallioinen, et al. (1998). "In vivo biocompatibility evaluation of nickel-titanium shape memory metal alloy: Muscle and perineural tissue responses and encapsule membrane thickness." Journal of Biomedical Materials Research 41(3): 481-488.

Ryhanen, J. (2000). "Biocompatibility of Nitinol." Minimally Invasive Therapy and Allied Technologies 9(2): 99-105.

Ryhanen, J., E. Niemi, et al. (1997). "Biocompatibility of nickel-titanium shape memory metal and its corrosion behavior in human cell cultures." Journal of Biomedical Materials Research 35: 451-457.

Shabalovskaya, S. A. (1996). "On the nature of the biocompatibility and on medical applications of NiTi shape memory and superelastic alloys." Bio-Medical Materials and Engineering 6(4): 267-289.

Thierry, B., Y. Merhi, et al. (2002). "Nitinol versus stainless steel stents: acute thrombogenicity study in an ex vivo porcine model." Biomaterials 23(14): 2997-3005.

Viceconti, M., R. Muccini, et al. (2000). "Large-sliding contact elements accurately predict levels of bone-implant micromotion relevant to osseointegration." Journal of Biomechanics 33(12): 1611-1618.

Wever, D., J. Elstrodt, et al. (2002). "Scoliosis correction with shape-memory metal: Results of an experimental study." European Spine Journal 11(2): 100-106.

Williams, D. F. (2008). "On the mechanisms of biocompatibility." Biomaterials 29(20): 2941-2953.

Yan, W. (2006). "Theoretical investigation of wear-resistance mechanism of superelastic shape memory alloy NiTi." Materials Science and Engineering A 427(1-2): 348-355.

Zhu, S. L., X. J. Yang, et al. (2008). "Effect of porous NiTi alloy on bone formation: A comparative investigation with bulk NiTi alloy for 15 weeks in vivo." Materials Science and Engineering C 28(8): 1271-1275.